VOLUME SIX HUNDRED AND TWENTY ONE

METHODS IN ENZYMOLOGY

Chemical and Synthetic Biology Approaches To Understand Cellular Functions - Part A

METHODS IN ENZYMOLOGY

Editors-in-Chief

ANNA MARIE PYLE
Departments of Molecular, Cellular and Developmental Biology and Department of Chemistry
Investigator, Howard Hughes Medical Institute
Yale University

DAVID W. CHRISTIANSON
Roy and Diana Vagelos Laboratories
Department of Chemistry
University of Pennsylvania
Philadelphia, PA

Founding Editors

SIDNEY P. COLOWICK and NATHAN O. KAPLAN

VOLUME SIX HUNDRED AND TWENTY ONE

METHODS IN ENZYMOLOGY

Chemical and Synthetic Biology Approaches To Understand Cellular Functions - Part A

Edited by

ARUN K. SHUKLA
Associate Professor, Joy Gill Chair Professor
and EMBOYoung Investigator,
Intermediate Fellow, Wellcome Trust DBT India Alliance,
Department of Biological Sciences and Bioengineering,
Indian Institute of Technology, Kanpur 208016, India

ELSEVIER

ACADEMIC PRESS
An imprint of Elsevier

Academic Press is an imprint of Elsevier
50 Hampshire Street, 5th Floor, Cambridge, MA 02139, United States
525 B Street, Suite 1650, San Diego, CA 92101, United States
The Boulevard, Langford Lane, Kidlington, Oxford OX5 1GB, United Kingdom
125 London Wall, London, EC2Y 5AS, United Kingdom

First edition 2019

ISBN: 978-0-12-818117-1
ISSN: 0076-6879

For information on all Academic Press publications
visit our website at https://www.elsevier.com/books-and-journals

Publisher: Zoe Kruze
Acquisition Editor: Ashlie M. Jackman
Editorial Project Manager: Shellie Bryant
Production Project Manager: James Selvam
Cover Designer: Christian Bilbow

Typeset by SPi Global, India

Contents

Contributors

David Albesa-Jové
Structural Biology Unit, CIC bioGUNE, Derio; IKERBASQUE, Basque Foundation for Science, Bilbao, Spain

Abdullah Aldehaiman
Division of Biological and Environmental Sciences and Engineering (BESE), Computational Bioscience Research Center (CBRC), King Abdullah University of Science and Technology (KAUST), Thuwal, Saudi Arabia

Stefan T. Arold
Division of Biological and Environmental Sciences and Engineering (BESE), Computational Bioscience Research Center (CBRC), King Abdullah University of Science and Technology (KAUST), Thuwal, Saudi Arabia

Joshua T. Atkinson
Systems, Synthetic, and Physical Biology Graduate Program, Rice University, Houston, TX, United States

Taysir Bader
Department of Chemistry, University of Minnesota, Minneapolis, MN, United States

Leonard Breindel
Department of Chemistry, University at Albany, State University of New York, Albany, NY, United States

Keri M. Bryson
Chemical Biology Laboratory, Center for Cancer Research, National Cancer Institute, National Institutes of Health, Frederick, MD, United States

David S. Burz
Department of Chemistry, University at Albany, State University of New York, Albany, NY, United States

Javier O. Cifuente
Structural Biology Unit, CIC bioGUNE, Derio, Spain

Kyle H. Cole
Department of Molecular Biology and Biochemistry, University of California, Irvine, CA, United States

Samuel J. Danishefsky
Chemical Biology Program, Sloan Kettering Institute, Memorial Sloan Kettering Cancer Center, New York, NY, United States

M. Escarlet Díaz Galicia
Division of Biological and Environmental Sciences and Engineering (BESE), Computational Bioscience Research Center (CBRC), King Abdullah University of Science and Technology (KAUST), Thuwal, Saudi Arabia

Bryan C. Dickinson
Department of Chemistry, The University of Chicago, Chicago, IL, United States

Mark Distefano
Department of Chemistry, University of Minnesota, Minneapolis, MN, United States

Nobuo Fukuda
Biomedical Research Institute, National Institute of Advanced Industrial Science and Technology (AIST), Tsukuba, Japan

John E. Gadbery
Biochemistry and Structural Biology Graduate Program, Stony Brook University, New York, NY, United States

Tata Gopinath
Department of Biochemistry, Molecular Biology, and Biophysics, University of Minnesota, Minneapolis, MN, United States

Anwesha Goswami
Division of Chemical Biology and Medicinal Chemistry, UNC Eshelman School of Pharmacy, University of North Carolina, Chapel Hill, NC, United States

Raik Grünberg
Division of Biological and Environmental Sciences and Engineering (BESE), Computational Bioscience Research Center (CBRC), King Abdullah University of Science and Technology (KAUST), Thuwal, Saudi Arabia

Marcelo E. Guerin
Structural Biology Unit, CIC bioGUNE, Derio; IKERBASQUE, Basque Foundation for Science, Bilbao, Spain

Erik K.R. Hanko
BBSRC/EPSRC Synthetic Biology Research Centre (SBRC), Centre for Biomolecular Sciences, School of Life Sciences, The University of Nottingham, Nottingham, United Kingdom

Shinya Honda
Biomedical Research Institute, National Institute of Advanced Industrial Science and Technology (AIST), Tsukuba, Japan

SeungBeom Hong
Division of Biological and Environmental Sciences and Engineering (BESE), Computational Bioscience Research Center (CBRC), King Abdullah University of Science and Technology (KAUST), Thuwal, Saudi Arabia

Masayuki Iwamoto
Department of Molecular Physiology and Biophysics, University of Fukui, Faculty of Medical Sciences, Fukui, Japan

Erik K. Larsen
Department of Chemistry, University of Minnesota, Minneapolis, MN, United States

David S. Lawrence
Department of Chemistry; Department of Pharmacology, University of North Carolina, Chapel Hill, NC, United States

Yi-Zong Lee
Instrument Center; Institute of Bioinformatics and Structural Biology, National Tsing Hua University, Hsinchu, Taiwan

Yaohao Li
Institute of Materia Medica, Chinese Academy of Medical Sciences and Peking Union Medical College, Beijing, China; Department of Biochemistry and BioFrontiers Institute, University of Colorado, Boulder, CO, United States

Andrej Lupták
Department of Molecular Biology and Biochemistry; Department of Pharmaceutical Sciences; Department of Chemistry, University of California, Irvine, CA, United States

Naglis Malys
BBSRC/EPSRC Synthetic Biology Research Centre (SBRC), Centre for Biomolecular Sciences, School of Life Sciences, The University of Nottingham, Nottingham, United Kingdom

Jordan L. Meier
Chemical Biology Laboratory, Center for Cancer Research, National Cancer Institute, National Institutes of Health, Frederick, MD, United States

Nigel P. Minton
BBSRC/EPSRC Synthetic Biology Research Centre (SBRC), Centre for Biomolecular Sciences, School of Life Sciences, The University of Nottingham, Nottingham, United Kingdom

Bradley S. Moore
Center for Marine Biotechnology and Biomedicine, Scripps Institution of Oceanography; Skaggs School of Pharmacy and Pharmaceutical Sciences, University of California at San Diego, La Jolla, San Diego, CA, United States

Colin P. O'Banion
Division of Chemical Biology and Medicinal Chemistry, UNC Eshelman School of Pharmacy, University of North Carolina, Chapel Hill, NC, United States

Shigetoshi Oiki
Department of Molecular Physiology and Biophysics, University of Fukui, Faculty of Medical Sciences, Fukui, Japan

Simone Rauch
Department of Biochemistry and Molecular Biology; Department of Chemistry, The University of Chicago, Chicago, IL, United States

Nicole S. Sampson
Biochemistry and Structural Biology Graduate Program; Department of Chemistry, Stony Brook University, New York, NY, United States

Laura Segatori
Department of Biosciences; Department of Chemical and Biomolecular Engineering; Department of Bioengineering, Rice University, Houston, TX, United States

Alexander Shekhtman
Department of Chemistry, University at Albany, State University of New York, Albany, NY, United States

Jonathan J. Silberg
Department of Biosciences; Department of Chemical and Biomolecular Engineering; Department of Bioengineering, Rice University, Houston, TX, United States

Shih-Che Sue
Institute of Bioinformatics and Structural Biology; Department of Life Science, National Tsing Hua University, Hsinchu, Taiwan

Zhongping Tan
Institute of Materia Medica, Chinese Academy of Medical Sciences and Peking Union Medical College, Beijing, China; Department of Biochemistry and BioFrontiers Institute, University of Colorado, Boulder, CO, United States

Xiaoyu Tang
Center for Marine Biotechnology and Biomedicine, Scripps Institution of Oceanography, University of California at San Diego, La Jolla, San Diego, CA, United States

Justin M. Thomas
Chemical Biology Laboratory, Center for Cancer Research, National Cancer Institute, National Institutes of Health, Frederick, MD, United States

Amy H. Tran
Department of Biochemistry and BioFrontiers Institute, University of Colorado, Boulder, CO, United States

Beatriz Trastoy
Structural Biology Unit, CIC bioGUNE, Derio, Spain

Kazuhiro Urakubo
Department of Molecular Physiology and Biophysics, University of Fukui, Fukui, Japan

Gianluigi Veglia
Department of Biochemistry, Molecular Biology, and Biophysics; Department of Chemistry, University of Minnesota, Minneapolis, MN, United States

Songlin Wang
Department of Biochemistry, Molecular Biology, and Biophysics, University of Minnesota, Minneapolis, MN, United States

Yu Wang
State Key Laboratory of Stem Cell and Reproductive Biology, Institute of Zoology; Institute for Stem Cell and Regeneration, Chinese Academy of Sciences; University of Chinese Academy of Sciences, Beijing, China

Daniel K. Weber
Department of Biochemistry, Molecular Biology, and Biophysics, University of Minnesota, Minneapolis, MN, United States

Shixian Wei
University of Chinese Academy of Sciences, Beijing, China

Bingyan Wu
Biochemistry and Cell Biology Graduate Program; Department of Biosciences, Rice University, Houston, TX, United States

Kazuya Yamanaka
Department of Life Science and Technology, Kansai University, Osaka, Japan

Jia Jia Zhang
Center for Marine Biotechnology and Biomedicine, Scripps Institution of Oceanography, University of California at San Diego, La Jolla, San Diego, CA, United States

Chen Zhao
State Key Laboratory of Stem Cell and Reproductive Biology, Institute of Zoology, Chinese Academy of Sciences, Beijing, China

Preface

Investigating the basic structure, function, and dynamics of cellular proteins is crucial to develop a complete understanding of cellular and physiological processes. In this context, development of novel methods and tools, and improvement of existing ones is of immense importance and broad interest. Over the last few years, a series of developments in the area of chemical and synthetic biology have surfaced that have allowed us to generate novel insights into a wide range of cellular processes. Many of these recent advances are now pushing the boundaries of our knowledge in various domains of cellular biology not only at the molecular and systems level.

In these three volumes of *Methods in Enzymology*, we cover a broad spectrum of topics including protein design and engineering, novel labeling approaches, emerging biochemical and biophysical methods, design of novel genetic circuits, genome engineering, and computational techniques. Some of the chapters in these volumes provide step-by-step experimental protocols that can be directly utilized in different laboratories, while others provide an overview of a specific method or a particular topic for a comprehensive understanding.

I take this opportunity to express my sincere gratitude to all the authors who readily agreed to contribute to this volume and delivered their work in stipulated time frame. I also want to thank the members of the editorial and production team for doing a wonderful job on getting these chapters ready for publication and efficiently coordinating the entire publication process.

On behalf of the entire *Methods in Enzymology* team, I present to you these three volumes of *Methods in Enzymology* on Chemical and Synthetic Biology and sincerely hope that you find the topics covered in these volumes very useful for your research. As always, we look forward to your comments and feedback.

Arun K. Shukla, PhD
Associate Professor, Joy Gill Chair Professor and EMBO Young Investigator
Intermediate Fellow, Wellcome Trust DBT India Alliance
Department of Biological Sciences and Bioengineering
Indian Institute of Technology, Kanpur 208016, India

CHAPTER ONE

Targeted m^{6}A reader proteins to study the epitranscriptome

Simone Rauch[a,b], Bryan C. Dickinson[b,*]

[a]Department of Biochemistry and Molecular Biology, The University of Chicago, Chicago, IL, United States
[b]Department of Chemistry, The University of Chicago, Chicago, IL, United States
*Corresponding author: e-mail address: dickinson@uchicago.edu

Contents

Abstract

Posttranscriptional regulation of RNA has emerged as an important regulator of genetic information flow in eukaryotic systems. In particular, chemical modifications of RNA have recently been established as key regulatory marks that affect the lifetime, location, trafficking, and function of messenger RNA (mRNA). In mammalian systems, *N*^{6}-methyladenosine (m^{6}A) is the most prevalent mRNA modification, and the writer, eraser, and reader proteins that install, remove, or recognize m^{6}A have been rapidly uncovered and studied at the whole cell level. Understanding the effects of specific m^{6}A modifications and their regulation at the single transcript level is the key next step to understanding the mechanism and consequences of epitranscriptomic regulation. We recently developed programmable m^{6}A reader proteins to study the effects of epitranscriptomic regulatory factors at individual RNA transcripts. In this chapter,

Methods in Enzymology, Volume 621
ISSN 0076-6879
https://doi.org/10.1016/bs.mie.2019.02.035

we discuss the application of targeted m^{6}A readers to study RNA regulation at single endogenous sites. We briefly introduce what is currently known about the *N*^{6}-methyltranscriptome and the Cas13 RNA-targeting family of proteins before detailing our protocol to study RNA modifications with targeted reader proteins.

1. Introduction

RNA that has been transcribed in the nucleus undergoes a wide variety of modifications until it reaches the site of translation in the cytoplasm. Well-known RNA regulatory processes, including capping, splicing, polyadenylation, and trafficking, occur in the nucleus and can regulate the lifetime and function of RNA (Anderson & Kedersha, 2009; Chen & Manley, 2009; Filipowicz, Bhattacharyya, & Sonenberg, 2008; Glisovic, Bachorik Jennifer, Yong, & Dreyfuss, 2008; Lee & Rio, 2015; Lim & Allada, 2013). More recently, a diverse set of chemical modifications has been found to alter stability, transport, protein binding, and translation of RNA inside the cell (Karijolich, Kantartzis, & Yu, 2010; Li & Mason, 2014; Roundtree, Evans, Pan, & He, 2017). In mammals, *N*^{6}-methyladenosine (m^{6}A) is the most prevalent messenger RNA (mRNA) modification (Roundtree et al., 2017). Regulatory functions of such chemical modifications are commonly mediated by m^{6}A reader proteins that preferentially bind to *N*^{6}-methylated adenosines in RNA and link the chemical modification to a known cellular pathway such as RNA degradation or cap-dependent translation activation (Du et al., 2016; Ivanova et al., 2017; Wang et al., 2013, 2015).

Currently, m^{6}A in RNA can be detected at single-nucleotide resolution using antibody-based mRNA enrichment followed by next generation sequencing (Dominissini et al., 2012; Linder et al., 2015). However, studies investigating the cellular effects of these mRNA modifications still rely on whole-cell approaches. m^{6}A effector proteins are either overexpressed or knocked down/out using genetic manipulation and the resulting transcriptome and cellular changes are assessed (Du et al., 2016; Ivanova et al., 2017; Li et al., 2017; Schwartz et al., 2014; Tirumuru et al., 2016; Xu et al., 2015; Zheng et al., 2013). Considering that there are over 7000 m^{6}A sites in mammalian mRNA (Dominissini, Moshitch-Moshkovitz, Salmon-Divon, Amariglio, & Rechavi, 2013), these functional studies can only provide a general, averaged assessment of modification-based RNA regulation, and could potentially miss the divergent regulatory systems acting on different RNAs. To truly understand the intricacies of m^{6}A-mediated RNA regulation in live cells, more specific, programmable tools are needed. In this chapter, we discuss the application of our recently developed targeted m^{6}A-reader proteins to study modifications on individual transcripts at single m^{6}A sites.

The remainder of this chapter is organized as follows: An introduction to RNA modifications, m^{6}A biology, RNA-targeting Cas systems, and the design of programmable reader proteins are presented in Section 2. In Section 3, we describe representative experiments deploying programmable RNA reader proteins in mammalian cells, measuring both RNA and protein level outputs on reporter genes and endogenous transcripts.

2. Programmable RNA reader proteins to study the epitranscriptome

2.1 RNA modifications

Over the last 50 years, more than 100 different RNA modifications have been uncovered (Cantara et al., 2011; Roundtree et al., 2017). Some of these modifications, including the 5′cap and poly(A) tail in eukaryotic systems, have been extensively studied and linked to the regulation of RNA metabolism (Glisovic et al., 2008), while the roles of the other chemical modifications are less well understood. Diverse chemical RNA modifications are found in ribosomal, transfer, and other cellular RNAs (rRNA, tRNA and mRNA/lncRNA) (Karijolich et al., 2010; Roundtree et al., 2017; Song & Yi, 2017; Yu, Shu, & Steitz, 1997). It is well known that rRNA is commonly modified by pseudouridines and 2′-*O*-methylation for increased stability (Wang & He, 2014; Yu et al., 1997). tRNA is the most prevalent modified RNA species and is frequently modified to ensure tRNA discrimination, translation fidelity and stability (Kirchner & Ignatova, 2015; Lorenz, Lunse, & Morl, 2017). The third major class of modifications are eukaryotic mRNA/lncRNA modifications. Most commonly, N^6-methyladenosine, N^1-methyladenosine, pseudouridine, and 2′-*O*-methylations are found within transcribed oligonucleotides (Song & Yi, 2017). For the purpose of this chapter, we will focus on how to target m^{6}A-modified transcripts site-specifically as the regulatory "reader" proteins for m^{6}A are the most studied.

2.2 N^6-methyladenosine (m^{6}A)

In eukaryotes, N^6-methyladenosine is the most prevalent and well-studied internal mRNA modification, averaging three modifications per transcript (Zhao, Roundtree, & He, 2017). m^{6}A has been shown to play a crucial role in several key developmental steps of the cell (Aguilo et al., 2015; Ivanova et al., 2017; Wen et al., 2018), as well as numerous other cellular processes including RNA processing, and has regulatory functions in viral infection (Gokhale et al., 2016; Lichinchi et al., 2016; Tirumuru et al., 2016) and cancer (Li et al., 2017; Wang et al., 2017; Zhang et al., 2017). m^{6}A commonly occurs at a DR(m^{6}A)CH consensus site (Dominissini et al., 2013) and is installed by a METTL3/METTL14 heterodimeric methyltransferase complex that is further stabilized by WTAP and KIAA1429 (Patil, Pickering, & Jaffrey, 2018; Yue et al., 2018). Additionally, m^{6}A can be removed by the demethylases FTO and ALKBH5, permitting dynamic alterations to the transcriptome through the competitive interplay between the "writer" and "eraser" proteins (Li et al., 2017; Zheng et al., 2013) (Fig. 1).

2.3 RNA "reader" proteins

In addition to the "writers" and "erasers" that install and remove m^{6}A, a variety of reader proteins have been discovered. Reader proteins link individual m^{6}A sites on target transcripts to a downstream functional or regulatory change. There are three classes of m^{6}A binding proteins: the YT521-B homology domain-containing proteins, the m^{6}A switch proteins, and the common RNA binding domain proteins including the IGF2BP family (Zhou & Pan, 2018). The YTHDF protein family was the first

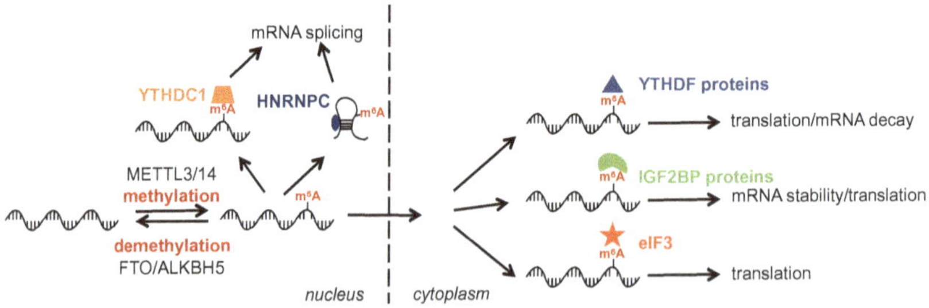

Fig. 1 Regulation of the *N6*-methyltranscriptome. m^{6}A is initially co-transcriptionally installed by the METTL3/METTL14 methyltransferase complex. It can furthermore be actively removed by ALKBH5 or FTO, the two characterized m^{6}A demethylases. After installment of the chemical mark, it can be read by an array of m^{6}A-specific reader proteins in the nucleus as well as the cytoplasm. The cytoplasmic YTH domain family (YTHDF) proteins were the first discovered m^{6}A binding protein family and includes members that are reported to influence both translation as well as mRNA degradation.

identified m^6A-binding protein family and remains to be the most well-studied to date. YTHDF1-3 are cytoplasmic reader proteins that bind to m^6A at the commonly identified DRACH consensus site (Dominissini et al., 2012; Du et al., 2016; Shi et al., 2017; Wang et al., 2013, 2015).

We have focused our initial studies on YTHDF1 (Y1) and YTHDF2 (Y2), as these are well-characterized proteins with potentially different cellular functions. YTHDF2 has been found to recruit the CCR4-NOT deadenylation machinery to N^6-methylated transcripts in cells, leading to a decrease in mRNA stability and initiation of RNA degradation (Du et al., 2016; Wang et al., 2013). YTHDF1, on the other hand, increases the translation efficiency of a target transcript through interactions with the translation initiation machinery and the ribosome directly (Wang et al., 2015). While these are the primarily reported functions of YTDHF1 and YTHDF2, there is some debate in the field whether all YTHDF proteins have identical functions in all contexts. For example, YTHDF1 and YTHDF3 have also been found to induce deadenylation of targeted RNA transcripts in tethering experiments (Du et al., 2016). Additionally, studies of viruses concluded that all YTHDF proteins can induce translation upregulation or RNA degradation and do not have different roles within cells (Gokhale et al., 2016; Kennedy et al., 2016). These findings provide further motivation for the development of tools to study individual transcripts, which are necessary to elucidate a full mechanistic and functional understanding of RNA regulation by chemical modifications.

2.4 Site-specific RNA targeting with CRISPR/Cas systems

The discovery and characterization of the prokaryotic antiviral defense system Clustered regularly interspaced short palindromic repeats (CRISPR)/Cas9 has revolutionized the genome editing field in recent years (O'Connell et al., 2014; Strutt, Torrez, Kaya, Negrete, & Doudna, 2018; Wiedenheft, Sternberg, & Doudna, 2012). In addition to providing a programmable nuclease, a nuclease-inactivated ("dead") version of Cas9 (dCas9) can be used as a delivery tool for a variety of cargoes to specific locations in the cell. For example, epigenetic effectors, fluorescent proteins, or transcription regulators can be delivered to any DNA location of interest using dCas9 delivery systems (Hilton et al., 2015; Kearns et al., 2015; Strutt et al., 2018). In 2016, Abudayyeh et al. reported the first single protein component RNA-targeting Cas protein of the Cas13 protein family (Abudayyeh et al., 2016). Analogous to Cas9, Cas13 can be inactivated by mutating the HEPN domains and easily targeted to different transcriptome locations using a guide RNA (gRNA), which like DNA-targeting systems, relies on Watson-Crick-Franklin base-pair interactions with target sequences. It has since been shown that Cas13 systems can be used to deliver fluorescent proteins, RNA editing enzymes, or splicing modulators to individual transcripts in cells (Abudayyeh et al., 2017; Cox et al., 2017; Konermann et al., 2018). Using this dCas13 delivery system, we have fused m^6A reader proteins to develop the first targeted m^6A reader proteins for epitranscriptome studies (Rauch, He, & Dickinson, 2018).

2.5 Targeted m^6A reader proteins

To provide the field with a tool to study the biological consequences of m^6A reader proteins, we leveraged the dCas13 RNA-targeting system to deliver the N-terminal, functional output domains of YTHDF1 and YTHDF2. The C-terminal domain of both these proteins contain the YTH-domain that recognized m^6A. By omitting this region, we aimed to simplify studies by developing methylation-independent tools. We reasoned that a targeted reader protein should be as small as possible to minimally perturb the cell. We furthermore showed that the N-terminal domain of the YTHDF proteins retain their function independent of the YTH domain. We therefore opted to fuse it to dCas13b as its sole delivery moiety. Additionally, while we fused the entire N-terminal domain of YTHDF1 to dCas13, we used a truncated version of YTHDF2 100–200 for our dCas13-Y2 construct because it had been previously identified as the subdomain interacting with the CCR4-NOT deadenylation machinery (Du et al., 2016). In our initial studies, we also identified a Y1 truncation that maintained its key translational activation properties (Rauch et al., 2018) (Fig. 2).

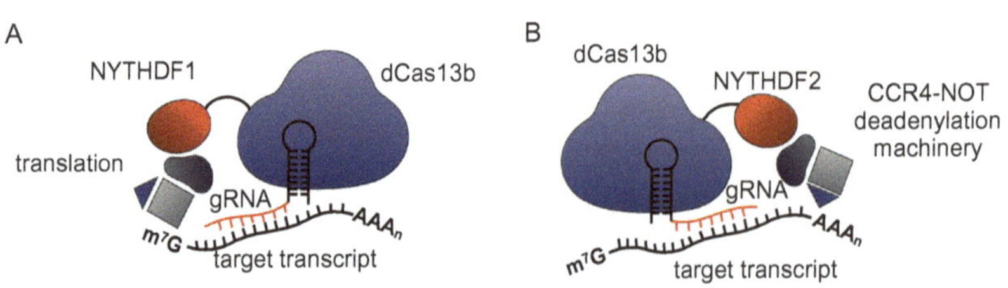

Fig. 2 Schematic of the targeted m^6A-reader proteins. (A) dCas13b-Y1 can be targeted to endogenous transcripts to interact with the translation machinery. (B) dCas13b-Y2 can be programmed to target RNA to induce mRNA degradation through the recruitment of the CCR4-NOT deadenylation machinery.

3. Design and applications of targeted RNA reader proteins

3.1 Equipment

Cell culture incubator (37 °C, 5% CO_2)
1. Cell culture biosafety cabinet
2. Plate reader (e.g., Biotek Synergy neo2 multimode or equivalent)
3. Benchtop centrifuge (e.g., Eppendorf or equivalent)
4. RT-qPCR machine (e.g., Roche LightCycler 96 or equivalent)

3.2 Materials

HEK293T cells (ATCC)
DMEM (L-glutamine, high glucose, sodium pyruvate, phenol red; Corning) supplemented with 10% fetal bovine serum (FBS, Gemini Benchmark), Penicillin/Streptomycin (P/S, Gibco/Life Technologies)
Opti-MEM I Reduced Serum Media (ThermoFisher Scientific)
Lipofectamine 2000 (Invitrogen)
Mammalian 96-well plates (Corning)
Mammalian 48-well plates (Corning)
Black 96-well plates (Corning)
Dual-Glo Luciferase Assay System (Promega)
RNeasy Mini kit (Qiagen)
PrimeScript RT Reagent Kit (Takara)
FastStart Essential DNA Green Master (Roche)
LightCycler 480 Multiwell Plate 96 (Roche)

3.3 Design of guide RNAs (gRNAs)

Designing gRNAs for the dCas13-Y1/2 proteins initially depends on the dCas13b protein's ability to bind to and access an RNA transcript. In our proof-of-principle experiments we chose substrates that have been previously targeted by different Cas13 proteins and identified gRNAs that have worked in the literature (Abudayyeh et al., 2017; Cox et al., 2017; Konermann et al., 2018). However, we found that our YTHDF fusions did not always have the same substantial effect at the sites the active Cas13 nuclease showed the largest effect. We therefore chose to pick several of the most promising RNA regions and rescreen our constructs.

In our experience, using a guiding region of 30 nucleotides was sufficient for successful RNA targeting. To target methylated sites in endogenous RNA, we suggest designing several gRNAs flanking the region of interest and testing those gRNAs both with active Cas13 nuclease as well as the YTHDF fusion proteins to ensure that the changes observed are a result of dCas13 targeting.

3.4 Mammalian dual luciferase assay to monitor protein levels

As a proof-of-concept and positive control experiment, we will first discuss the delivery and study of reader proteins to reporter genes. For these experiments, we will utilize the Promega dual luciferase reporter system, which permits facile analysis of protein levels by normalized luciferase activity measurements of the target firefly luciferase protein, relative to Renilla luciferase as a control. Firefly and Renilla luciferase have distinct evolutionary origins and share no sequence homology. Additionally, RT-qPCR analysis can be used to assess the relative RNA levels. For all experiments, multiple controls are needed. To assess on-target vs off-target activity, a control gRNA is needed in addition to the targeting gRNA. To assess effector-specific activity, an on-targeted gRNA is used in conjunction with a control fusion protein, such as the dCas13b system without any fusion protein.

3.4.1 *Transfection of targeted m^6A reader proteins*

All tissue culture steps are to be performed in a sterile biosafety cabinet. Cells are maintained in DMEM supplemented with 10% FBS and 1% penicillin/streptomycin and plated in DMEM with 10% FBS for transfections. Prior to transfection, all required DNA plasmids have to purified carefully to avoid any bacterial contamination in the tissue culture work.

1. To conduct a dual luciferase plate reader assay, plate HEK293T cells at 40% confluency in 100 μL media on a 96-well plate.
2. After 16–24 h, transfect each well with 150 ng dCas13b-Y1/Y2, 100 ng gRNA, and 12.5 ng luciferase reporter plasmids. First, dilute 0.5 μL of lipofectamine 2000 in 10 μL Opti-MEM. Then, dilute all plasmids for transfection in 10 μL Opti-MEM before combining the diluted DNA with the diluted transfection reagent.
3. Incubate the combined transfection mixture for 15 min at room temperature before gently pipetting onto the cells. To avoid experimental

variability, we recommend making a master mix for all three biological replicates before dividing it up into individual wells.

4. After 48 h of incubation at 37 °C with 5% CO_2, the cells are ready for the luciferase assays.

3.4.2 *Dual luciferase assay analysis of protein levels*

We found that the amount of dual luciferase reporter added to the system greatly influences the quality and reproducibility of the experiments. Therefore, some optimization might be required if cell lines other than HEK or HeLa cells are used. Verify that your cell culture plates are compatible with the plate reader first.

1. After 48 h of incubation, remove cells from the incubator.
2. For dual luciferase reads, we use the DualGlo Luciferase Assay System from Promega that enables the measurement of luciferase directly in the growth media. In the first step, firefly luciferase signal is measured followed by an addition of a combined lysis/firefly substrate buffer. Then, the substrate for the second, control Renilla luciferase is diluted in Stop&Glo buffer and added to the mixture to stop further luminescence readout from firefly and induce Renilla luminescence. Because an equal volume of growth media: substrate is required in the first step, we remove 60 μL prior to the start of this assay.
3. To each well, add 40 μL of reconstituted DualGlo Luciferase Reagent, mix well, and incubate for 30–40 min in the dark. Make sure to include control wells with no added luciferase reagent or luciferase reporter construct.
4. Measure the firefly luciferase readout on a plate reader (shake 5 s, then take read on luminescence setting).
5. Dilute the Renilla substrate in Stop&Glo buffer (1:100) and mix well. Add 40 μL of the Renilla substrate mixture to each well, mix by pipetting up and down, and incubate for at least 40 min. We found that in HEK293T cells, the Renilla signal is weaker than the firefly signal and takes longer to accumulate. Incubation times of less than 40 min yielded noisy, often inconclusive data for us.
6. Take the Renilla luciferase readout on the plate reader (shake 5 s, then take read on luminescence setting).
7. To analyze the dual luciferase reporter assay, take the ratio between the firefly luciferase luminescence and Renilla luminescence. Standard error can be calculated by dividing the standard deviation by the square root of the sample number. Changes in protein expression can be found by comparing a non-targeting gRNA and a targeting gRNA that has been co-transfected with the fusion protein of interest (Fig. 3).

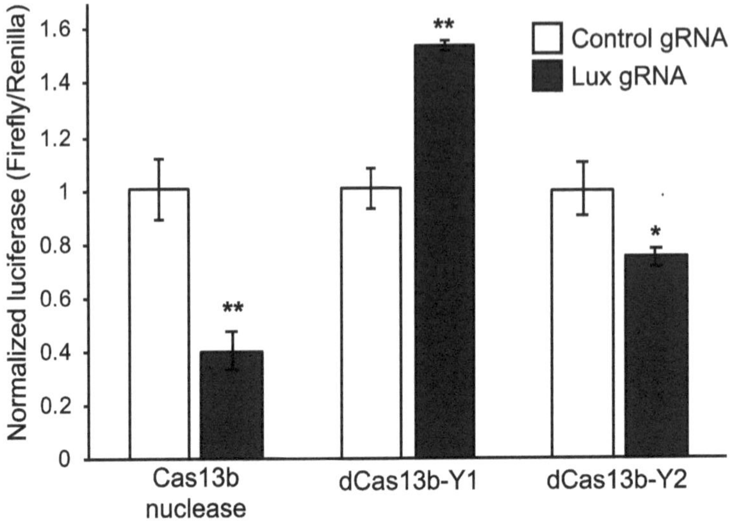

Fig. 3 Dual luciferase assay: HEK293T were transfected with Cas13b nuclease/dCas13b-Y1/dCas13b-Y2 fusion protein, off-target or firefly-luciferase targeting gRNA, and the dual luciferase reporter plasmid. Active Cas13b nuclease was used as an assay control. The luminescence readout of firefly luciferase was divided by the luminescence readout of Renilla luciferase. The off-target condition was normalized to 1 for easier visualization. Transfection with dCas13b-Y1 yielded an increase in luciferase signal, whereas transfection of cells dCas13b-Y2 resulted in decreased luciferase signal.

3.5 RT-qPCR to measure RNA levels

All previously described cell culture techniques should still be followed.

3.5.1 Transfection

1. To quantify mRNA levels after treatment with the fusion proteins, plate HEK293T cells at 40% confluency in 500 μL media on a 48-well plate.
2. After 16–24 h, transfect each well with 500 ng dCas13b-Y1/Y2, and 300 ng gRNA, (and 30 ng luciferase reporter plasmids where applicable). Dilute 2 μL of lipofectamine 2000 in 20 μL Opti-MEM and all plasmids together in 20 μL Opti-MEM before combining the diluted DNA with the diluted transfection reagent.
3. Incubate the combined transfection mixture for 15 min at room temperature before addition it to the cells.
4. After 48 h of incubation at 37 °C with 5% CO_2, the cells can be harvested for total RNA extraction.

3.5.2 RNA isolation

1. Purify total RNA using the Qiagen RNeasy Plus kit. In brief, lyse with 350 μL of RLT Plus reagent supplemented with 10 μL β-mercaptoethanol/1 mL lysis buffer.
2. Apply lysate to gDNA eliminator columns and spin for 30 s at 13,000 rpm.
3. Save the supernatant and add 350 μL of 70% EtOH before adding the mixture to the provided RNeasy spin columns. Centrifuge for 30 s at 13,000 rpm.
4. Wash the column with 700 μL RW1 buffer and centrifuge again. Discard the flow through.
5. Wash the column with 500 μL RPE buffer, centrifuge and repeat the wash step. Discard the flow through.
6. Spin at 13,000 rpm for an additional 2 min to dry the column completely.
7. Elute with 40 μL of water and continue to the reverse transcription step.

3.5.3 Reverse transcription (RT)

For our experiments, we generally use the PrimeScript RT Reagent Kit. The following steps are for the PrimeScript kit. However, other RT kits can easily be substituted by following the manufacturer's protocol to obtain cDNA libraries.

1. Measure the RNA concentration obtained from purification. Use 250 ng of purified total RNA for each reverse transcription reaction.
2. Follow the manufacturer's protocol to perform the RT reaction by adding:

   2 μL 5 × PrimeScript™ Buffer (1)
   0.5 μL PrimeScript™ RT Enzyme Mix I (2)
   0.5 μL Oligo dT Primers (3)
   0.5 μL Random 6mers (4)
   250 ng purified total RNA
   water to make a final volume of 10 μL

3. Run the RT reaction for 15 min at 37 °C, followed by a 5 s enzyme deactivation step at 85 °C

3.5.4 RT-qPCR

To conduct our RT-qPCR reactions, we use the Roche LightCycler 96 system. Make sure you carefully pick the appropriate housekeeping gene that should remain unchanged for your experiment to simplify subsequent analysis (Table 1).

Table 1 qPCR primers.

Gene name	Forward	Reverse
GAPDH	GTCTCCTCTGACTTCAACAGCG	ACCACCCTGTTGCTGTAGCCAA
KRAS	TGGTGGCTGATGCTTTGA	CACTGGATAGGGTTCTGTCTATTC
PPIB	AACGCAGGCAAAGACACCAACG	TCTGTCTTGGTGCTCTCCACCT

1. To prepare for the qPCR reaction, dilute the reverse-transcribed cDNA library 25-fold for all qPCR reactions. Dilute qPCR primers in pairs to a final concentration of 5 μM.
2. For each qPCR reaction, combine 3 μL of water, 2 μL of diluted primer pairs, and 10 μL of FastStart Essential DNA Green Master mix (Roche). Distribute across a 96-well Roche qPCR plate.
3. Add 5 μL of the diluted cDNA library to each reaction and run the qPCR using a standard qPCR protocol (Preincubation: 95 °C for 600 s; 3 Step Amplification: 95 °C for 20 s, 60 °C for 20 s, 72 °C for 20 s; Melting: 95 °C for 10 s, 65 °C for 60 s, 97 °C for 1 s (continuous)).
4. To analyze the resultant qPCR data, export the C_t values for each individual well. For every sample condition performed, a control qPCR for a housekeeping gene (e.g., GAPDH or ACTB) should be run. Note that for endogenous gene targeting, multiple housekeeping genes may be needed to accurately assess the effects of targeted readers on a given transcript, if the regulation of that transcript effects other RNAs, for example.
   - **a.** Subtract the C_t value of the housekeeping gene from the C_t value of the gene of interest (goi): $\Delta C_t = C_t(\text{goi}) — C_t(\text{housekeeping gene})$. This first step normalizes the RNA input level and will permit the future quantification of changes between different experimental conditions (for example, in this case, the comparison would likely be on- vs. off-target gRNA).
   - **b.** Average the biological replicates of the off-target condition. Then, calculate ΔΔCt = ΔCt (on-target condition)—ΔCt (off-target condition). ΔΔCt allows the comparison of RNA levels between on-target and off-target experimental conditions. At this point, the off-target condition is normalized to 1 and the on-target condition yields the relative change in RNA level when treated with the fusion protein (Fig. 4).

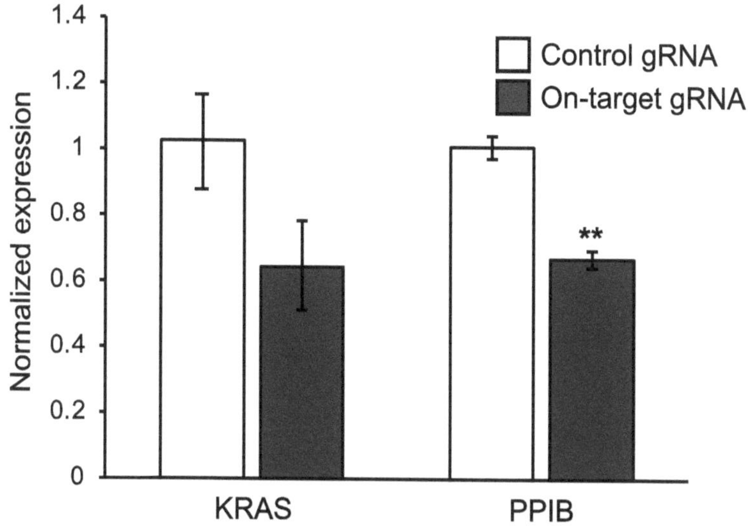

Fig. 4 Representative qPCR data after transfection with dCas13b-Y2 and either off-target or on-target gRNA targeting an endogenous RNA transcript (KRAS or PPIB). Normalized expression was calculated by taking a $\Delta\Delta C_t$. We first calculated ΔC_t of our target gene and GAPDH, followed by calculating the $\Delta\Delta C_t$ between the on-target and off-target conditions. We find that both the KRAS and PPIB RNA levels decrease after transfection with the dCas13b-Y2 construct and the on-target gRNA.

4. Summary and conclusions

Understanding RNA regulation of individual chemical modifications by the interplay between reader proteins is crucial to understanding how RNA modifications regulate a vast array of processes throughout the mammalian transcriptome. To provide investigators in the field with the tools to conduct site-specific experiments, we described the experimental procedure for RNA-guided RNA-targeted m^{6}A readers. This chapter describes the design rationale and the use of targeted m^{6}A-reader proteins to assess expression changes at both the RNA as well as protein level of individual transcripts. We describe the use of our targeted reader proteins and demonstrate how to use them to investigate consequences of reader proteins at individual modification sites in live cells. Due to Cas13's tolerance to many different fusion proteins, analogous protocols could be easily adapted to deploy a vast variety of targeted m^{6}A effector proteins including different readers, as well as writers and erasers in the future.

Acknowledgments

This work was supported by the University of Chicago, the National Institute of General Medical Sciences (R35 GM119840) and the National Human Genome Research Institute (RM1 HG008935) of the National Institutes of Health.

References

Abudayyeh, O. O., Gootenberg, J. S., Essletzbichler, P., Han, S., Joung, J., Belanto, J. J., et al. (2017). RNA targeting with CRISPR–Cas13. *Nature*, *550*, 280.

Abudayyeh, O. O., Gootenberg, J. S., Konermann, S., Joung, J., Slaymaker, I. M., Cox, D. B. T., et al. (2016). C2c2 is a single-component programmable RNA-guided RNA-targeting CRISPR effector. *Science*, *353*(6299), aaf5573.

Aguilo, F., Zhang, F., Sancho, A., Fidalgo, M., Di Cecilia, S., Vashisht, A., et al. (2015). Coordination of m6A mRNA methylation and gene transcription by ZFP217 regulates pluripotency and reprogramming. *Cell Stem Cell*, *17*(6), 689–704.

Anderson, P., & Kedersha, N. (2009). RNA granules: Post-transcriptional and epigenetic modulators of gene expression. *Nature Reviews. Molecular Cell Biology*, *10*, 430.

Cantara, W. A., Crain, P. F., Rozenski, J., McCloskey, J. A., Harris, K. A., Zhang, X., et al. (2011). The RNA modification database, RNAMDB: 2011 update. *Nucleic Acids Research*, *39*, D195–D201. Database issue.

Chen, M., & Manley, J. L. (2009). Mechanisms of alternative splicing regulation: Insights from molecular and genomics approaches. *Nature Reviews. Molecular Cell Biology*, *10*, 741.

Cox, D. B. T., Gootenberg, J. S., Abudayyeh, O. O., Franklin, B., Kellner, M. J., Joung, J., et al. (2017). RNA editing with CRISPR-Cas13. *Science*, *358*(6366), 1019–1027.

Dominissini, D., Moshitch-Moshkovitz, S., Salmon-Divon, M., Amariglio, N., & Rechavi, G. (2013). Transcriptome-wide mapping of N(6)-methyladenosine by m(6) A-seq based on immunocapturing and massively parallel sequencing. *Nature Protocols*, *8*(1), 176–189.

Dominissini, D., Moshitch-Moshkovitz, S., Schwartz, S., Salmon-Divon, M., Ungar, L., Osenberg, S., et al. (2012). Topology of the human and mouse m6A RNA methylomes revealed by m6A-seq. *Nature*, *485*, 201.

Du, H., Zhao, Y., He, J., Zhang, Y., Xi, H., Liu, M., et al. (2016). YTHDF2 destabilizes m6A-containing RNA through direct recruitment of the CCR4–NOT deadenylase complex. *Nature Communications*, 7, 12626.

Filipowicz, W., Bhattacharyya, S. N., & Sonenberg, N. (2008). Mechanisms of post-transcriptional regulation by microRNAs: Are the answers in sight? *Nature Reviews. Genetics*, *9*(2), 102–114.

Glisovic, T., Bachorik Jennifer, L., Yong, J., & Dreyfuss, G. (2008). RNA-binding proteins and post-transcriptional gene regulation. *FEBS Letters*, *582*(14), 1977–1986.

Gokhale, N. S., McIntyre, A. B. R., McFadden, M. J., Roder, A. E., Kennedy, E. M., Gandara, J. A., et al. (2016). N6-methyladenosine in flaviviridae viral RNA genomes regulates infection. *Cell Host & Microbe*, *20*(5), 654–665.

Hilton, I. B., D'Ippolito, A. M., Vockley, C. M., Thakore, P. I., Crawford, G. E., Reddy, T. E., et al. (2015). Epigenome editing by a CRISPR-Cas9-based acetyltransferase activates genes from promoters and enhancers. *Nature Biotechnology*, *33*, 510.

Ivanova, I., Much, C., Di Giacomo, M., Azzi, C., Morgan, M., Moreira, P. N., et al. (2017). The RNA m6A reader YTHDF2 is essential for the post-transcriptional regulation of the maternal transcriptome and oocyte competence. *Molecular Cell*, *67*(6), 1059–1067.e4.

Karijolich, J., Kantartzis, A., & Yu, Y.-T. (2010). RNA modifications: A mechanism that modulates gene expression. In M. Sioud (Ed.), *RNA therapeutics: function, design, and delivery* (pp. 1–19). Totowa, NJ: Humana Press.

Kearns, N. A., Pham, H., Tabak, B., Genga, R. M., Silverstein, N. J., Garber, M., et al. (2015). Functional annotation of native enhancers with a Cas9–histone demethylase fusion. *Nature Methods*, *12*, 401.

Kennedy, E. M., Bogerd, H. P., Kornepati, A. V. R., Kang, D., Ghoshal, D., Marshall, J. B., et al. (2016). Posttranscriptional m6A editing of HIV-1 mRNAs enhances viral gene expression. *Cell Host & Microbe*, *19*(5), 675–685.

Kirchner, S., & Ignatova, Z. (2015). Emerging roles of tRNA in adaptive translation, signalling dynamics and disease. *Nature Reviews. Genetics*, *16*(2), 98–112.

Konermann, S., Lotfy, P., Brideau, N. J., Oki, J., Shokhirev, M. N., & Hsu, P. D. (2018). Transcriptome engineering with RNA-targeting type VI-D CRISPR effectors. *Cell*, *173*(3), 665–676.e14.

Lee, Y., & Rio, D. C. (2015). Mechanisms and regulation of alternative pre-mRNA splicing. *Annual Review of Biochemistry*, *84*(1), 291–323.

Li, S., & Mason, C. E. (2014). The pivotal regulatory landscape of RNA modifications. *Annual Review of Genomics and Human Genetics*, *15*(1), 127–150.

Li, Z., Weng, H., Su, R., Weng, X., Zuo, Z., Li, C., et al. (2017). FTO plays an oncogenic role in acute myeloid leukemia as a N6-Methyladenosine RNA demethylase. *Cancer Cell*, *31*(1), 127–141.

Lichinchi, G., Zhao, B. S., Wu, Y., Lu, Z., Qin, Y., He, C., et al. (2016). Dynamics of human and viral RNA methylation during Zika virus infection. *Cell Host & Microbe*, *20*(5), 666–673.

Lim, C., & Allada, R. (2013). Emerging roles for post-transcriptional regulation in circadian clocks. *Nature Neuroscience*, *16*, 1544.

Linder, B., Grozhik, A. V., Olarerin-George, A. O., Meydan, C., Mason, C. E., & Jaffrey, S. R. (2015). Single-nucleotide-resolution mapping of m6A and m6Am throughout the transcriptome. *Nature Methods*, *12*, 767.

Lorenz, C., Lunse, C. E., & Morl, M. (2017). tRNA modifications: Impact on structure and thermal adaptation. *Biomolecules*, 7(2), E35.

O'Connell, M. R., Oakes, B. L., Sternberg, S. H., East-Seletsky, A., Kaplan, M., & Doudna, J. A. (2014). Programmable RNA recognition and cleavage by CRISPR/Cas9. *Nature*, *516*, 263.

Patil, D. P., Pickering, B. F., & Jaffrey, S. R. (2018). Reading m6A in the transcriptome: m6A-binding proteins. *Trends in Cell Biology*, *28*(2), 113–127.

Rauch, S., He, C., & Dickinson, B. C. (2018). Targeted m(6)a reader proteins to study epi-transcriptomic regulation of single RNAs. *Journal of the American Chemical Society*, *140*(38), 11974–11981.

Roundtree, I. A., Evans, M. E., Pan, T., & He, C. (2017). Dynamic RNA modifications in gene expression regulation. *Cell*, *169*(7), 1187–1200.

Schwartz, S., Mumbach, M. R., Jovanovic, M., Wang, T., Maciag, K., Bushkin, G. G., et al. (2014). Perturbation of m6A writers reveals two distinct classes of mRNA methylation at internal and 5′ sites. *Cell Reports*, *8*(1), 284–296.

Shi, H., Wang, X., Lu, Z., Zhao, B. S., Ma, H., Hsu, P. J., et al. (2017). YTHDF3 facilitates translation and decay of N(6)-methyladenosine-modified RNA. *Cell Research*, *27*(3), 315–328.

Song, J., & Yi, C. (2017). Chemical modifications to RNA: A new layer of gene expression regulation. *ACS Chemical Biology*, *12*(2), 316–325.

Strutt, S. C., Torrez, R. M., Kaya, E., Negrete, O. A., & Doudna, J. A. (2018). RNA-dependent RNA targeting by CRISPR-Cas9. *eLife*, 7, e32724.

Tirumuru, N., Zhao, B. S., Lu, W., Lu, Z., He, C., & Wu, L. (2016). N6-methyladenosine of HIV-1 RNA regulates viral infection and HIV-1 gag protein expression. *eLife*, *5*, e15528.

Wang, X., & He, C. (2014). Dynamic RNA modifications in posttranscriptional regulation. *Molecular Cell*, *56*(1), 5–12.

Wang, X., Lu, Z., Gomez, A., Hon, G. C., Yue, Y., Han, D., et al. (2013). N6-methyladenosine-dependent regulation of messenger RNA stability. *Nature*, *505*, 117.

Wang, S., Sun, C., Li, J., Zhang, E., Ma, Z., Xu, W., et al. (2017). Roles of RNA methylation by means of N6-methyladenosine (m6A) in human cancers. *Cancer Letters*, *408*, 112–120.

Wang, X., Zhao, B. S., Roundtree, I. A., Lu, Z., Han, D., Ma, H., et al. (2015). N6-methyladenosine modulates messenger RNA translation efficiency. *Cell*, *161*(6), 1388–1399.

Wen, J., Lv, R., Ma, H., Shen, H., He, C., Wang, J., et al. (2018). Zc3h13 regulates nuclear RNA m6A methylation and mouse embryonic stem cell self-renewal. *Molecular Cell*, *69*(6), 1028–1038.e6.

Wiedenheft, B., Sternberg, S. H., & Doudna, J. A. (2012). RNA-guided genetic silencing systems in bacteria and archaea. *Nature*, *482*, 331.

Xu, C., Liu, K., Ahmed, H., Loppnau, P., Schapira, M., & Min, J. (2015). Structural basis for the discriminative recognition of N6-methyladenosine RNA by the human YT521-B homology domain family of proteins. *Journal of Biological Chemistry*, *290*(41), 24902–24913.

Yu, Y. T., Shu, M. D., & Steitz, J. A. (1997). A new method for detecting sites of 2'-O-methylation in RNA molecules. *RNA*, *3*(3), 324–331.

Yue, Y., Liu, J., Cui, X., Cao, J., Luo, G., Zhang, Z., et al. (2018). VIRMA mediates preferential m6A mRNA methylation in 3′UTR and near stop codon and associates with alternative polyadenylation. *Cell Discovery*, *4*(1), 10.

Zhang, S., Zhao, B. S., Zhou, A., Lin, K., Zheng, S., Lu, Z., et al. (2017). m6A demethylase ALKBH5 maintains tumorigenicity of glioblastoma stem-like cells by sustaining FOXM1 expression and cell proliferation program. *Cancer Cell*, *31*(4), 591–606.e6.

Zhao, B. S., Roundtree, I. A., & He, C. (2017). Post-transcriptional gene regulation by mRNA modifications. *Nature Reviews. Molecular Cell Biology*, *18*(1), 31–42.

Zheng, G., Dahl, J. A., Niu, Y., Fedorcsak, P., Huang, C. M., Li, C. J., et al. (2013). ALKBH5 is a mammalian RNA demethylase that impacts RNA metabolism and mouse fertility. *Molecular Cell*, *49*(1), 18–29.

Zhou, K. I., & Pan, T. (2018). An additional class of m6A readers. *Nature Cell Biology*, *20*(3), 230–232.

CHAPTER TWO

Synthetic gene expression circuits regulating sexual reproduction

Nobuo Fukuda*, Shinya Honda
Biomedical Research Institute, National Institute of Advanced Industrial Science and Technology (AIST), Tsukuba, Japan
*Corresponding author: e-mail address: nob-fukuda@aist.go.jp

Contents

Abstract

The budding yeast *Saccharomyces cerevisiae* has been widely utilized in fermentative production since ancient times. Several approaches for modification of yeast traits have been developed, including mutagenesis, protoplast fusion, and genetic modification. Crossbreeding provides an attractive means to improve and combine strain traits based on sexual reproduction. Common crossbreeding strategies require the isolation of *MAT**a*** and *MAT*α haploids via sporulation, as most of parental yeasts are *MAT**a**/*α diploids and unable to mate directly. Unfortunately, many yeast strains used in industry exhibit low sporulation rates resulting in limited crossbreeding efficiency and numerous technical challenges. Here, we review the construction of synthetic gene expression circuits as a means to provide alternative methods for sporulation for yeast crossbreeding. These methods enable researchers to convert the sequence of the *MAT* locus and subsequently acquire crossbreds via mating of isolated yeast strains. The purpose of this chapter is to provide a basic guide for researchers who are attempting to expand the variety of yeast resources using the sexual reproduction machinery of yeast.

Methods in Enzymology, Volume 621
ISSN 0076-6879
https://doi.org/10.1016/bs.mie.2019.02.036

1. Introduction

Fermentation of carbohydrate sources into ethanol using yeast is one of the oldest human technologies. The earliest evidence for fermented beverage production dates back to the Neolithic period (Duan et al., 2018), and currently yeasts are still used for many essential biotechnological processes including the production of beer, wine, and biofuels. *Saccharomyces cerevisiae* is one of the most extensively characterized yeast species, and its fast growth, well-developed genetics, and robustness in large scale fermentations have made it the preferred host for many bioprocesses (Fukuda & Honda, 2018). Although it has been approximately two decades since the genome of the S288C laboratory strain (Brachmann et al., 1998) of *S. cerevisiae* was sequenced, the complex regulation of gene expression underlying alcoholic fermentation remains poorly understood (Dashko, Zhou, Compagno, & Piškur, 2014). To aid in this understanding, crossbreeding provides an attractive approach that allows researchers to use sexual reproduction to generate novel yeast strains exhibiting combined preferred characteristics of multiple parental strains. Unfortunately, however, classical crossbreeding can be problematic as many yeast strains exhibiting excellent fermentation characteristics possess poor sexual reproduction ability. Here, we highlight improved crossbreeding technologies and provide a practical guide that will allow researchers to cross any available yeast strains.

2. Control of sexual reproduction

2.1 Mating type and gene regulation machinery

Yeasts exhibit asexual and sexual reproductive cycles, and the most common mode of vegetative growth is asexual reproduction by budding. While haploid cells will die under high-stress conditions such as nutrient starvation, diploid cells can enter sexual reproduction to endure harsh environmental conditions (Fig. 1). A variety of haploid spores are produced during meiosis, and these cells mate with each other to reform diploid cells following spore germination.

Haploid cells exist as one of two mating types, **a** or α, due to differences in the genetic composition at the *MA*ting *T*ype (*MAT*) locus (Herskowitz, 1983). Mating of opposite mating-type haploid cells (*MAT**a*** and *MATα*) results in the formation of a new and heterozygous diploid strain (*MAT**a**/α*).

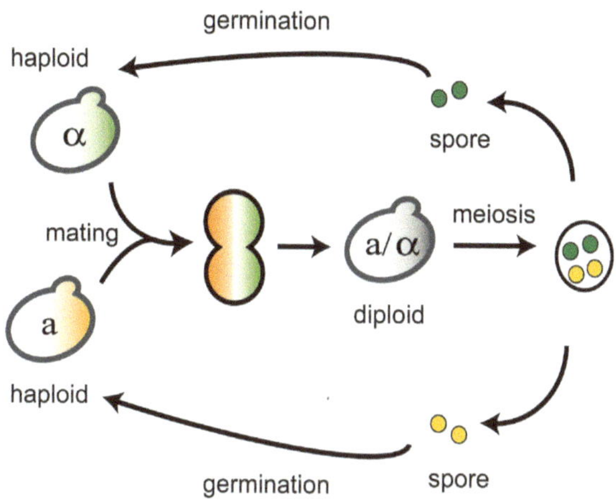

Fig. 1 Sexual reproduction within the life cycle of budding yeasts.

Similar to the phenotypic differences between haploid **a** and α cells, different gene expression regulation patterns are responsible for the phenotypic differences observed between haploid and *MAT**a**/α* diploid cells (Fig. 2). *MAT**a*** haploids express the genes ***a**1* from the *MAT* locus whereas *MATα* haploids express the genes *α1* and *α2* from that same locus (Haber, 1998). Genes involved in mating are divided into three categories that include **a**-type-specific genes (***a**sg*), α-type-specific genes (*αsg*), and haploid-specific genes (*hsg*). As shown in Fig. 2, the synthesized protein α1 functions as the activator of *αsg*, and α2 functions as a repressor of *asg*. While the synthesized **a**1 protein does not influence the expression of genes involved in mating, the formation of the **a**1-α2 complex does repress the expression of *α1* and *hsg* in *MAT**a**/α* diploid cells.

*MAT**a**/α* diploid cells can undergo meiosis under starvation conditions; however, *MAT**a*** and *MATα* haploid cells cannot undergo meiosis when exposed to an identical environment. This difference is a result of differential gene regulation by the **a**1-α2 complex (Fig. 3). Entry into the meiotic is regulated by the expression of the *I*nducer of *ME*iosis 1 (*IME1*) gene, commonly known as the master initiator of meiosis (Gelfand et al., 2011). *IME1* expression is activated by starvation signaling and repressed by the gene product of *R*egulator of *ME*iosis 1 (*RME1*). In both *MAT**a*** and *MATα* haploid cells, the absence of the **a**1-α2 complex induces expression of the *RME1* gene. As expression of the *IME1* gene is repressed by the gene product of *RME1*, meiosis is not induced by starvation signaling in haploid cells. In contrast, the **a**1-α2 complex directly represses expression of the

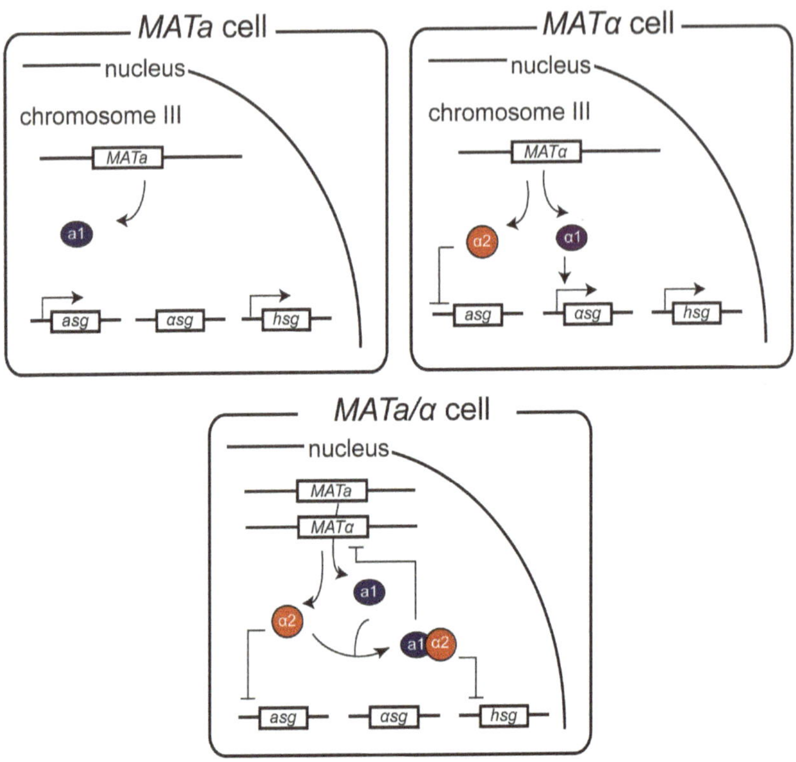

Fig. 2 Mating type-dependent expression of genes required for yeast mating.

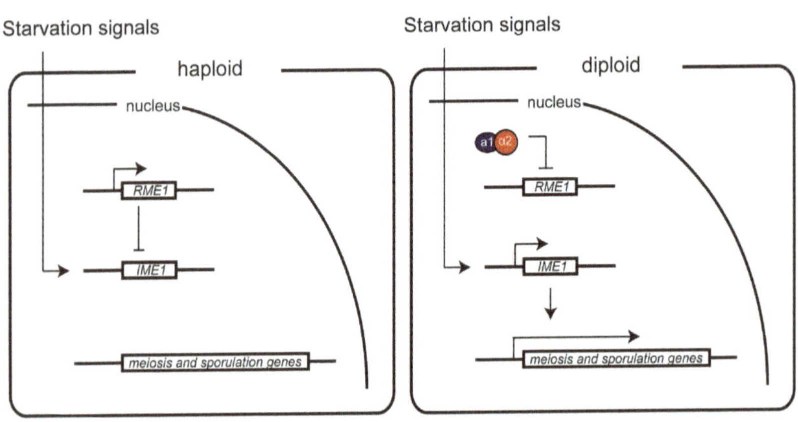

Fig. 3 Regulation of meiosis and sporulation genes.

RME1 gene, and the gene product of *IME1* activates the meiosis pathway in **a**/α-type diploid cells in response to starvation. Again, it should be noted that the **a**1-α2 complex is the most important factor distinguishing haploid (**a** and α) from diploid (**a**/α) cells.

2.2 Irregular means of sexual reproduction

Many industrial *S. cerevisiae* strains are known to exhibit poor sexual reproduction ability. Previously, it was reported (Snoek et al., 2015) that >80% of the lager, >60% of sake, and >50% of ale strains did not undergo sporulation, indicating that these strains are not suitable for classical crossbreeding. Additionally, the majority of the strains exhibiting sporulation ability were found to be homothallic yeasts, where the Ho endonuclease mediates bidirectional replacement of DNA sequences at the *MAT* locus (Fig. 4). In haploid cells, mating type switching is induced by the *HO* gene encoding Ho endonuclease belonging to the *hsg* which is repressed by the **a**1-α2 complex. After cleavage by Ho endonuclease, DNA repair proceeds using mating type sequences derived from one of two silent donor loci, *HMR**a*** and *HMLα*. When using the classical approach to crossbreeding of homothallic yeasts, undesirable inbreeding between mother and daughter cells (autopolyploidization) inevitably occurs even after tetrad dissection. This then impedes subsequent crossbreeding.

Interestingly, there are several irregular sexual reproduction routes that are thought to be derived from chromosomal aberrations occurring during mitotic division (Fig. 5). The *MAT* genes are located on chromosome III

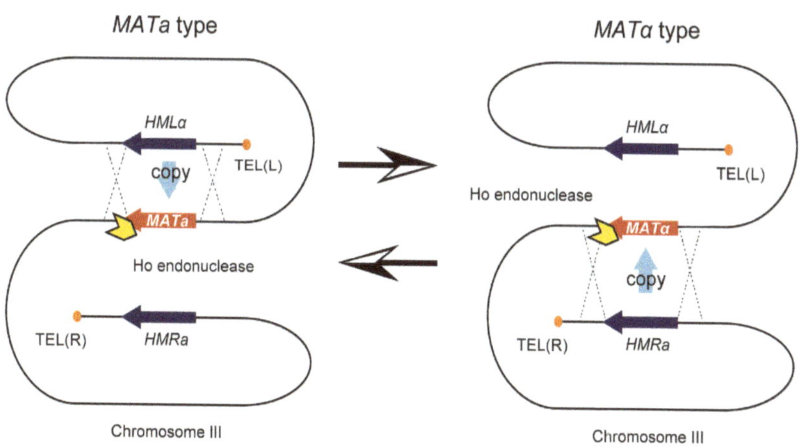

Fig. 4 Mating type switching triggered by Ho endonuclease.

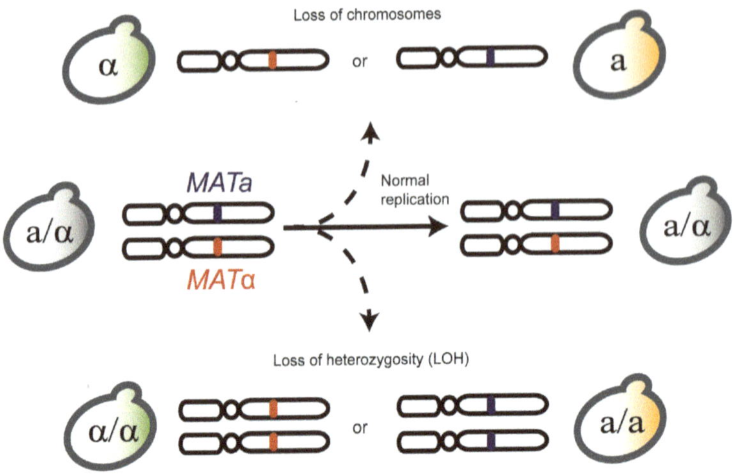

Fig. 5 Chromosomal aberrations occurring via mitotic division.

that is the most unstable of the 16 *S. cerevisiae* chromosomes (Kumaran, Yang, & Leu, 2013). There are several kinds of chromosomal aberration such as loss of heterozygosity (LOH) and mitotic chromosome loss. LOH is a naturally occurring process that generates homozygous loci via chromosomal rearrangement of heterozygous loci (Andersen, Nelson, Hetrick, & Gottschling, 2008). LOH occurring at the *MAT* loci within *MAT**a**/α* cells produces either a *MAT**a**/**a*** or a *MATα/α* cell, and these cells possess the same mating ability as either *MAT**a*** and *MATα* haploids generated via sporulation, respectively. LOH occurs at a spontaneous frequency of less than 1×10^{-4} (Hiraoka, Watanabe, Umezu, & Maki, 2000). Mitotic chromosome loss is also a naturally-occurring event where polyploid cells lose single or multiple chromosomes (Mayer & Aguilera, 1990). Loss of chromosome III frequency in yeast diploid cells has been reported at 5×10^{-5} (Kumaran et al., 2013), and yeast cells having lost one of the two copies of chromosome III containing the *MAT**a*** or *MATα* gene acquire either **a**- or α-type mating ability. While the spontaneous frequencies of such events are quite low, polyploids and aneuploids can be naturally generated not by sporulation but by chromosomal aberration during mitotic division.

2.3 Synthetic gene expression circuit for crossbreeding

Here, we describe two methods for crossbreeding of yeasts exhibiting sporulation deficiency. To convert yeast mating types, we utilize Ho endonuclease or chromosomal aberration in **a**/α-type polyploids.

Expression of the *HO* gene in *MAT***a**/α diploid cells induces replacement of **a** or α DNA sequences at the *MAT* loci and generates *MAT***a**/**a** and *MAT*α/α diploid cells. These cells possess the ability to mate with each other in nature. To avoid autopolyploidization and to isolate these mating-competent cells, the **a**1-α2 complex is artificially formed by introducing synthetic gene expression circuits (Fukuda, Matsukura, & Honda, 2013) into parental diploid cells (Fig.6). *MAT***a**/**a** or *MAT*α/α diploids serve as target cells and are easily identified and isolated using fluorescence intensity. The introduced circuits can be rendered non-functional in yeast cells by the removal of episomal vectors. Prior to employing this method, researchers must note "stuck" (uncuttable) mutations (Ray, White, & Haber, 1991). If these mutations exist at the *MAT* loci of parental diploid cells, Ho endonuclease cannot mediate mating type conversion, and researchers must utilize an alternative approach.

Spontaneous chromosomal aberrations occurring during cultivation result in low numbers of both **a**- and α-type mating-competent cells within cultures containing large numbers of parental *MAT***a**/α cells. The inclusion of the **a**1-α2 complex suppresses mating of the rare **a**- and α-type cells, and this allows these cells to grow independently in the cell mixture without forming nonhybrid (autopolyploid) cells (Fig. 7). Here, the expression of the marker gene is induced in a mating type-specific manner (Fukuda & Honda, 2013). This permits the efficient selection of **a**- or α-type cells from within a mixed cell population. Unlike auxotrophic laboratory yeasts, industrially used yeasts are generally prototrophic, which prevents the use of auxotrophic marker genes. Given this, antibiotic resistance marker genes should be prepared for the transformation of parental cells and subsequent isolation of **a**- and α-type derivative and hybrid cells (Fukuda, Kaishima, Ishii, Kondo, & Honda, 2016). Synthetic gene expression circuits can be inactivated by the removal of episomal vectors.

Industrial yeast strains possess remarkably low genetic transformation efficiencies compared to those of laboratory strains. For example, sake yeasts display a transformation efficiency of 10^1–10^2 cfu/μg-DNA while this value is 10^4–10^5 cfu/μg-DNA in laboratory yeast (Ogata, Okumura, Tadenuma, & Tamura, 1993). As a result of these known differences, we recommend the use of 1–10 μg of expression vectors for transforming industrial yeasts. It should be noted that throughout the entire genome of crossbreeds there is no detection of exogenous DNA after removal of episomal vectors. This allows many valuable yeast resources to be efficiently utilized for the generation of new strains exhibiting desirable properties for industrial applications.

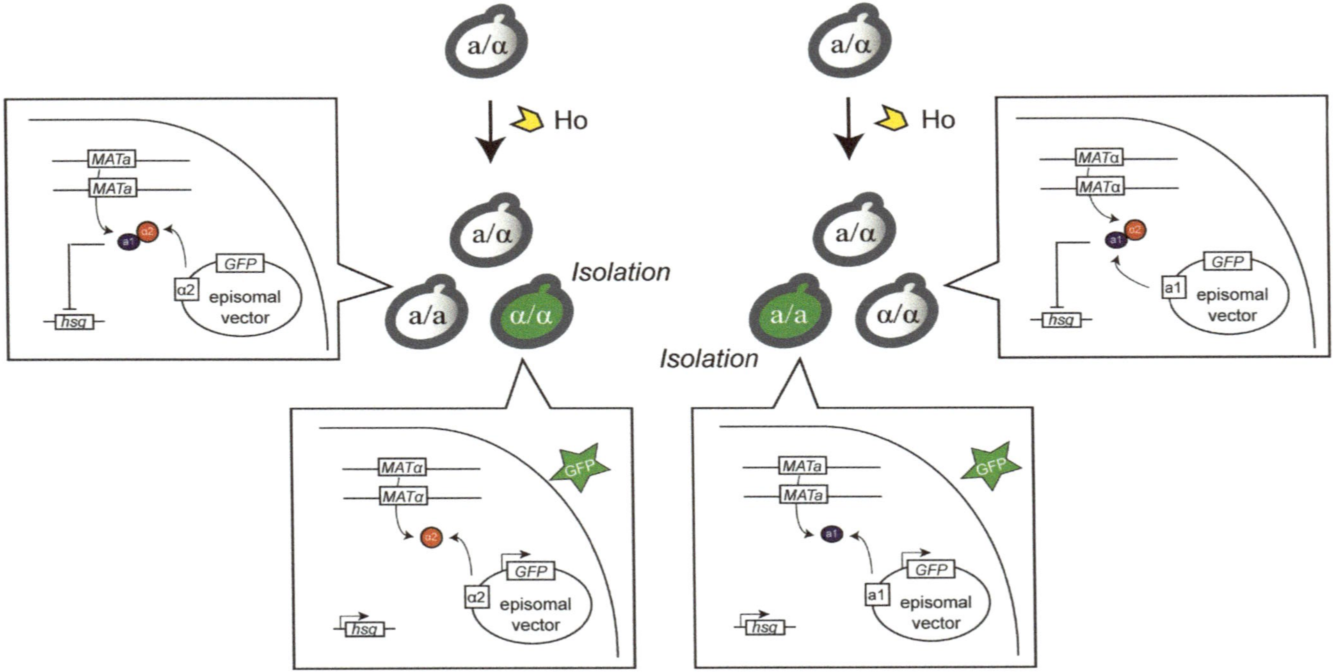

Fig. 6 Schematic outline of artificial mating type conversion using Ho endonuclease.

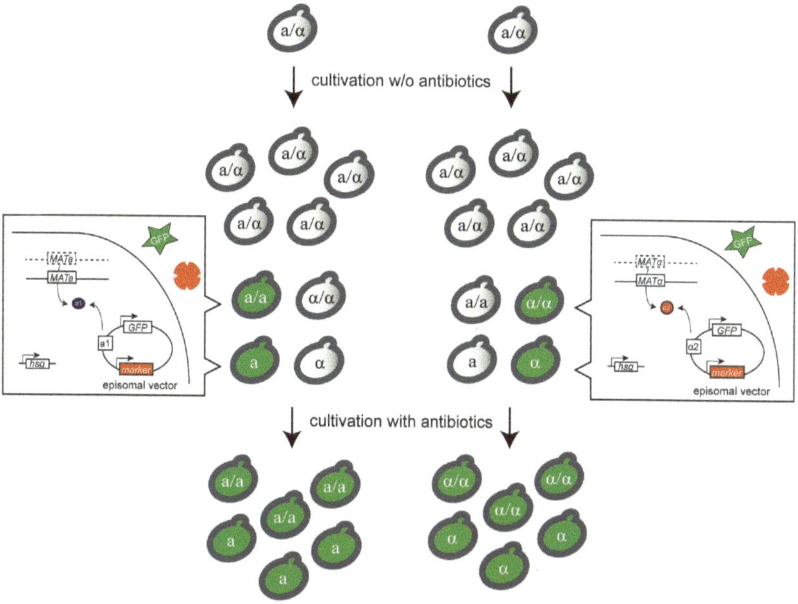

Fig. 7 Schematic outline of screening of mating competent cells from mixed cell populations.

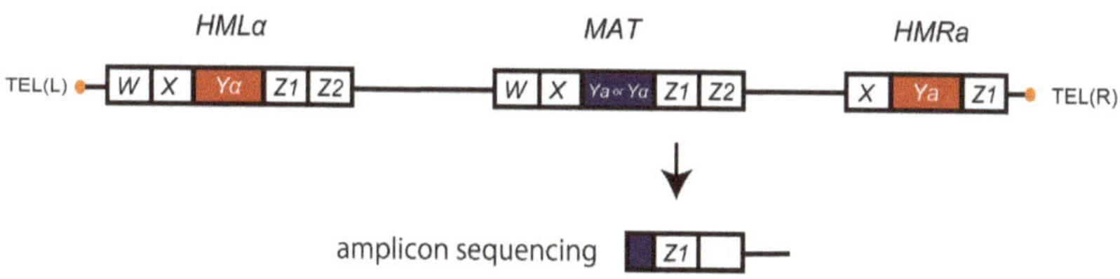

Fig. 8 The Z1 region of the *MAT* loci targeted by Ho endonuclease.

3. Experimental design and protocol

3.1 Before you begin: Search for *MAT* gene sequences

First, method-1 (Fig. 6) or method-2 (Fig. 7) must be selected for mating type conversion of parental yeast cells. To confirm the presence or absence of "stuck" mutations, the sequence of the *MAT* gene sequences of target cells should be confirmed informatically or experimentally. Amplicon sequencing is a useful means for experimentally determining the *MAT* gene sequences. As shown in Fig. 8, the *MAT* locus contains W, X, Y (**a** or α), Z1, and Z2 regions as well as the *HMLα* locus. Conversely, *HMR**a*** contains the X, Y**a,** and Z1 regions (Strathern et al., 1982). The nucleotide sequences recognized by Ho endonuclease reside within the Z1 regions of the *MAT* loci (Ray et al., 1991), and DNA fragments should be amplified with primer sets annealing to the Y (**a** or α) region and downstream of the *MAT* loci. Method-1 is preferable when the start of the Z1 region can be cleaved by the Ho endonuclease (sequence: CGCAACAGTATAATTTTATAA), otherwise method-2 must be utilized. If the initial confirmation process will not be used, we recommend using method-2 for converting the mating type of parental yeast cells.

3.2 Design, construction, and introduction of episomal vectors

In both method-1 and -2, selection marker genes are required for introduction of the constructed episomal vectors into yeast cells (Fig. 9). For transformation of prototrophic yeast strains, *kanMX4*, *hygro*, and *AUR-1C* genes can be used as antibiotic resistance selection markers (Fukuda et al., 2016). Additionally, a selection marker gene regulated by the ***a**sg* ($P_{\mathbf{a}sg}$) or *αsg* ($P_{\alpha sg}$) promotor is required for isolation of **a**- or α-type mating-competent cells using method-2 (Fig. 7). The promoters of the *STE2* and *STE3* genes provide examples of $P_{\mathbf{a}sg}$ and $P_{\alpha sg}$, respectively (Fukuda & Honda, 2013). Additional information regarding ***a**sg* and *αsg* can be obtained from a previous report (Botstein, 2004) and the Saccharomyces Genome Database (https://www.yeastgenome.org/).

For method-1 (Fig. 6), *HO* gene expression should be regulated by an inducible promoter such as P_{GAL1} to strictly control activation and inactivation of gene expression (Fukuda, Kaishima, Ishii, & Honda, 2017). To prevent autopolyploidization in mating type conversion triggered by Ho

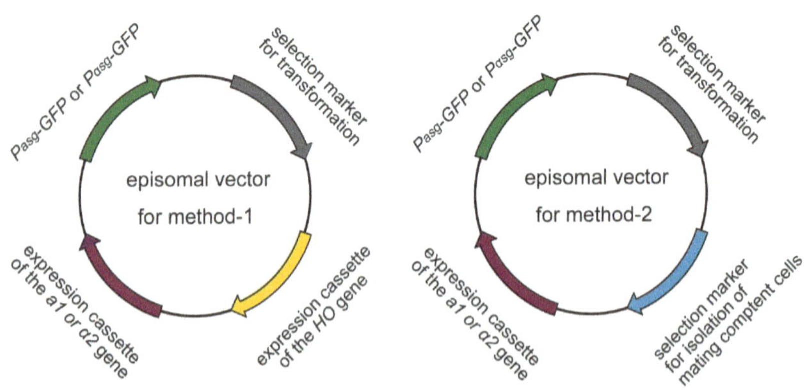

Fig. 9 Constitution of episomal vectors.

endonuclease, the **a**1-α2 complex must be formed in non-target yeast cells. The expression of the *a1* gene must be regulated by a strong constitutive promoter such as P_{PGK1}, whereas P_{asg} should be utilized for expression of the *α2* gene (Fukuda et al., 2013). Expression cassettes containing the *GFP* gene under the control of P_{asg} or $P_{\alpha sg}$ are not required, but their use is recommended to distinguish yeast mating types. Similarly, for method-2, expression of the **a**1 or α2 genes and use of GFP is optional but recommended.

Constructed vectors can be introduced into yeast cells (Fig. 10) using the lithium acetate method (Gietz, St. Jean, Woods, & Schiestl, 1992), which is one of most popular yeast transformation methods. Before selection on solid medium, we strongly recommend short cultivation of the treated cells in the absence of selection pressure to improve the viability of transformed cells.

3.3 Techniques for acquiring mating-competent yeast cells

In method-1, parental yeast cells are grown at 30 °C overnight in cultivation medium spiked with inducers of *HO* gene expression (Fig. 6). The

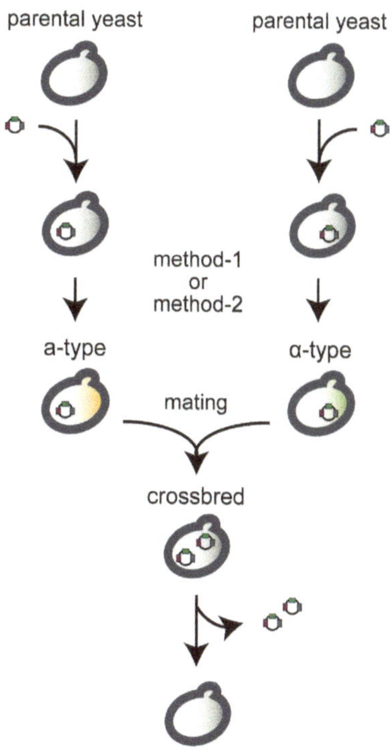

Fig. 10 Schematic overview of yeast crossbreeding described in the current chapter.

cultivated cells are then harvested, washed, and resuspended in distilled water to an optical density of 1.0×10^{-5} at 600 nm ($OD_{600} = 1.0 \times 10^{-5}$). Cell suspensions (approximately 500 μL) are spread on solid medium lacking inducers to terminate mating type conversion. After incubation at 30 °C for 2 days, colonies exhibiting green fluorescence are isolated as the target mating-competent cells. We recommend selecting colonies exhibiting high fluorescence intensity. If required, single colony isolation can be repeated to exclude off-target cells generated after removal of inducers.

For method-2, parental yeast cells are grown at 30 °C overnight in cultivation medium in the absence of antibiotics (Fig. 7). The cultivated cells are then harvested, washed, and resuspended in distilled water to an OD_{600} of 1.0. Cell suspensions (ranging from 100 to 500 μL) are spread on solid medium containing antibiotics to screen for mating-competent cells. After incubation at 30 °C for 2 days, fluorescence intensity of the generated colonies should be measured for confirmation of mating type.

3.4 Generation and isolation of crossbreds

For the isolation of crossbreds, **a**- and α-type cells must express different selection marker genes. If they do not, episomal vector exchange must be performed (Fukuda et al., 2016) as described in "3.5 Removal of Episomal Vectors." After preparation, **a**- and α-type cells are mixed and grown in cultivation medium at 30 °C for 1.5 h, to an initial OD_{600} of 0.1 for each cell type. After cultivation, yeast cells are harvested, washed, and resuspended in distilled water to an OD_{600} of 0.01. Cell suspensions (ranging from 100 to 500 μL) are spread on solid medium containing two types of antibiotics corresponding to selection marker genes. Crossbreds that have acquired two types of episomal vectors can be selected from this media (Fig. 10).

3.5 Removal of episomal vectors

Episomal vectors can be removed from yeast cell populations by repeated cultivation in medium lacking selective agents. Yeast cells containing episomal vectors are grown in cultivation medium at 30 °C, and subsequently a portion of the culture fluid is transferred into fresh medium. Cultivation time to reach growth plateau is predicable based on the doubling time (<2 h) of yeast cells (Salari & Salari, 2017). Researchers can perform four culture passages in 2 days by modulating initial OD_{600}. After cultivation in this manner, yeast cells are harvested, washed and resuspended in distilled water to an OD_{600} of 1.0×10^{-5}. Cell suspensions (approximately 500 μL) are spread on solid medium lacking antibiotics and incubated at 30 °C for 2 days. To confirm removal of episomal vectors, replicas of the generated colonies should be monitored for growth inhibition in medium containing each antibiotic. Absence of episomal vectors can also be confirmed by PCR specifically targeting selection marker genes (Fukuda et al., 2016).

4. Conclusions

There are numerous valuable yeast resources that cannot be fully exploited for crossbreeding. In this chapter, we have demonstrated methods to generate mating-competent cells from parental yeasts exhibiting severe deficiencies in sporulation. Importantly, all genetic modifications described in these methods can be introduced into yeast cells using episomal vectors. This permits complete removal of the modifications following isolation of the crossbreds. Crossbreeding induces genomic shuffling based on mechanisms of sexual reproduction, and the resulting derivative strains may therefore acquire phenotypes different from those of the parental strains. These methods will allow researcher to utilize unused resources by facilitating the generation of new yeast strains exhibiting desirable properties for industrial applications.

Acknowledgments

This work was supported by JSPS KAKENHI Grant Number 25820406, 16K14497 and 18K05424.

References

Andersen, M. P., Nelson, Z. W., Hetrick, E. D., & Gottschling, D. E. (2008). A genetic screen for increased loss of heterozygosity in *Saccharomyces cerevisiae*. *Genetics*, *179*, 1179–1195.

Botstein, D. (2004). Ira herskowitz: 1946-2003. *Genetics*, *166*, 653–660.

Brachmann, C. B., Davies, A., Cost, G. J., Caputo, E., Li, J., Hieter, P., et al. (1998). Designer deletion strains derived from *Saccharomyces cerevisiae* S288C: A useful set of strains and plasmids for PCR-mediated gene disruption and other applications. *Yeast*, *14*, 115–132.

Dashko, S., Zhou, N., Compagno, C., & Piškur, J. (2014). Why, when, and how did yeast evolve alcoholic fermentation? *FEMS Yeast Research*, *14*, 826–832.

Duan, S. F., Han, P. J., Wang, Q. M., Liu, W. Q., Shi, J. Y., Li, K., et al. (2018). The origin and adaptive evolution of domesticated populations of yeast from Far East Asia. *Nature Communications*, *9*, 2690.

Fukuda, N., & Honda, S. (2013). Development of growth selection systems to isolate a-type or α-type of yeast cells spontaneously emerging from *MAT***a**/α diploids. *Journal of Biological Engineering*, 7, 27.

Fukuda, N., & Honda, S. (2018). Artificial mating-type conversion and repetitive mating for polyploid generation. *ACS Synthetic Biology*, *7*, 1413–1423.
Fukuda, N., Kaishima, M., Ishii, J., & Honda, S. (2017). Positive detection of GPCR antagonists using a system for inverted expression of a fluorescent reporter gene. *ACS Synthetic Biology*, *6*, 1554–1562.
Fukuda, N., Kaishima, M., Ishii, J., Kondo, A., & Honda, S. (2016). Continuous crossbreeding of sake yeasts using growth selection systems for a-type and α-type cells. *AMB Express*, *6*, 45.
Fukuda, N., Matsukura, S., & Honda, S. (2013). Artificial conversion of the mating-type of *Saccharomyces cerevisiae* without autopolyploidization. *ACS Synthetic Biology*, *2*, 697–704.
Gelfand, B., Mead, J., Bruning, A., Apostolopoulos, N., Tadigotla, V., Nagaraj, V., et al. (2011). Regulated antisense transcription controls expression of cell-type-specific genes in yeast. *Molecular and Cellular Biology*, *31*, 1701–1709.
Gietz, D., St. Jean, A., Woods, R. A., & Schiestl, R. H. (1992). Improved method for high efficiency transformation of intact yeast cells. *Nucleic Acids Research*, *20*, 1425.
Haber, J. E. (1998). Mating-type gene switching in *Saccharomyces cerevisiae*. *Annual Review of Genetics*, *32*, 561–599.
Herskowitz, I. (1983). Cellular differentiation, cell lineages, and transposable genetic cassettes in yeast. *Current Topics in Developmental Biology*, *18*, 1–14.
Hiraoka, M., Watanabe, K., Umezu, K., & Maki, H. (2000). Spontaneous loss of heterozygosity in diploid *Saccharomyces cerevisiae* cells. *Genetics*, *156*, 1531–1548.
Kumaran, R., Yang, S. Y., & Leu, J. Y. (2013). Characterization of chromosome stability in diploid, polyploid and hybrid yeast cells. *PLoS One*, *8*, e68094.
Mayer, V. W., & Aguilera, A. (1990). High levels of chromosome instability in polyploids of *Saccharomyces cerevisiae*. *Mutation Research*, *231*, 177–186.
Ogata, T., Okumura, Y., Tadenuma, M., & Tamura, G. (1993). Improving transformation method for industrial yeasts: Construction of *ADH1-APT2* gene and using electroporation. *The Journal of General and Applied Microbiology*, *39*, 285–294.
Ray, B. L., White, C. I., & Haber, J. E. (1991). Heteroduplex formation and mismatch repair of the stuck mutation during mating-type switching in *Saccharomyces cerevisiae*. *Molecular and Cellular Biology*, *11*, 5372–5380.
Salari, R., & Salari, R. (2017). Investigation of the best *Saccharomyces cerevisiae* growth condition. *Electronic Physician*, *9*, 3592–3597.
Snoek, T., Picca, N. M., Van den Bremt, S., Mertens, S., Saels, V., Verplaetse, A., et al. (2015). Large-scale robot-assisted genome shuffling yields industrial *Saccharomyces cerevisiae* yeasts with increased ethanol tolerance. *Biotechnology for Biofuels*, *8*, 32.
Strathern, J. N., Klar, A. J., Hicks, J. B., Abraham, J. A., Ivy, J. M., Nasmyth, K. A., et al. (1982). Homothallic switching of yeast mating type cassettes is initiated by a double-stranded cut in the *MAT* locus. *Cell*, *31*, 183–192.

CHAPTER THREE

Nucleotide resolution sequencing of N4-acetylcytidine in RNA

Justin M. Thomas, Keri M. Bryson, Jordan L. Meier*
Chemical Biology Laboratory, Center for Cancer Research, National Cancer Institute, National Institutes of Health, Frederick, MD, United States
*Corresponding author: e-mail address: jordan.meier@nih.gov

Contents

Abstract

Posttranscriptional modifications of RNA represent an emerging class of regulatory elements in human biology. Improved methods for studying how these elements are controlled and where they occur has the potential to transform our understanding of gene expression in development and disease. Here we describe a chemical method for nucleotide resolution sequencing of N4-acetylcytidine (ac4C), a highly conserved modified nucleobase whose formation is catalyzed by the essential cytidine acetyltransferase enzyme NAT10. This approach enables the sensitive, PCR-amplifiable detection of individual ac4C sites from nanograms of unfractionated cellular RNA. The sensitive and quantitative nature of this assay provides a powerful tool to understand how cytidine acetylation is targeted, profile RNA acetyltransferase dynamics, and validate the sites and stoichiometry of ac4C in novel RNA species.

Methods in Enzymology, Volume 621
ISSN 0076-6879
https://doi.org/10.1016/bs.mie.2019.02.022

1. Introduction

To date more than 100 modified nucleotides have been identified in RNA across all domains of life (Boccaletto et al., 2018). The majority of these posttranscriptional modifications have been characterized in abundant ribosomal RNAs (rRNAs) and transfer RNAs (tRNAs), where they often are found at highly conserved sites and can influence properties such as ribosome maturation and tRNA stability (Agris, 2004; Sloan et al., 2017). More recently, advances in next-generation sequencing technologies have enabled the mapping of several posttranscriptional modifications in messenger RNAs (mRNAs) as well as in functional non-coding RNAs (e.g., microRNAs) (Alarcon, Lee, Goodarzi, Halberg, & Tavazoie, 2015; Gilbert, Bell, & Schaening, 2016). In these species the presence of posttranscriptional modifications, referred to as the "epitranscriptome," is capable of influencing diverse functions including translational efficiency, mRNA stability, and RNA-protein interactions (Frye, Harada, Behm, & He, 2018; Schwartz, 2016).

1.1 Cytidine acetylation

The regulatory potential of RNA modifications has often been viewed as conceptually analogous to the epigenome, where dynamic changes in histone acetylation serve as a hallmark of active transcription (He, 2010; Meier, 2013). Acetylation of RNA also exists, in the form of three modified nucleobases: N6-acetyladenosine and N4-acetyl-2′-O-methylcytidine (ac4Cm), which are found in thermophilic archaea; and N4-acetylcytidine (ac4C), which is found in all domains of life (Boccaletto et al., 2018) (Fig. 1). Historically, ac4C was first identified in the 1960s during the initial

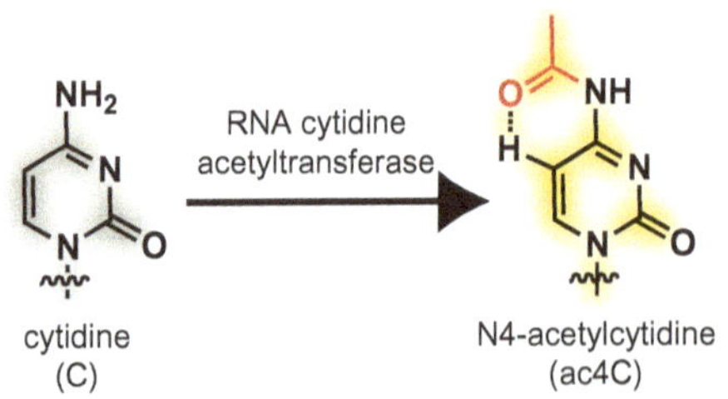

Fig. 1 Posttranscriptional acetylation of RNA cytidine residues forms N4-acetylcytidine (ac4C), shown in the more stable proximal conformation which preserves the native Watson-Crick edge of cytidine.

sequence determination of yeast and mammalian transfer RNA (tRNAs) (Kowalski, Yamane, & Fresco, 1971; Staehelin, Rogg, Baguley, Ginsberg, & Wehrli, 1968; Zachau, Dutting, & Feldmann, 1966). Subsequently, this modification was also discovered to occur in eukaryotic ribosomal RNA (Thomas, Gordon, & Rogg, 1978) and bacterial tRNAs (Oashi et al., 1972). Recent reports suggest ac4C may also be found in small RNAs and messenger RNAs in human cell lines (Arango et al., 2018; Lan et al., 2018; Tardu, Lin, & Koutmou, 2018), as well as viral RNA (McIntyre et al., 2018). The function of ac4C has been most well studied in bacterial tRNA, where it occurs at the wobble position of the initiator $tRNA^{Met}$ and improves translational fidelity by favoring pairing with the AUG codon, which specifies methionine, over the AUA codon, which specifies isoleucine (Stern & Schulman, 1978; Taniguchi et al., 2018). This effect has been attributed to the ability of ac4C's preferred pseudo-bicyclic conformation, in which the acetyl group is oriented proximal to C5 of cytidine, to influence Watson-Crick base pairing and mismatch discrimination (Fig. 1) (Kumbhar, Kamble, & Sonawane, 2013; Parthasarathy, Ginell, De, & Chheda, 1978). Less is known about the role of ac4C in eukaryotic systems, although studies in yeast have pointed to a role in rRNA biogenesis and tRNA stability (Ito et al., 2014; Johansson & Bystrom, 2004; Sharma et al., 2015).

Regulation of ac4C is accomplished via RNA cytidine acetyltransferase enzymes. Thus far, all organisms in which ac4C has been characterized harbor a single RNA acetyltransferase gene. These enzymes can be broadly grouped into three general types (Fig. 2A):

(1) TmcA-type RNA acetyltransferases are found in bacteria where they catalyze acetylation of the wobble base of $tRNA^{Met}$ (Chimnaronk et al., 2009; Ikeuchi, Kitahara, & Suzuki, 2008) (Fig. 2B). These are multi-domain enzymes approximately 600–700 amino acids in length which contain both a Walker-type ATPase/helicase domain and a Gcn5-related *N*-acetyltransferase (GNAT) domain. Structural and functional studies of these enzymes favor a reaction mechanism in which cytidine acetylation occurs following unwinding of the tRNA anticodon stem by the helicase domain. Genetic studies in *E. coli* and *H. volcanii* have found that TmcA is a non-essential gene whose deletion does not cause overt growth defects unless combined with other mutations (El Yacoubi et al., 2009; Ikeuchi et al., 2008).

(2) Nat10/Kre33-type RNA acetyltransferases are eukaryotic enzymes (Nat10, human; Kre33, yeast) that mediate cytidine acetylation in

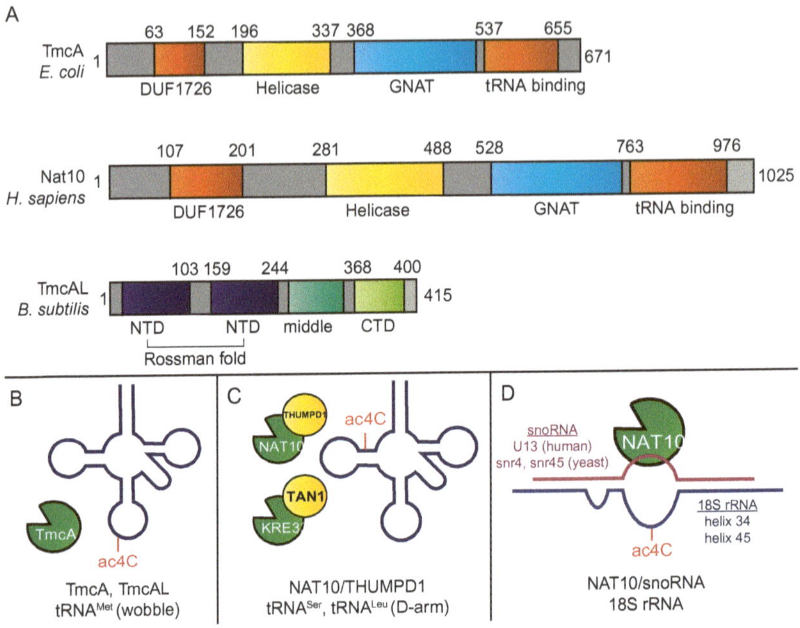

Fig. 2 Cytidine acetylation enzymes, substrates, and adapter molecules. (A) Cytidine acetylation enzymes. (B) Known substrates of TmcA and TmcAL cytidine acetylation enzymes. (C) Known tRNA substrates of Nat10/Kre33-type cytidine acetyltransferases. tRNA acetylation in eukaryotes requires an adapter protein (THUMPD1, human; Tan1, yeast). (D) Known ribosomal RNA substrates of Nat10/Kre33-type cytidine acetyltransferases. rRNA Acetylation requires short nuclear RNA (snoRNA) adapters (human helix 45, U13; yeast helix 45, snr45; yeast helix 34, snr4). Note: Identify of snoRNA guiding Human helix 34 acetylation has not yet been identified.

the stem of the D arm of $tRNA^{Leu}$ and $tRNA^{Ser}$, as well as in helix 34 and helix 45 of 18S ribosomal RNA (Ikeuchi et al., 2008; Ito et al., 2014; Sharma et al., 2015). These are ~1000 amino acid proteins comprised of a helicase domain, a GNAT domain, and multiple predicted RNA-binding domains. Nat10 and its homologs are essential in humans, mice, and yeast, although mutational studies indicate the acetyltransferase activity of the latter may be dispensable (Sharma et al., 2015). Relatively little biochemical data exists for these enzymes. However, all sites of ac4C that have been localized in organisms with Nat10-type acetyltransferases occur within base-paired 5′-CCG-3′ sequences. Furthermore, cellular studies indicate deposition of these sites requires either an adapter protein for tRNA acetylation (Thumpd1, human; Tan1, yeast) (Johansson & Bystrom, 2004) (Fig. 2C), or an adapter antisense snoRNA to guide rRNA acetylation (U13, human; snr4/snr45, yeast) (Sharma et al., 2015; Sharma et al., 2017) (Fig. 2D). Chemical proteomic profiling indicates Nat10 may be among the most abundant and active acetyltransferases in human cell lines (Montgomery, Sorum, Guasch, Nicklaus, & Meier, 2015; Montgomery, Sorum, & Meier, 2014).

(3) TmcAL-type enzymes are a third class of cytidine acetylation catalysts which were recently discovered (Taniguchi et al., 2018) (Fig. 2A–B). These enzymes were found through the study of bacteria that utilize ac4C at the wobble base of $tRNA^{Met}$ but do not harbor TmcA homologs in their genome. TmcAL is distinguished from TmcA and Nat10 in that it does not contain a helicase domain, a GNAT domain, or use acetyl-CoA as a cofactor. Instead, TmcAL uses a two-step mechanism in which acetate and ATP react to form a high energy acetyl adenylate intermediate, which is subsequently acetylates the exocyclic amine of cytidine (Taniguchi et al., 2018). This mechanism is similar to that used by aminoacyl-tRNA synthetases. Bacteria lacking TmcAL display a cold-sensitive phenotype and mistranslate AUA codons, which specify isoleucine, as AUG, which specifies methionine.

Dysregulation of the only known human cytidine acetyltransferase, Nat10, has recently been linked to many diseases including Hutchinson-Gilford progeria syndrome (Larrieu, Britton, Demir, Rodriguez, & Jackson, 2014), ovarian cancer (Tan et al., 2013), colorectal cancer (Zhang et al., 2014), and liver cancer (Tschida et al., 2017). The pathophysiological associations and essential role of Nat10 in eukaryotes, as well as recent proposals of pervasive cytidine acetylation in coding RNAs (Arango et al., 2018), have spurred new interest in the development of methods to study the sites, dynamics, and regulation of ac4C.

1.2 Detection, localization, and quantification of N4-acetylcytidine in RNA

Early methods used to identify ac4C in tRNA and rRNA involved partial enzymatic digestions and two-dimensional paper chromatography (Kowalski et al., 1971; Staehelin et al., 1968; Zachau et al., 1966). More recently, ac4C has been localized at precise sites in eukaryotic rRNA using two methods: (1) LC-MS analysis of RNase-digests, and (2) UV-HPLC analysis of nucleosides derived from mung-bean nuclease assays (Ito, Akamatsu, et al., 2014; Sharma et al., 2015; Sharma, Marchand, Motorin, & Lafontaine, 2017). While these digest-based methods have greatly advanced our understanding of ac4C, they suffer from poor sensitivity due to a lack of signal amplification. This leads to significant material requirements which can limit utility and throughput, and has prompted us to explore alternative methods for the analysis of cytidine acetylation (Sinclair et al., 2017), including nucleotide resolution sequencing (Motorin, Muller, Behm-Ansmant, & Branlant, 2007; Schwartz & Motorin, 2017).

Sequencing-based detection relies on the ability of RNA modifications to interfere with reverse transcription. For example, many RNA modifications including N1-methyadenosine (m1A) and N3-methycytidine (m3C) block Watson-Crick hydrogen bonding (Harcourt, Kietrys, & Kool, 2017). This causes reverse transcriptase (RT) enzymes to prematurely terminate or misincorporate non-cognate deoxynucleoside triphosphates (dNTPs) during cDNA synthesis, the latter of which can be detected by PCR amplification and sequencing. In contrast to m1A and m3C, many RNA modifications including ac4C, 5-methylcytidine (5mC), and pseudouridine do not alter the Watson Crick edge, rendering them nearly "invisible" during reverse transcription (Ryvkin et al., 2013) (Fig. 1). These modifications can be instead be detected by antibody-based methods, or by exploiting differences in chemical reactivity to distinguish their RT profile from that of canonical nucleobases. A prototypical example of the latter approach is bisulfite sequencing of 5mC in DNA and RNA (Hayatsu, 2008; Schaefer, 2015).

Recently we reported a chemical reaction to enable sequencing-based detection of ac4C in RNA (Thomas et al., 2018). Analogous to bisulfite sequencing, our approach exploits the unique electronic structure of ac4C, which renders its 5,6-double bond susceptible to nucleophilic attack by sodium borohydride (Fig. 3) (Cerutti & Miller, 1967; Miller & Cerutti, 1967). The addition of two equivalents of hydride forms the reduced nucleotide N4-acetyl-3,4,5,6-tetrahydrocytidine. Analysis of model substrates as well as physiological sites in rRNA revealed that chemical reduction of ac4C causes premature termination and misincorporation during reverse transcription (Fig. 3). The ratio of termination (RT stop) to misincorporation was found to vary based on the specific polymerase used, with the thermostable group II intron reverse transcriptase (TGIRT) showing most efficient read through (Thomas et al., 2018) (Fig. 4). This approach is sensitive, requiring only a few hundred nanograms RNA for assessment of individual ac4C sites by Sanger sequencing. Furthermore, the extent of misincorporation was found to vary linearly with ac4C content, indicating the utility of this approach to study RNA acetylation stoichiometry (Fig. 5). Here we describe

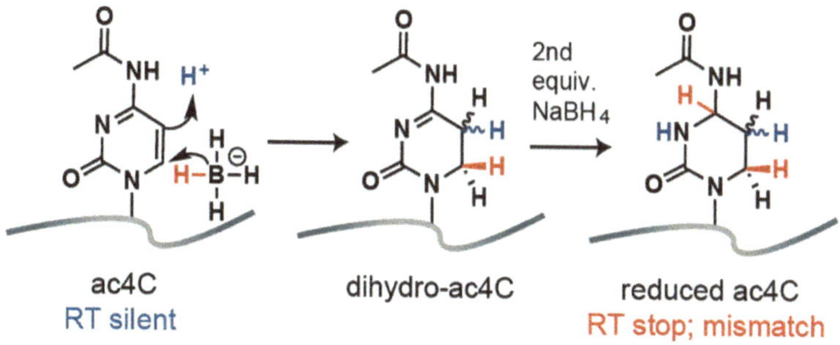

Fig. 3 Chemical reduction of N4-acetylcytidine in RNA to N4-acetyl-3,4,5,6-tetrahydrocytidine (referred to here as "reduced ac4C"). Reduction of ac4C can cause termination (RT stop) or dNTP misincorporation (mismatch) during reverse transcription. *Figure is adapted with permission from Thomas, J. M., Briney, C. A., Nance, K. D., Lopez, J. E., Thorpe, A. L., Fox, S. D., et al. (2018). A chemical signature for cytidine acetylation in RNA. Journal of the American Chemical Society.* 140, *12667–12,70. Copyright American Chemical Society, 2018.*

detailed protocols for the sequencing-based detection of ac4C in RNA. We specify key steps and controls important for assessing ac4C in RNA, and highlight future opportunities for applying this chemistry to better understand the role of cytidine acetylation in fundamental biology and human disease.

2. Technical aspects

2.1 Sample preparation and RNA isolation

This protocol utilizes total RNA extracted from tissues or cultured cells by TRIzol extraction. High quality total RNA isolated using alternative methods may also be used. Treatment with DNase is essential to avoid downstream amplification of ribosomal DNA. All reagents used should be free of RNases and care should be taken at all times to avoid contamination of samples with RNases which come from human skin and are prevalent in laboratory environments. It is recommended to wear gloves during all procedures and avoid contact between gloves and the sealing surfaces of microcentrifuge tube lids which can occur when opening.

2.1.1 Materials

TRIzol Reagent (Thermofisher,15596018)

PBS (Thermofisher, 10010)

Chloroform:isoamyl alcohol 24:1, suitable for nucleic acid purification (Sigma, CO549)

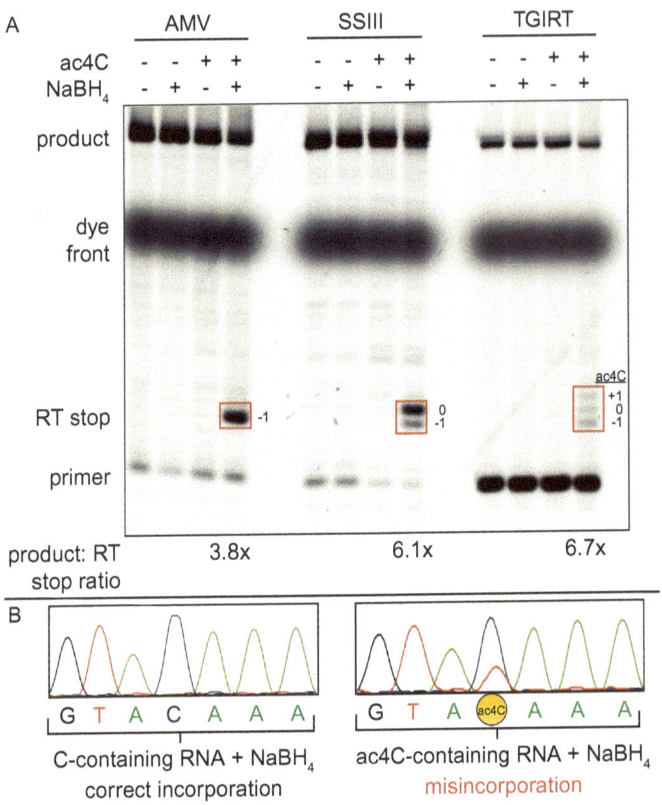

Fig. 4 Monitoring reduction of ac4C in RNA by RT stop and misincorporation. (A) Comparison of primer extension following $NaBH_4$ reduction of an ac4C-containing RNA. A model in vitro transcribed RNA containing a single site of ac4C was treated with $NaBH_4$ (100 mM, 37 °C, 1 h). The product of this reaction was incubated with a Cy5 primer, dNTPs, and the specified reverse transcriptase according to the manufacturer's recommended conditions. Substantial RT stop was observed with AMV, while slightly more product was formed using SuperScript III and TGIRT RT. The position of the RT stop product is specified relative to the ac4C site (+1, 0, −1). (B) Analysis of PCR-amplified full-length cDNAs (sense strand) by Sanger sequencing. Misincorporation of dNTPs is observed only for the ac4C-containing model RNA. All cDNAs were produced via TGIRT RT. *Figure is adapted with permission from Thomas, J. M., Briney, C. A., Nance, K. D., Lopez, J. E., Thorpe, A. L., Fox, S. D., et al. (2018). A chemical signature for cytidine acetylation in RNA. Journal of the American Chemical Society.* 140, *12667–12670. Copyright American Chemical Society, 2018.*

Isopropyl Alcohol (Macron Fine Chemicals, 3032-16)
TURBO DNase kit (2 U/μL) (Invitrogen, AM2238)
UltraPure Phenol:Chloroform:isoamyl Alcohol 24:24:1 (Thermofisher, 15593-049)

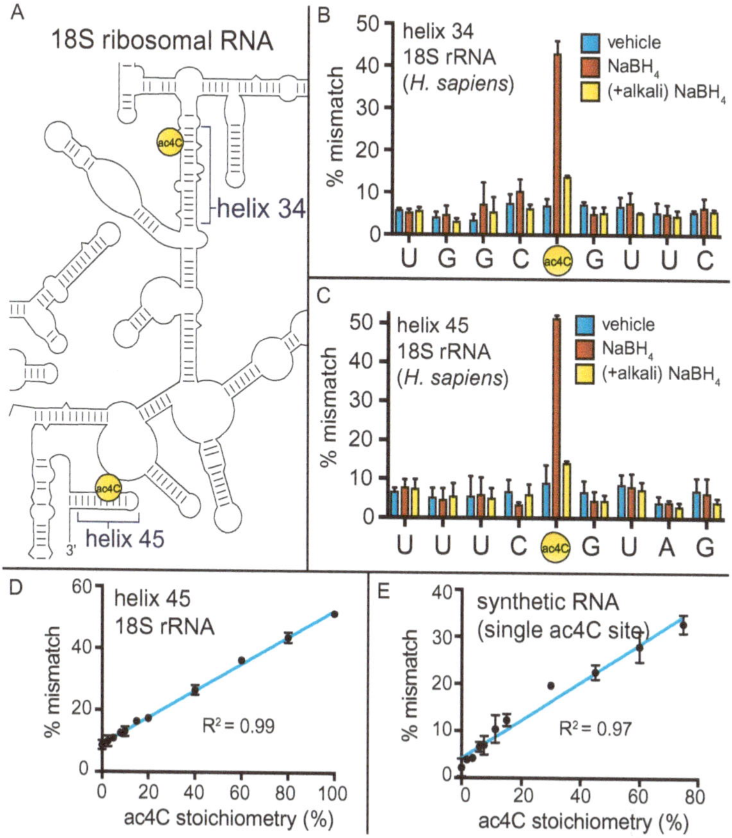

Fig. 5 Quantitative sequencing-based detection of ac4C in model substrates and 18S rRNA. (A) Schematic of human rRNA ac4C sites. (B) Sequencing-based detection of ac4C-dependent mismatches in cDNAs generated from human rRNA helix 34 and (C) helix 45 following borohydride treatment and RT. Sequence corresponds to the cDNA sense strand. Vehicle = water; $NaBH_4$ = 100 mM sodium borohydride; (+alkali) $NaBH_4$ = alkali pre-treatment (100 mM $NaCO_3$, pH 10, 60 °C, 1 h), precipitation, then borohydride. (D) Relationship between misincorporation signal and stoichiometry of ac4C in rRNA and (E) a synthetic RNA harboring ac4C in an "ACA" sequence context. For all data, error bars indicate the standard deviation (n = 3). *Figure is reprinted with permission from Thomas, J. M., Briney, C. A., Nance, K. D., Lopez, J. E., Thorpe, A. L., Fox, S. D., et al. (2018). A chemical signature for cytidine acetylation in RNA. Journal of the American Chemical Society.* 140, *12667–12670. Copyright American Chemical Society, 2018.*

Molecular biology grade Ethanol, Absolute (200 Proof) (Fisher BioReagents, BP2818500)
Nuclease free ultra-pure water (Invitrogen 10977)
Sodium Acetate (3 M) pH 5.5 (Invitrogen, AM9740)
Thermo Scientific NanoDrop 2000 Spectrophotometer

Centrifuge capable accepting 15 mL falcon tubes, refrigeration to 4 °C and >12,000 × *g*
Tabletop microcentrifuge capable of refrigeration to 4 °C and >18,000 × *g*.

2.1.2 RNA isolation

1. Homogenize freshly washed cell pellets using TRIzol reagent. For homogenization of frozen cell pellets, initially resuspend cells in ice-cold PBS, pellet by centrifugation (200 × *g*, 4 °C, 4 min), and discard supernatant. Next, add TRIzol reagent (approximately 1 mL TRIzol per 15 cm^2 dish of cells harvested) directly to the freshly washed cell pellet. Vortex and incubate 5 min at room temperature.
2. Add chloroform:isoamyl alcohol solution such that the final proportion of the sample is four parts TRIzol, 1 part chloroform:isoamyl alcohol (4:1 ratio). Once the chloroform-containing solution is added, the sample should become cloudy.
3. Vortex for 15 s, incubate at room temperature for 5 min, and centrifuge (12,000 × *g*, 4 °C, 15 min).
4. Transfer top aqueous layer, which contains RNA, to a new tube. Care should be taken to avoid touching the middle/interface layer, which contains DNA.
5. To precipitate RNA, add isopropyl alcohol such that the final proportion of the sample is 1:1 isopropanol:aqueous layer. Mix by vortexing for 15 s, incubate at room temperature for 10 min, and centrifuge (12,000 × *g*, 4 °C, 15 min).
6. Discard the supernatant and add a volume of 75% ethanol equal to the amount of TRIzol used during homogenization (step #1).
7. Vortex for 5 s and centrifuge (7500 × *g*, 4 °C, 5 min).
8. Discard supernatant and dry pellet for up to 10 min.
9. Dissolve pellet in 100 μL of ultra-pure water and quantify using the NanoDrop 2000 Spectrophotometer

2.1.3 DNase-treatment of total RNA

1. Dilute RNA to 200 μg/mL with ultra-pure water and add 10 × Turbo DNase buffer to 1 × concentration.
2. Add 1 μL Turbo DNase for every 10 μg RNA. Mix gently and incubate at 37 °C for 1 h.

3. If DNase treatment reaction is <200 μL, add sufficient ultra-pure water to bring the volume to 200 μL.
4. Next you will phenol: chloroform extract and ethanol precipitate the total RNA in order to remove DNase. First, add equal volume phenol: chloroform:isoamyl alcohol to the sample. Vortex and centrifuge at 12,000 × *g* at room temperature for 5 min using a tabletop ultracentrifuge.
5. Collect top aqueous layer and transfer to a clean tube.
6. Add a 1:1 volume of chloroform to the aqueous layer. Vortex and centrifuge at 12,000 × *g* at room temperature for 5 min using a tabletop ultracentrifuge.
7. Collect top aqueous layer and transfer to a clean tube.
8. To ethanol precipitate DNase-treated RNA, Add 1:9 volume of 3 M sodium acetate and 3:1 volume of 100% Ethanol to the sample. Mix thoroughly.
9. Store in −80 °C for ≥1 h.
10. After −80 °C incubation, spin in a microcentrifuge at 18,000 × *g* at 4 °C for 30 min.
11. Decant the supernatant, and wash pellet with 500 μL of ice-cold 70% ethanol. Spin in a microcentrifuge for an additional 10 min at 18,000 × *g* at 4 °C.
12. Decant supernatant and air-dry pellet for 10–15 min.
13. Resuspend pellet in 50 μL ultra-pure water per 15 cm^2 dish of cells harvested and heat/incubate at 55 °C for 15 min.
14. Quantify RNA using the NanoDrop 2000 Spectrophotometer.

2.2 Chemical manipulation of ac4C in total cellular RNA

As little as 10 pg of total RNA can be used to successfully generate cDNA for nucleotide resolution ac4C sequencing; however, for convenient sample handling when material is not limiting, we recommend beginning with approximately 3 μg of total RNA from each biological condition to be analyzed.

2.2.1 *Materials*

Sodium Borohydride (Sigma-Aldrich, 632287)
Hydrochloric acid (Sigma-Aldrich, H1758)
Sodium Bicarbonate (Sigma Aldrich, S5761)
UltraPure 1 M Tris-HCI Buffer, pH 7.5 (Invitrogen, 15567027)

2.2.2 Reductive RNA treatment

This procedure details the parallel preparation of samples subjected to three sets of conditions: Borohydride treated RNA (+borohydride), borohydride treated deacetylated RNA (+borohydride +alkali) and mock treated RNA (−borohydride). The latter two conditions serve as controls. Sodium borohydride selectively reduces ac4C, converting it to tetrahydroacetycytidine while leaving cytidine unmodified (Fig. 3). Caution should be used when preparing and handling solutions containing $NaBH_4$ as borohydride decomposes in aqueous solutions to produce hydrogen gas which can pressurize reaction vessels. We recommend preparing $NaBH_4$ stocks immediately before use and taking care to vent reactions during or after chemical treatments.

1. Prepare fresh 1 M $NaBH_4$ solution immediately before using in ice cold nuclease-free water. Once dissolved the solution will begin bubbling (release of hydrogen gas). Reactions should be vented in order to prevent pressurization of vessels.
2. Prepare three separate reactions per sample: (1) For the +borohydride RNA, bring 1 μg total RNA to 90 μL with nuclease-free water. Add 10 μL 1 M $NaBH_4$. (2) For the +borohydride +alkali RNA, bring alkali-treated total RNA (prepared in Section 2.2.3) to 90 μL with nuclease-free water and add 10 μL 1 M $NaBH_4$. (3) For the −borohydride RNA, bring 1 μg RNA to 100 μL in nuclease-free water.
3. Incubate at 37 °C for 60 min.
4. Add 15 μL 1 M HCl to each sample to quench residual $NaBH_4$. Addition of acid will cause borohydride containing samples to bubble vigorously.
5. Once bubbling has ceased or 1 min, add 30 μL 1 M Tris pH 7.5 to neutralize residual HCL.
6. Adjust volume to 200 μL and ethanol precipitate as described above in Section 2.1.3 steps 8–12.
7. Resuspend pellet in 25 μL ultra-pure water and vortex.
8. Quantify RNA using the NanoDrop 2000 spectrophotometer

2.2.3 Chemical deacetylation control

Chemical deacetylation via base-pretreatement serves as a convenient control for ac4C dependent misincorporation. A portion of the sample (1 μg) is subjected to treatment in mild alkali. This causes hydrolysis of ac4C to cytidine, rendering it non-reactive with borohydride and limiting borohydride-dependent dNTP misincorporation during RT.

1. To chemically deacetylate ac4C, incubate 1 μg of RNA sample in 100 μL of 100 mM sodium bicarbonate pH 9.5 (prepared in nuclease-free water)

at 60 °C for 1 h. This condition promotes substantial conversion (>80%) of ac4C to cytidine without causing excessive degradation of the RNA backbone as assessed by Bioanalyzer.

2. Adjust the volume of the RNA reaction to 200 μL with nuclease-free water. Ethanol precipitate the sample as described above in Section 2.1.2, steps 8–12.
3. Resuspend pellet in 25 μL ultra-pure water and vortex.
4. Quantify RNA using the NanoDrop 2000 spectrophotometer

2.3 Nucleotide resolution sequencing of individual ac4C sites

2.3.1 Materials

TGIRT-III reverse transcription enzyme (INGEX)
5 × TGIRT reaction buffer (user prepared with nuclease-free reagents, 2.25 M NaCl, 100 mM Tris-HCL pH 7.5)
RNasin plus, 40 units/μL (Promega, N2611)
Optimized dNTP mix [10 mM dTTP, 10 mM dCTP, 10 mM dATP, 5 mM dGTP] (prepared from 100 mM dNTPs set, NEB N0446S)
Phusion High-Fidelity PCR Kit (New England Biolabs, E0553S)
Agarose LE, Molecular Biology Grade (Thomas Scientific, C996H59)
100 bp DNA Ladder (New England Biolabs, N3231S)
SYBR Safe DNA Gel Stain (Invitrogen, S33102)
UltraPure TBE Buffer, 10 × (Invitrogen, 15581044)
UV transilluminator (UVP MultiDoc-It Imager Benchtop UV Transilluminator)
6 × DNA gel loading dye (NEB, B7021S)
QIAquick Gel extraction kit (Qiagen, 28704)
Thermocycler
Agarose gel electrophoresis equipment.

2.3.2 Reverse transcription of borohydride-treated RNA

Reverse transcription of sodium borohydride-treated ac4c results in partial incorporation of adenosine in the cDNA strand opposite the reduced ac4C. It should also be noted that reduced ac4C results in partial termination of reverse transcription at or adjacent to its location, often referred to as RT-stops (Fig. 4A). Of the reverse transcriptase enzymes screened by our lab TGIRT-III RT (Ingex) produces the highest proportion of full-length to stop product, as well as the most robust C-to-T signature of the RT enzymes screened by our lab (Fig. 4B). The following protocol describes the use of TGIRT RT for sequence-defined cDNA generation, which we have used to profile ac4C in model substrates and human 18S rRNA (Thomas et al., 2018).

1. Dilute RNA from Section 2.2.2 to 100 pg/μL in water for use as template.
2. Perform TGIRT RT reactions. We typically carry out RT in 20 μL volumes in PCR tubes. Exemplary protocol: combine 4 μL 5 × TGIRT reaction buffer with 200 pg (2 μL) of template, 4 pM DNA primer, 2 μL 50 mM $MgCl_2$ and sufficient ultra-pure water to bring the mixture to 17 μL. Note 1: TGIRT reactions are incubated at 57 °C, higher than many other commercially available reverse transcription enzymes. Therefore, it is essential to design the reverse transcription primer to form a stable duplex with the target sequence at this temperature. Note 2: Fluorophore or radiolabeled primers can be substituted at this step to assess RT stop via primer extension assay (Fig. 4)
3. Heat to 75 °C for 3 min and placed on ice for 1 min to anneal the primer and RNA.
4. After cooling, add 0.5 μL TGIRT-III, 0.5 μL Rnasin Plus and 100 μL 100 mM DTT. Incubate 20 min at room temperature. Initiate the reverse transcription reaction by adding 1 μL of optimized dNTP mix. In optimization studies we have found that decreasing the dGTP concentration from 500 to 250 μM increases the sensitivity of C-to-T misincorporation. Incubate the reaction for one hour at 57 °C and proceed to PCR amplification step (Section 2.3.3).
5. Note on additional controls: We recommend performing a reverse transcription reaction in the absence of template to ensure the reagents being used are free from contaminating RNA/DNA. It is also essential to include control reactions that lack RT enzyme for each biological replicate, which allows confirmation that the PCR reaction (Section 2.3.3) is amplifying only cDNA generated through reverse transcription and not contaminating gDNA from other sources.

2.3.3 PCR amplification and purification of cDNAs

1. In order to PCR amplify cDNA, first prepare 50 μL PCR reactions using Phusion Hot Start Flex DNA polymerase according to the manufacturers recommended protocol using 1 μL TGIRT reaction product as template. Final reaction conditions used are 1 × HF buffer, 200 μM dNTPs, 0.5 μM forward primer, 0.5 μM reverse primers, 1 unit polymerase, and 1 μL TGIRT reaction product. Note: A PCR reaction

lacking template is recommended to check that PCR reagents are not contaminated with amplifiable genomic DNA. Annealing temperature and number of amplification cycles should be determined empirically for each cDNA to be analyzed. To amplify cDNAs derived from the ac4C-continaing helix 45 of human 18S rRNA, we utilize an annealing temperature of 67.4 °C and amplify for 33 cycles (Thomas et al., 2018).

2. Purifying the PCR product by agarose gel purification/extraction is recommended prior to submission of PCR product for sequencing for best results. Mix PCR product with 6 × loading buffer and run on 2% agarose gel containing 1 × SYBRsafe DNA gel stain until bromophenol blue marker dye runs approximately two-thirds of the way across the gel. 1 μg of 100 bp dsDNA ladder should be run alongside PCR reaction products.
3. Visualize PCR product on UV transilluminator and excise bands of expected size with a clean razor blade. Control reactions (minus RT and minus template) should be run alongside and visualized to ensure no PCR product is present in these samples.
4. Process agarose gel slices using commercially available gel extraction kit, elute DNA and submit for Sanger sequencing with an appropriate primer.

2.3.4 Sanger sequencing-based analysis of cytidine acetylation

1. To calculate ac4C-dependent misincorporation, open processed Sanger sequencing traces using sequence trace viewing software. Free software which allow peak heights to be calculated include 4peaks (Mac) and Applied Biosystems Sequence Scanner 2.0 (Windows). Only high-quality sequence traces with low background noise and little or no peak ghosting should be analyzed.
2. Identify sites of ac4C by locating bases that appear as a mixture of C and T in borohydride treated samples (+borohydride). Sites of high ac4C stoichiometry may produce more intense signal for T than C in sequencing chromatograms and may be assigned as T in base called sequences. Sites of ac4C exhibit substantially decreased C-to-T mutation in alkali pretreated controls (+borohydride +alkali) and only background levels of misincorporation (<10%) in untreated controls (−borohydride) (Fig. 5).
3. Quantify misincorporation at ac4C sites by measuring the height of C and T peaks in the sequencing traces and calculate misincorporation using the equation: percent misincorporation = 100*(T-peak)/((C-peak) + (T-peak)).

4. To estimate ac4C stoichiometry at a given site, fit calculated misincorporation to a standard curve. A standard curve for ac4C dependent misincorporation can be generated by plotting the average fraction of C-to-T conversion at the site of interest across a series of RNA mixtures prepared from defined ratios of wild-type and Nat10 knockout cells (100:0, 80:20, 60:40, etc.; Fig. 5).

3. Discussion

3.1 Critical parameters and troubleshooting

A critical aspect in sequencing-based detection of ac4C is the use of proper genetic and/or chemical controls. At least three other modified RNA nucleobases are susceptible to reaction with sodium borohydride: dihydrouridine, 7-methylguanosine and N3-methylcytidine (m3C) (Cerutti, Kondo, Landis, & Witkop, 1968; Marchand et al., 2018; Wintermeyer & Zachau, 1970). One way in which ac4C can be distinguished from these other borohydride-reactive modifications is by monitoring the influence of deletion or mutation of cellular cytidine acetyltransferase on misincorporation (Arango et al., 2018). Cytidine acetyltransferases are highly conserved, and can be rapidly identified by searching genomes for homologs to TmcA, Kre33, and Nat10. The selection of genetic controls is further simplified by the fact that in all organisms known to harbor ac4C, a single cytidine acetylation enzyme is known. One caveat to this approach is the fact that in eukaryotes Nat10/Kre33-type cytidine acetyltransferases have been found to be essential for life (Sharma et al., 2015; Wang et al., 2015). In these cases, siRNA knockdown, targeted mutation of the acetyltransferase domain, or conditional knockout may serve as useful controls. Genetic ablation may also be used to target non-essential protein and snoRNA adapters required for tRNA and rRNA acetylation (Fig. 3C and D) (Sharma, Yang, et al., 2017).

In the absence of genetic approaches, chemical deacetylation provides a valuable alternative for attributing a borohydride-dependent misincorporation event to ac4C (Sinclair et al., 2017). These controls take advantage of the fact that the electron-deficient ac4C acetamide is susceptible to cleavage by nucleophiles and alkali conditions, which convert ac4C sites to unmodified cytidine. This property is unique compared to other borohydride-reactive RNA modifications, and defines base-sensitive borohydride-dependent misincorporation as a unique chemical signature of ac4C in RNA (Thomas et al., 2018). An important aspect of these chemical controls, as well as borohydride-treatment itself, is that they create basic conditions that can cause RNA strand scission. This may contribute to background in RT stop assays (Marchand et al., 2018), and is an important consideration for PCR amplification and next-generation sequencing library preparation strategies. Other features of ac4C's reductive chemistry that distinguish it from other modifications include that its reduction proceeds best in unbuffered solutions of aqueous borohydride, in contrast to the buffered solutions used to reduce dihydrouridine (Cerutti & Miller, 1967). Finally, for Sanger sequencing studies of ac4C in abundant species such as 18S rRNA we have found that purity of the borohydride-treated amplified cDNA is critical for obtaining reproducible quantitative data. Controls that can be used to assess contamination include performing RT in the absence of an RNA template, ensuring the dependence of PCR amplification on the presence RT enzyme, and performing PCR in the absence of template cDNA. In addition, we recommend purifying the final PCR amplicon by agarose gel to obtain the highest quality sequencing results.

3.2 Applications and future directions

Here we have described a sensitive protocol for the nucleotide resolution sequencing-based detection of the RNA modification ac4C. The method described here is immediately applicable to the analysis ac4C's known rRNA and pre-rRNA sites. The ability to quantitatively assess the acetylation of these sites has several applications, including studying how helix 34 and helix 45 ac4C is affected by stimuli such as small molecules (Larrieu et al., 2014), the metabolic state of the cell (Montgomery et al., 2016), and disease states. In addition, similar to how primer extension assays have facilitated the understanding of snoRNA-dependent ribose methylation (Bachellerie & Cavaille, 1997), the availability of a sensitive and site-specific assay for ac4C should be useful in defining how snoRNAs target cytidine acetylation for deposition. In addition to ac4C, the chemistry described here should also be useful for studies of ac4Cm, which is found in some Archaea (Bruenger et al., 1993; Kawai et al., 1991). One limitation of our method is its ability to interrogate ac4C in more densely-functionalized modification landscapes. The known ac4C sites in eukaryotic $tRNA^{Ser}$ and $tRNA^{Leu}$ lie directly adjacent to dihydrouridine, whose coordinate reduction may severely limit RT readthrough. In the future we anticipate this limitation may be addressed by integrating borohydride-based ac4C detection with recently reported tRNA sequencing methods, exploring the use of more processive polymerases such as HIV and Marathon RT, and by integrating our protocol with deep sequencing methods (Sexton, Wang, Rutenberg-Schoenberg, & Simon, 2017; Zheng et al., 2015; Zhou et al., 2018). Finally, we anticipate nucleotide resolution sequencing will form an essential component of strategies to identify, validate, and assess the stoichiometry of novel ac4C sites in coding RNAs, whose existence is supported by recent RIP-Seq and LC-MS studies in humans and yeast (Arango et al., 2018; Tardu et al., 2018). Overall, our methods provide a foundation for the quantitative and site-specific analysis of ac4C, and as such have the potential to expand our understanding of acetylation as a component of the emerging epitranscriptome.

Acknowledgments

We are grateful to members of the Meier lab for providing helpful discussions and critical feedback on this chapter. This work was supported by the Intramural Research Program of the NIH, National Cancer Institute, Center for Cancer Research (ZIA-BC011488-06).

References

Agris, P. F. (2004). Decoding the genome: A modified view. *Nucleic Acids Research*, *32*, 223–238.

Alarcon, C. R., Lee, H., Goodarzi, H., Halberg, N., & Tavazoie, S. F. (2015). N6-methyladenosine marks primary microRNAs for processing. *Nature*, *519*, 482–485.

Arango, D., Sturgill, D., Alhusaini, N., Dillman, A. A., Sweet, T. J., Hanson, G., et al. (2018). Acetylation of cytidine in mRNA promotes translation efficiency. *Cell*, *175*, 1872–1886.e24.

Bachellerie, J. P., & Cavaille, J. (1997). Guiding ribose methylation of rRNA. *Trends in Biochemical Sciences*, *22*, 257–261.

Boccaletto, P., Machnicka, M. A., Purta, E., Piatkowski, P., Baginski, B., Wirecki, T. K., et al. (2018). MODOMICS: A database of RNA modification pathways. 2017 update. *Nucleic Acids Research*, *46*, D303–D307.

Bruenger, E., Kowalak, J. A., Kuchino, Y., McCloskey, J. A., Mizushima, H., Stetter, K. O., et al. (1993). 5S rRNA modification in the hyperthermophilic archaea Sulfolobus solfataricus and Pyrodictium occultum. *The FASEB Journal*, *7*, 196–200.

Cerutti, P., Kondo, Y., Landis, W. R., & Witkop, B. (1968). Photoreduction of uridine and reduction of dihydrouridine with sodium borohydride. *Journal of the American Chemical Society*, *90*, 771–775.

Cerutti, P., & Miller, N. (1967). Selective reduction of yeast transfer ribonucleic acid with sodium borohydride. *Journal of Molecular Biology*, *26*, 55–66.

Chimnaronk, S., Suzuki, T., Manita, T., Ikeuchi, Y., Yao, M., Suzuki, T., et al. (2009). RNA helicase module in an acetyltransferase that modifies a specific tRNA anticodon. *The EMBO Journal*, *28*, 1362–1373.

El Yacoubi, B., Phillips, G., Blaby, I. K., Haas, C. E., Cruz, Y., Greenberg, J., et al. (2009). A gateway platform for functional genomics in Haloferax volcanii: Deletion of three tRNA modification genes. *Archaea*, *2*, 211–219.

Frye, M., Harada, B. T., Behm, M., & He, C. (2018). RNA modifications modulate gene expression during development. *Science*, *361*, 1346–1349.

Gilbert, W. V., Bell, T. A., & Schaening, C. (2016). Messenger RNA modifications: Form, distribution, and function. *Science*, *352*, 1408–1412.

Harcourt, E. M., Kietrys, A. M., & Kool, E. T. (2017). Chemical and structural effects of base modifications in messenger RNA. *Nature, 541*, 339–346.
Hayatsu, H. (2008). Discovery of bisulfite-mediated cytosine conversion to uracil, the key reaction for DNA methylation analysis–a personal account. *Proceedings of the Japan Academy. Series B, Physical and Biological Sciences, 84*, 321–330.
He, C. (2010). Grand challenge commentary: RNA epigenetics? *Nature Chemical Biology, 6*, 863–865.
Ikeuchi, Y., Kitahara, K., & Suzuki, T. (2008). The RNA acetyltransferase driven by ATP hydrolysis synthesizes N4-acetylcytidine of tRNA anticodon. *The EMBO Journal, 27*, 2194–2203.
Ito, S., Akamatsu, Y., Noma, A., Kimura, S., Miyauchi, K., Ikeuchi, Y., et al. (2014). A single acetylation of 18 S rRNA is essential for biogenesis of the small ribosomal subunit in Saccharomyces cerevisiae. *The Journal of Biological Chemistry, 289*, 26201–26212.
Ito, S., Horikawa, S., Suzuki, T., Kawauchi, H., Tanaka, Y., Suzuki, T., et al. (2014). Human NAT10 is an ATP-dependent RNA acetyltransferase responsible for N4-acetylcytidine formation in 18 S ribosomal RNA (rRNA). *The Journal of Biological Chemistry, 289*, 35724–35730.
Johansson, M. J., & Bystrom, A. S. (2004). The Saccharomyces cerevisiae TAN1 gene is required for N4-acetylcytidine formation in tRNA. *RNA, 10*, 712–719.
Kawai, G., Ue, H., Yasuda, M., Sakamoto, K., Hashizume, T., McCloskey, J. A., et al. (1991). Relation between functions and conformational characteristics of modified nucleosides found in tRNAs. *Nucleic Acids Symposium Series*, 49–50.
Kowalski, S., Yamane, T., & Fresco, J. R. (1971). Nucleotide sequence of the "denaturable" leucine transfer RNA from yeast. *Science, 172*, 385–387.
Kumbhar, B. V., Kamble, A. D., & Sonawane, K. D. (2013). Conformational preferences of modified nucleoside N(4)-acetylcytidine, ac4C occur at "wobble" 34th position in the anticodon loop of tRNA. *Cell Biochemistry and Biophysics, 66*, 797–816.
Lan, M. D., Xiong, J., You, X. J., Weng, X. C., Zhou, X., Yuan, B. F., et al. (2018). Existence of diverse modifications in small-RNA species composed of 16-28 nucleotides. *Chemistry, 24*, 9949–9956.
Larrieu, D., Britton, S., Demir, M., Rodriguez, R., & Jackson, S. P. (2014). Chemical inhibition of NAT10 corrects defects of laminopathic cells. *Science, 344*, 527–532.
Marchand, V., Ayadi, L., Ernst, F. G. M., Hertler, J., Bourguignon-Igel, V., Galvanin, A., et al. (2018). AlkAniline-Seq: Profiling of m(7) G and m(3) C RNA modifications at single nucleotide resolution. *Angewandte Chemie (International Ed. in English), 57*, 16785–16790.
McIntyre, W., Netzband, R., Bonenfant, G., Biegel, J. M., Miller, C., Fuchs, G., et al. (2018). Positive-sense RNA viruses reveal the complexity and dynamics of the cellular and viral epitranscriptomes during infection. *Nucleic Acids Research, 46*, 5776–5791.
Meier, J. L. (2013). Metabolic mechanisms of epigenetic regulation. *ACS Chemical Biology, 8*, 2607–2621.
Miller, N., & Cerutti, P. A. (1967). Synthesis of N4-acetyl-3,4,5,6-tetrahydrocytidine and copolymers of cytidylic acid and N4-acetyl-3,4,5,6-tetrahydrocytidylic acid. *Journal of the American Chemical Society, 89*, 2767.
Montgomery, D. C., Garlick, J. M., Kulkarni, R. A., Kennedy, S., Allali-Hassani, A., Kuo, Y. M., et al. (2016). Global profiling of acetyltransferase feedback regulation. *Journal of the American Chemical Society, 138*, 6388–6391.
Montgomery, D. C., Sorum, A. W., Guasch, L., Nicklaus, M. C., & Meier, J. L. (2015). Metabolic regulation of histone acetyltransferases by endogenous Acyl-CoA cofactors. *Chemistry & Biology, 22*, 1030–1039.
Montgomery, D. C., Sorum, A. W., & Meier, J. L. (2014). Chemoproteomic profiling of lysine acetyltransferases highlights an expanded landscape of catalytic acetylation. *Journal of the American Chemical Society, 136*, 8669–8676.

Motorin, Y., Muller, S., Behm-Ansmant, I., & Branlant, C. (2007). Identification of modified residues in RNAs by reverse transcription-based methods. *Methods in Enzymology*, *425*, 21–53.

Oashi, Z., Murao, K., Yahagi, T., Von Minden, D. L., McCloskey, J. A., & Nishimura, S. (1972). Characterization of C + located in the first position of the anticodon of Escherichia coli tRNA Met as N 4 -acetylcytidine. *Biochimica et Biophysica Acta*, *262*, 209–213.

Parthasarathy, R., Ginell, S. L., De, N. C., & Chheda, G. B. (1978). Conformation of N4-acetylcytidine, a modified nucleoside of tRNA, and stereochemistry of codon-anticodon interaction. *Biochemical and Biophysical Research Communications*, *83*, 657–663.

Ryvkin, P., Leung, Y. Y., Silverman, I. M., Childress, M., Valladares, O., Dragomir, I., et al. (2013). HAMR: High-throughput annotation of modified ribonucleotides. *RNA*, *19*, 1684–1692.

Schaefer, M. (2015). RNA 5-methylcytosine analysis by bisulfite sequencing. *Methods in Enzymology*, *560*, 297–329.

Schwartz, S. (2016). Cracking the epitranscriptome. *RNA*, *22*, 169–174.

Schwartz, S., & Motorin, Y. (2017). Next-generation sequencing technologies for detection of modified nucleotides in RNAs. *RNA Biology*, *14*, 1124–1137.

Sexton, A. N., Wang, P. Y., Rutenberg-Schoenberg, M., & Simon, M. D. (2017). Interpreting reverse transcriptase termination and mutation events for greater insight into the chemical probing of RNA. *Biochemistry*, *56*, 4713–4721.

Sharma, S., Langhendries, J. L., Watzinger, P., Kotter, P., Entian, K. D., & Lafontaine, D. L. (2015). Yeast Kre33 and human NAT10 are conserved 18S rRNA cytosine acetyltransferases that modify tRNAs assisted by the adaptor Tan1/THUMPD1. *Nucleic Acids Research*, *43*, 2242–2258.

Sharma, S., Marchand, V., Motorin, Y., & Lafontaine, D. L. J. (2017). Identification of sites of 2'-O-methylation vulnerability in human ribosomal RNAs by systematic mapping. *Scientific Reports*, 7, 11490.

Sharma, S., Yang, J., van Nues, R., Watzinger, P., Kotter, P., Lafontaine, D. L. J., et al. (2017). Specialized box C/D snoRNPs act as antisense guides to target RNA base acetylation. *PLoS Genetics*, *13*, e1006804.

Sinclair, W. R., Arango, D., Shrimp, J. H., Zengeya, T. T., Thomas, J. M., Montgomery, D. C., et al. (2017). Profiling cytidine acetylation with specific affinity and reactivity. *ACS Chemical Biology*, *12*, 2922–2926.

Sloan, K. E., Warda, A. S., Sharma, S., Entian, K. D., Lafontaine, D. L. J., & Bohnsack, M. T. (2017). Tuning the ribosome: The influence of rRNA modification on eukaryotic ribosome biogenesis and function. *RNA Biology*, *14*, 1138–1152.

Staehelin, M., Rogg, H., Baguley, B. C., Ginsberg, T., & Wehrli, W. (1968). Structure of a mammalian serine tRNA. *Nature*, *219*, 1363–1365.

Stern, L., & Schulman, L. H. (1978). The role of the minor base N4-acetylcytidine in the function of the Escherichia coli noninitiator methionine transfer RNA. *The Journal of Biological Chemistry*, *253*, 6132–6139.

Tan, T. Z., Miow, Q. H., Huang, R. Y., Wong, M. K., Ye, J., Lau, J. A., et al. (2013). Functional genomics identifies five distinct molecular subtypes with clinical relevance and pathways for growth control in epithelial ovarian cancer. *EMBO Molecular Medicine*, *5*, 1051–1066.

Taniguchi, T., Miyauchi, K., Sakaguchi, Y., Yamashita, S., Soma, A., Tomita, K., et al. (2018). Acetate-dependent tRNA acetylation required for decoding fidelity in protein synthesis. *Nature Chemical Biology*, *14*, 1010–1020.

Tardu, M., Lin, Q., & Koutmou, K. (2018). N4-acetylcytidine and 5-formylcytidine are present in Saccharomyces cerevisiae mRNAs. *biorxiv*. https://doi.org/10.1101/327585.

Thomas, J. M., Briney, C. A., Nance, K. D., Lopez, J. E., Thorpe, A. L., Fox, S. D., et al. (2018). A chemical signature for cytidine acetylation in RNA. *Journal of the American Chemical Society, 140*, 12667–12670.

Thomas, G., Gordon, J., & Rogg, H. (1978). N4-Acetylcytidine. A previously unidentified labile component of the small subunit of eukaryotic ribosomes. *The Journal of Biological Chemistry, 253*, 1101–1105.

Tschida, B. R., Temiz, N. A., Kuka, T. P., Lee, L. A., Riordan, J. D., Tierrablanca, C. A., et al. (2017). Sleeping beauty insertional mutagenesis in mice identifies drivers of steatosis-associated hepatic tumors. *Cancer Research, 77*, 6576–6588.

Wang, T., Birsoy, K., Hughes, N. W., Krupczak, K. M., Post, Y., Wei, J. J., et al. (2015). Identification and characterization of essential genes in the human genome. *Science, 350*, 1096–1101.

Wintermeyer, W., & Zachau, H. G. (1970). A specific chemical chain scission of tRNA at 7-methylguanosine. *FEBS Letters, 11*, 160–164.

Zachau, H. G., Dutting, D., & Feldmann, H. (1966). Nucleotide sequences of two serine-specific transfer ribonucleic acids. *Angewandte Chemie (International Ed. in English), 5*, 422.

Zhang, H., Hou, W., Wang, H. L., Liu, H. J., Jia, X. Y., Zheng, X. Z., et al. (2014). GSK-3beta-regulated N-acetyltransferase 10 is involved in colorectal cancer invasion. *Clinical Cancer Research, 20*, 4717–4729.

Zheng, G., Qin, Y., Clark, W. C., Dai, Q., Yi, C., He, C., et al. (2015). Efficient and quantitative high-throughput tRNA sequencing. *Nature Methods, 12*, 835–837.

Zhou, K. I., Clark, W. C., Pan, D. W., Eckwahl, M. J., Dai, Q., & Pan, T. (2018). Pseudouridines have context-dependent mutation and stop rates in high-throughput sequencing. *RNA Biology, 15*, 892–900.

CHAPTER FOUR

A guide for drug inducible genome editing with HIT systems

Chen Zhao[a,*], Shixian Wei[b], Yu Wang[a,b,c,*]

[a]State Key Laboratory of Stem Cell and Reproductive Biology, Institute of Zoology, Chinese Academy of Sciences, Beijing, China
[b]University of Chinese Academy of Sciences, Beijing, China
[c]Institute for Stem Cell and Regeneration, Chinese Academy of Sciences, Beijing, China
*Corresponding authors: e-mail address: zhaochen@ioz.ac.cn; yuwang@post.harvard.edu

Contents

Abstract

Technologies toward precise control of biological events are desired for biomedical research and potential clinical applications. Our recently reported HIT systems based on CRISPR/Cas9 and TAL effectors can achieve temporal and dose dependent drug control for genome editing. Methods are presented for the application of these optimized HIT systems in human cells in this chapter.

1. Introduction

Systematic interpretation of gene function requires the ability to manipulate genes in a robust and controlled inducible manner. Recent discoveries of TALEs (transcription activator-like effectors) and CRISPR

Methods in Enzymology, Volume 621
ISSN 0076-6879
https://doi.org/10.1016/bs.mie.2019.02.031

(clusters of regularly interspaced short palindromic repeats) have enabled targeted genetic modifications in cultured cells, as well as in whole organisms with high efficiency and specificity (Kim & Kim, 2014; Wei et al., 2013). Investigating gene functions under drug control may provide potential avenues toward safer gene therapies because it reduces undesired off-target events and allows genetic modification in greater precision. One way to achieve drug control is by utilizing estrogen receptor (ER), as exemplified by its successful coupling with Cre recombinase (Branda & Dymecki, 2004; Indra et al., 1999). Upon ligand binding, ER translocates from the cytoplasm to the nucleus. Consequently, fusion with an ER domain enables access of a protein to the nucleus, where genomic DNA resides in. ER^{T2}, an artificial ER mutant bearing three mutations G400V/M543A/L544A, shows high selectivity to the synthetic estrogen antagonist 4-OHT over endogenous ER ligand β-estradiol, reducing undesired perturbation from background activity (Feil, Wagner, Metzger, & Chambon, 1997). Inspired by the success of $CreER^{T2}$, we developed and optimized a series of chemical inducible devices based on the fusion of ER^{T2} to CRISPR and TALE systems, which we named hybrid inducible technologies (HIT). Herein, we describe the methods for drug inducible genome editing with our HIT systems.

1.1 Introduction to HIT-Cas9

The CRISPR/Cas (clustered, regularly interspaced, short palindromic repeats/CRISPR associated proteins) systems are adaptive immune defense machineries in bacteria and archaea (Bhaya, Davison, & Barrangou, 2011). Among them, type II CRISPR/Cas9 system has been widely used in site-directed genome editing in eukaryotic cells because of its ease of use and high efficiency (Mali, Esvelt, & Church, 2013; Mali, Yang, et al., 2013). With the lead of guide RNA (gRNA), Cas9 protein binds to its complementary DNA adjacent to a protospacer-adjacent motif (PAM), a species specific requirement for Cas9 recognition (Cong et al., 2013; Jinek et al., 2012; Mali, Esvelt, et al., 2013; Mali, Yang, et al., 2013). Upon binding, the nuclease activity of Cas9 protein introduces DNA double-strand breaks, which induces targeted gene editing from cellular DNA-damage response, either through error prone non-homologous end-joining (NHEJ) or precise homology directed repair (HDR) (Cong et al., 2013; Jinek et al., 2012; Mali, Esvelt, et al., 2013; Mali, Yang, et al., 2013).

To achieve drug inducible genome editing using CRISPR/Cas9 technology, we envisioned engrafting ER^{T2} to Cas9, a method to control the access of Cas9 protein to genome. After a series of optimization, we concluded at a HIT-Cas9 construct (Zhao, Zhao, et al., 2018) fusing two nuclear export signal (NES) peptides and two tandem ER^{T2} domains on the C-terminus of Cas9 (Fig. 1). The examination of both NHEJ and HDR events induced by C2N2E demonstrated that tight and efficient genome editing was accomplished in a 4-OHT inducible fashion across multiple human cell lines including a human embryonic stem cell (hESC) line H9, adipose derived human mesenchymal stem cells (MSCs), and a liver cancer cell line HepG2 (Zhao, Zhao, et al., 2018). In addition, HIT-Cas9 did not introduce significant off-target activity at several known sites. We also demonstrated that drug induction is titratable, selective and rapid, which grants further precision.

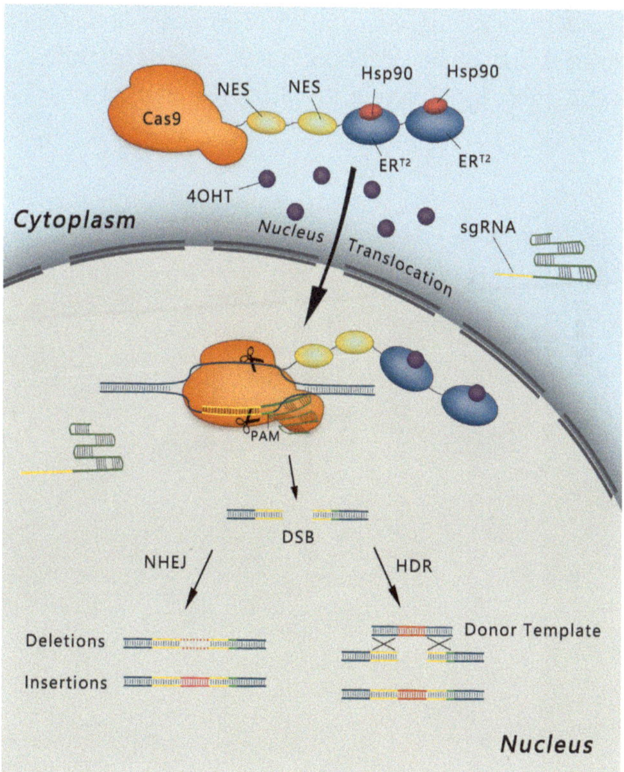

Fig. 1 Cartoon illuminating the mechanism of HIT-Cas9 system.

1.2 Introduction to HIT-TALEN

As transcriptional activators, TALEs found in plant pathogenic *Xanthomonas* can specifically bind and regulate genes during pathogenesis (Miller et al., 2011). TAL effectors are comprised of a central repeat domain mediating DNA recognition, nuclear localization signals (NLS), and an acidic transcriptional activation domain (Boch et al., 2009). Each repeat contains 33–35 amino acids specifying one target base due to its repeat variable diresidues (RVDs) in the middle (Boch et al., 2009; Moscou & Bogdanove, 2009). Fusion of a TAL effector DNA binding domain and a *Fok*I cleavage domain generates a TAL effector nuclease (TALEN), a powerful tool for targeted gene editing (Huang et al., 2011; Joung & Sander, 2013; Miller et al., 2011).

Similar to HIT-Cas9, Hybrid Inducible TALEN (HIT-TALEN) was generated based on fusion of ER^{T2} and TALENs (Fig. 2). TALEN constructs with two tandem ER^{T2} N-terminal fusions enabled chemical inducible genomic DNA cleavage despite background activity, whereas single fusion did not (Zhao, Zhang, et al., 2018). We further optimized the experimental protocol and observed background reduction while maintaining drug inducible effect by lowering the delivered amount of HIT-TALEN plasmid DNA.

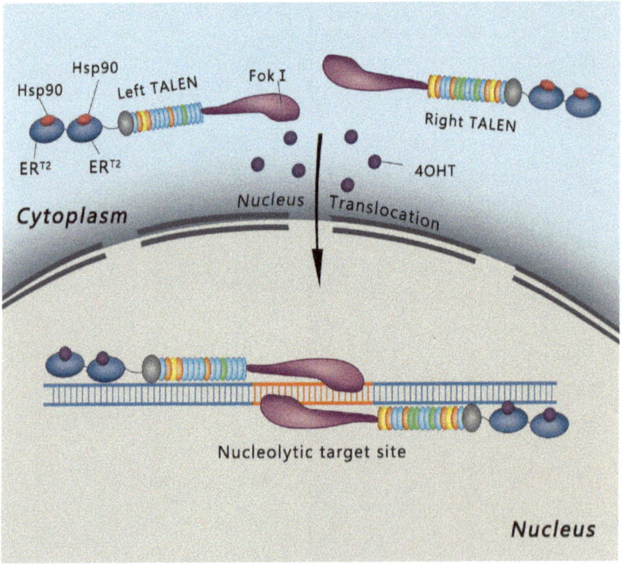

Fig. 2 Cartoon illuminating the mechanism of HIT-TALEN system.

2. Drug inducible genome editing by transient transfection

2.1 Transient transfection of HIT-systems into human cells

In the HIT-Cas9 system, effective sgRNA(s) targeting specific gene locus is required. Drug inducible genome editing can be achieved by co-transfecting C2N2E and sgRNA into target cells using transfection reagent. As for HIT-TALEN system, a pair of customized TALENs targeting specific gene sites is fused with two tandem ER^{T2} for controlled gene editing. We here described the transfection protocol for both HIT-systems using HEK293T cells as an example.

2.1.1 Equipment

- CO_2 incubator

2.1.2 Buffers and reagents

- Phosphate buffer saline (PBS)—137 mM NaCl, 2.7 mM KCl, 10 mM Na_2HPO_4, and 2 mM KH_2PO_4
- 0.25% Trypsin (Thermo Fisher)
- Dulbecco's Modified Eagle Medium (Thermo Fisher)
- Fetal Bovin Serum (FBS) (Thermo Fisher)
- GlutaMAX (Thermo Fisher)
- Penicillin-Streptomycin, 100 × (Thermo Fisher)
- Transfection reagent—Polyethylenimine (PEI), Linear, MW 25000, 1 mg/mL
- Opti-MEM (Thermo Fisher)
- 4-Hydroxytamoxifen (4-OHT)—Dissolve 10 mg of 4-OHT in 2.58 mL of DMSO to reach a stock concentration of 10 mM, then aliquot and store it at −20 °C

2.1.3 Procedure

This protocol is used to transfect HIT-Cas9 or HIT-TALEN systems into HEK293T cells in a 24-well plate. All amounts and volumes are given on a per well basis. Readers can scale up or down according to cell number and plate type.

1. HEK293T cells are maintained in Dulbecco's Modified Eagle's Medium (DMEM) supplemented with 10% FBS, 2 mM GlutaMAX (Thermo Fisher), 100 U/mL penicillin and 100 μg/mL streptomycin at 37 °C and with 5% CO_2 until the cultures are over 80% confluent.
2. The day before transfection, cells are washed by PBS and then dispersed using 0.25% trypsin. After counting the cell number, around $1.5–2 \times 10^5$ cells are pre-seeded per well in 500 μL of complete growth medium.
3. On the day of transfection, a total of 800 ng of HIT-Cas9 constructs (C2N2E and sgRNA at a molar ratio of 1:1.2) is used. As for HIT-TALEN system, 200 ng of each $2ER^{T2}$-TALEN in the pair are enough for low background activity under drug control. To achieve tight and effective genome editing using HIT-TALEN, readers can also decrease the amount of either one $2ER^{T2}$-TALEN construct under 100 ng while keeping 500 ng of another one. For each well, dilute both plasmids in 25 μL of Opti-MEM and incubate at room temperature for 5 min.
4. For each well, dilute 2.4 μL of PEI (1 mg/mL) in 25 μL of Opti-MEM and incubate at room temperature for 5 min.
5. Mix DNA and transfection reagent solutions gently and incubate 15–20 min at room temperature to form transfection complexes.
6. Transfer 50 μL of DNA-PEI complexes directly to each well and mix gently by rocking the plate back and forth.
7. Incubate the cells at 37 °C in a CO_2 incubator for 4–6 h, then replace the medium of each well with fresh complete growth medium supplemented with 500 nM 4-OHT.
8. Optional: Co-transfect a control plasmid expressing GFP or other fluorescent proteins with HIT-systems constructs to examine transfection efficiency, particularly for those cells with low transfection efficiency, such as MSCs, Hela and HepG2 cells.

2.1.4 Notes

1. The optimal cell confluence for transfection is approximately 60–70%. Higher or lower cell density may decrease the transfection efficiency.
2. All plasmids applied for the transfection must be endotoxin free.
3. The ratio of DNA (μg):PEI (μL) varies from the cells applied for the transfection. Basically, 1:3 ratio of DNA:PEI is suitable for most cell types such as HEK293T cells, Hela cells and HepG2 cells, while 1:1 ratio works better for mesenchymal stem cell (MSC) with lower cellular toxicity.
4. Transfection efficiency can be improved if the medium is changed prior to the incubation with DNA-PEI complexes.

2.2 Harvest of transfected cells and preparation of genomic DNA

For cells with high transfection efficiency such as HEK293T, genomic DNA can be prepared directly from transfected cells by treating with lysis buffer. Otherwise, fluorescence-activated cell sorting (FACS) is recommended for the enrichment of the transfected cells.

2.2.1 Equipment

- Centrifuge (Eppendorf)
- Digital Dry Bath Incubator

2.2.2 Buffers and reagents

- Phosphate buffer saline (PBS)—137 mM NaCl, 2.7 mM KCl, 10 mM Na_2HPO_4, and 2 mM KH_2PO_4
- 0.25% Trypsin (Thermo Fisher)
- PCR Lysis Buffer—10 mM Tris-Cl (pH 8.0), 2 mM EDTA (pH 8.0), 2.5% (vol/vol) Tween-20, 2.5% (vol/vol) Triton-X 100, 100 μg/mL of Proteinase K

2.2.3 Procedure

1. After 48 h induction of 4-OHT, cells are washed by PBS and then dispersed using 0.25% trypsin.
2. Stop digestion by adding complete growth medium and collect cells by centrifugation at 1000 rpm for 5 min.
3. The pellets are washed with PBS and centrifuged again at 1000 rpm for 5 min.
4. For cells with relative low transfection efficiency, it is recommended to co-transfect a plasmid expressing fluorescent proteins such as GFP or RFP, and thus the transfected cells can be enriched by sorting fluorescent cells using flow cytometry. Collect all selected cells by centrifuging at 1000 rpm for 5 min.
5. Remove the supernatant and add the lysis buffer at a concentration of no less than 1000 cells/μL to resuspend the pellets followed by a lysis procedure: 50 °C for 60 min, 95 °C for 15 min. Store the cell lysate at −20 °C until use. Typically, 1 μL of prepared lysate is used in a 25 μL PCR reaction system.

2.2.4 Notes

1. Cell confluence after 48 h of incubation with 4-OHT should be at least over 80% for HEK293T cells and proximately 100% for MSCs, Hela and HepG2 cells.
2. Lysis buffer can be directly added to the cells without trypsination if fluorescence-activated cell sorting process is not required. Usually, 100 μL of lysis buffer is enough for one well in a 24-well plate.
3. Homogenize cell lysate and PCR reaction by pipetting up and down.
4. Dry bath incubator can be replaced by a PCR instrument to perform lysis procedure or add corn oil to avoid the evaporation if a small sample volume (below 20 μL) is applied.

2.3 Examination of genome editing under drug control

In this section, we describe the protocol to examine genome editing of an EMX1 site by HIT-Cas9 in HEK293T cells. Two paralleled methods, Surveyor assay and TIDE assay, are used to evaluate the editing efficiency.

2.3.1 Equipment

- PCR system (Thermo Fisher)
- Vertical & Horizontal electrophoresis System
- Digital Dry Bath Incubator
- Gel imaging system (BioRad)

2.3.2 Buffer and reagents

- AccuPrime Taq DNA Polymerase System (Thermo Fisher)
- 1 × TAE—40 mM Tris base, 20 mM acetic acid and 1 mM EDTA
- 1% Agarose gel—Dissolve 1 g agarose in 100 mL distilled water
- DNA dye—10 mg/mL Ethidium Bromide (EB)
- 6 × Loading buffer (NEB)
- SURVEYOR Mutation Detection Kit (Idtdan Transgenomics)
- 10 × TBE—0.9 M Tris base, 0.9 M boric acid, 20 mM EDTA (pH 8.0)
- 20% Ammonium persulfate (APS)—Dissolve 2 g of APS in 10 mL distilled water, then aliquot and store it at −20 °C until use
- 3.5% Polyacrylamide gel (stacking gel)—850 μL of 30% Acrylamide-Bisacrylamide (29:1), 750 μL of 10 × TBE, 25 μL of 20% Ammonium persulfate (APS), 5 μL of *N,N,N′,N′*-tetramethylethylenediamine (TEMED), 5.9 mL of distilled water
- 15% Polyacry-lamide gel (separating gel)—9 mL of 30% Acrylamide-Bisacrylamide (29:1), 1.8 mL of 10 × TBE, 75 μL of 20% Ammonium persulfate (APS), 7.5 μL of *N,N,N′,N′*-tetramethylethylenediamine (TEMED), 7.2 mL of distilled water

2.3.3 Procedure

1. Prepare a 25 μL of PCR reaction mix in a 0.2 mL PCR tube to amplify DNA fragment containing edited gene sequences using AccuPrime Tag DNA Polymerase. The mixture includes 500 nM of each primer, 1 × AccuPrime™ PCR Buffer II, 1 μL of genomic DNA from Section 2.2, 1 μL of AccuPrime™ Taq DNA Polymerase.
2. Place PCR tube in the PCR system and perform the following program: pre-denaturalized at 94 °C for 4 min, then 35 cycles of 94 °C for 20 s, 59 °C for 20 s and 68 °C for 35 s, finally 5 min at 68 °C. The annealing temperature varies depending on the primers designed for the amplification.
3. Pipette 5 μL of PCR products for agarose gel electrophoresis to determine successful amplification.
4. Keep 5 μL of PCR products for SURVEYOR examination and leave the rest of PCR products for Sanger sequencing.
5. SURVEYOR examination:
   - Mix 5 μL of PCR products with 4.5 μL of 1 × AccuPrime™ PCR Buffer II in a 0.2 mL PCR tube and perform the following program: denatured by heating to 95 °C and slowly reannealed from 95 °C to 85 °C by −2 °C/s followed by −0.1 °C/s from 85 °C to 25 °C for rehybridization using PCR system.
   - Add 0.5 μL of Transgenomic's SURVEYOR Nuclease into the reannealed products. Mix it gently and incubate at 42 °C for 20–30 min.
   - Add 1 × loading buffer to stop the reaction and apply all products for polyacrylamide gel (PAGE) electrophoresis. The gel is run usually at 100 V for 2 h and then stained by EB to visualize in Gel imaging system.
6. The sequencing results can be analyzed using TIDE web tool (https://tide.nki.nl/), while the results from PAGE electrophoresis can be analyzed by imageJ software.

2.3.4 Notes

1. To reduce non-specific cleavage by SURVEYOR nuclease, the PCR primers must be highly specific to produce a single robust DNA band. PCR primers with high T_m values often produce optimal results according to our experiences.
2. Pre-run the PAGE gel for half an hour at 80 V for clearer gel image.
3. Duration to incubate the mixture of reannealed products and SURVEYOR Nuclease at 42 °C is recommended not to be over 40 min to avoid non-specific cleavage.
4. Design additional primers flanking the cleavage site at a distance of 150 bp for Sanger sequencing and TIDE analysis.

3. Drug inducible genome editing by lentiviral transduction

In certain circumstances such as scaled applications, it is desired to generate monoclonal stable cell lines expressing C2N2E and sgRNA(s) or a pair of 2ER^{T2} fused TALEN via lentiviral transduction. All plasmids introduced in this chapter are constructed using lentiviral expression vectors, which contain antibiotic resistant genes for enriching successfully infected cells. This protocol is an example to perform drug inducible editing of EMX1 site in human embryonic stem cell line H9.

3.1 Lentiviral packaging and transduction

We use three plasmid system to transfect 293T cells for lentivirus production. The packaged viruses need to be further concentrated by Amicon Ultra-15 centrifugal filter device before the infection of cells.

3.1.1 Equipment

- Biosafety cabinet (Thermo Fisher)
- CO_2 incubator (Thermo Fisher)
- Refrigerated centrifuge with horizontal rotor (Eppendorf)

3.1.2 Buffer and reagents

- Buffer and reagents used for transfection assay (see Section 2.1.2)
- Essential 8 Medium (Thermo Fisher)
- 0.5 M EDTA (Thermo Fisher)
- Polybrene (hexadimethrine bromide, Sigma)—Dissolve 10 mg of polybrene in 1 mL of ultrapure water, sterilize by passage through a 0.2 μm filter and store at −20 °C until use.

3.1.3 Procedure

HIT-Cas9 device includes two constructs, C2N2E and sgRNA, containing different antibiotic resistant genes (zeocin and puromycin, respectively). HIT-TALEN system contains a pair of 2ER^{T2} fused TALEN targeting specific gene loci. The antibiotic resistant genes within the two constructs are different (zeocin and neomycin), too. All plasmids are constructed using lentiviral transfer plasmid. Co-transfect each transfer plasmid and packaging and envelope plasmids into the HEK293T cells for virus production.

1. The procedure for transfection is the same as what described in Section 2.1.3. The molar ratio of three plasmids (transfer plasmid:packaging plasmid:envelope plasmid) is 8:6:3 in our system.
2. Transfection assay is performed using HEK293T cells pre-seeded in a 10 cm cell culture dish. The cell culture supernatant is collected twice, 48 and 72 h after transfection. The total amount of supernatant is 15 mL:10 mL of medium from the first replacement 6 h post transfection and 5 mL from the second collection of supernatant.
3. Collect all the cell culture supernatant into a 50 mL sterile centrifuge tube and centrifuge at 2000 rpm in a refrigerated centrifuge for 5 min at 4 °C to pellet the cellular debris.
4. Filter the cell supernatant through a 0.45 μm filter to remove any remaining cell debris and large particles.
5. Transfer the filtrate into the Amicon Ultra-15 centrifugal filter device and then centrifuge at 4000 rpm for 20–30 min at 4 °C. The volume of concentrated cell supernatant is normally about 1 mL after the centrifugation.
6. Dispense the concentrated virus into 200 μL aliquots in 1.5 mL microcentrifuge tube and store at −80 °C until use.
7. Before the day of transduction, H9 cells by 80% confluence are passaged from one well of six-well plate to a matrigel pre-coated six-well plate at a dilution of 1:10 using 0.5 M EDTA. The cell confluence should be around 30–40% on the day of transduction.
8. Mix 1.8 mL of Essential 8 Medium, 2 μL of polybrene and 200 μL of concentrated viral solution in a 15 mL microcentrifuge tube. Change the medium with the mixture and incubate at 37 °C in a CO_2 incubator for 6–8 h followed by the replacement of fresh E8 medium. Change the medium every day and add antibiotic into the E8 medium after 2 days of transduction before screening monoclones.

3.1.4 Notes

1. The initial density of the HEK293T cells is critical for successful viral packaging. The cell confluence on the day of transfection should be between 60% and 80%.
2. Concentration of viral supernatant is necessary because the large size of the lentiviral constructs used in HIT-Cas9 and HIT-TALEN systems, which results in low titer.
3. It is not recommended that lentiviral supernatant is subjected to multiple freeze-thaw cycles. Do not use the viral supernatants if it has been stored at −80 °C over 6 months.

3.2 Generation of monoclonal stable cell line with HIT-systems

This protocol can be used to generate stable cell lines expressing all necessary constructs of HIT-systems. Optimal drug inducible genome editing can be achieved by carefully identifying monoclonal H9 cell line upon antibiotic selection.

3.2.1 *Equipment*

- Biosafety cabinet
- CO_2 incubator

3.2.2 *Buffer and reagents*

- Essential 8 Medium (Thermo Fisher)
- Zeocin (10 mg/mL, Invivogen)
- Puromycin (10 mg/mL, Invivogen)
- Rock inhibitor (Y-27632, 10 mM, Selleck)
- TryplE Express (Thermo Fisher)

3.2.3 *Procedure*

1. H9 ESCs are maintained in E8 medium supplemented with antibiotic and 10 μM rock inhibitor at 37 °C and with 5% CO_2 after 48 h of transduction. The screening concentration used for zeocin is 2.5 μg/mL when infected with C2N2E viral particles, while that for puromycin is 1 μg/mL after sgRNA viral infection. The concentration applied for neomycin is 100 μg/mL when establishing the HIT-TALEN stable cell line.
2. Observe the infected H9 cells every day until the cells in the untransduced well are dead completely. Prepare cell stocks at liquid nitrogen once the polyclonal populations of resistant cells are expanded sufficiently.
3. Isolate individual cells from the polyclonal resistant cells by incubating with Tryple-E for 2–3 min to form the homogenized cell solution. Quantitate the cell concentration by a hemocytometer and transfer cells into a 10 cm cell culture dish precoated with matrigel at a concentration of 50 cells/mL. Cells can also be passaged to matrigel coated 96-well plate at a concentration of 10 cells/mL for monoclonal selection. Single-cell sorting by flow cytometry is another method to generate monoclonal cell line.
4. Maintain the cells in E8 medium supplemented with antibiotic and rock inhibitor in a CO_2 incubator for 7–14 days. Change the conditional E8 medium every 2 days within the first 7 days of culture and scan the plate for cell growth after 7 days.

5. Remove rock inhibitor from E8 medium once the cells appear as colonies in the well.
6. Transfer the monoclonal lines to 12-well plate by incubating with 0.5 M EDTA for 5 min at 37 °C when the cells have expanded to be confluent. Freeze replicated cell stocks in liquid nitrogen before screening the transgene expression and its function in genome editing.

3.2.4 Notes

1. It is recommended to generate and screen monoclonal H9 cell lines in two steps: first expressing C2N2E construct at a level for efficient and tight drug inducible performance, and second with sgRNA viral delivery.
2. The antibiotic selection should last at least 2 weeks for zeocin and neomycin or 1 week for puromycin until the control cells completely die. Use the same concentration of antibiotic for the following monoclonal selection.
3. Rock inhibitor is required when screening the stable cell lines by antibiotic selection, particularly for the isolation and growth of a monoclonal cell population.

3.3 Examination of drug inducible genome editing for monoclonal H9 cell lines

The methods to examine drug inducible genome editing are the same as what described in Section 2.3. It is noted that the concentration of 4-OHT applied for H9 cells is lower than that used for HEK293T cells due to the high sensitivity of H9 cells to drug induced cytotoxicity.

In addition, it is important to verify the expression of Cas9 or TALE protein prior to examination of genome editing efficiency for monoclonal cell lines. This should be executed for those monoclonal H9 cells survived after antibiotic selection. Both quantitative real-time PCR and western blot can be used to evaluate transgene expression.

3.3.1 Equipment

- See Section 2.3.

3.3.2 Buffer and reagents

- See Section 2.3.

3.3.3 Procedure

1. Maintain monoclonal H9 cell lines survived after antibiotic selection in E8 medium. Plate each clone in duplicate at 24-well plate and treat the cells with or without 250nM of 4-OHT.
2. Extract genomic DNA from each well after 3 days of 4-OHT induction and perform PCR amplification as described in Section 2.3.
3. Examine the genome editing events by Sanger sequencing and SURVEYOR assay as described in Section 2.3 and compare the efficiency between 4-OHT treated and untreated samples. Select monoclones displaying high editing efficiency and low background activity for the future experiments.

4. Summary and conclusions

We have designed and vigorously optimized drug inducible systems for genome editing based on CRISPR/Cas9 and TAL effectors via transient transfection in 293T cells. These systems displayed high efficiency and low background activity, and open new opportunities for precise chemical control of dynamic biological processes. Here we discussed their application by lipid particle based transient transfection and lentiviral infection. Other scenarios exist, such as in vivo delivery, and the protocols will need customized optimization to achieve optimal performance in a drug inducible manner.

Acknowledgments

We are grateful for all the members of the Wang lab for helpful discussions and technical advices. We thank support from Strategic Priority Research Program of the Chinese Academy of Sciences (XDA16010500), the National Basic Research Program of China (2015CB964800), the National Natural Science Foundation of China (31571514 and 21701280), and State Key Laboratory of Stem Cell and Reproductive Biology.

References

Bhaya, D., Davison, M., & Barrangou, R. (2011). CRISPR-Cas systems in bacteria and archaea: Versatile small RNAs for adaptive defense and regulation. *Annual Review of Genetics*, *45*, 273–297.

Boch, J., Scholze, H., Schornack, S., Landgraf, A., Hahn, S., Kay, S., et al. (2009). Breaking the code of DNA binding specificity of TAL-type III effectors. *Science*, *326*(5959), 1509–1512.

Branda, C. S., & Dymecki, S. M. (2004). Talking about a revolution: The impact of site-specific recombinases on genetic analyses in mice. *Developmental Cell*, *6*(1), 7–28.

Cong, L., Ran, F. A., Cox, D., Lin, S., Barretto, R., Habib, N., et al. (2013). Multiplex genome engineering using CRISPR/Cas systems. *Science*, *339*(6121), 819–823.

Feil, R., Wagner, J., Metzger, D., & Chambon, P. (1997). Regulation of Cre recombinase activity by mutated estrogen receptor ligand-binding domains. *Biochemical and Biophysical Research Communications*, *237*(3), 752–757.

Huang, P., Xiao, A., Zhou, M., Zhu, Z., Lin, S., & Zhang, B. (2011). Heritable gene targeting in zebrafish using customized TALENs. *Nature Biotechnology*, *29*(8), 699–700.

Indra, A. K., Warot, X., Brocard, J., Bornert, J. M., Xiao, J. H., Chambon, P., et al. (1999). Temporally-controlled site-specific mutagenesis in the basal layer of the epidermis: Comparison of the recombinase activity of the tamoxifen-inducible Cre-ERT and Cre-ERT2 recombinases. *Nucleic Acids Research*, *27*(22), 4324–4327.

Jinek, M., Chylinski, K., Fonfara, I., Hauer, M., Doudna, J. A., & Charpentier, E. (2012). A programmable dual-RNA-guided DNA endonuclease in adaptive bacterial immunity. *Science*, *337*(6096), 816–821.

Joung, J. K., & Sander, J. D. (2013). TALENs: A widely applicable technology for targeted genome editing. *Nature Reviews. Molecular Cell Biology*, *14*(1), 49–55.

Kim, H., & Kim, J. S. (2014). A guide to genome engineering with programmable nucleases. *Nature Reviews. Genetics*, *15*(5), 321–334.

Mali, P., Esvelt, K. M., & Church, G. M. (2013). Cas9 as a versatile tool for engineering biology. *Nature Methods*, *10*(10), 957–963.

Mali, P., Yang, L., Esvelt, K. M., Aach, J., Guell, M., DiCarlo, J. E., et al. (2013). RNA-guided human genome engineering via Cas9. *Science*, *339*(6121), 823–826.

Miller, J. C., Tan, S., Qiao, G., Barlow, K. A., Wang, J., Xia, D. F., et al. (2011). A TALE nuclease architecture for efficient genome editing. *Nature Biotechnology*, *29*(2), 143–148.

Moscou, M. J., & Bogdanove, A. J. (2009). A simple cipher governs DNA recognition by TAL effectors. *Science*, *326*(5959), 1501.

Wei, C., Liu, J., Yu, Z., Zhang, B., Gao, G., & Jiao, R. (2013). TALEN or Cas9—Rapid, efficient and specific choices for genome modifications. *Journal of Genetics and Genomics*, *40*(6), 281–289.

Zhao, C., Zhang, Y., Zhao, Y., Ying, Y., Ai, R., Zhang, J., et al. (2018). Multiple chemical inducible Tal effectors for genome editing and transcription activation. *ACS Chemical Biology*, *13*(3), 609–617.

Zhao, C., Zhao, Y., Zhang, J., Lu, J., Chen, L., Zhang, Y., et al. (2018). HIT-Cas9: A CRISPR/Cas9 genome-editing device under tight and effective drug control. *Molecular Therapy. Nucleic Acids*, *13*, 208–219.

Further reading

Certo, M. T., Ryu, B. Y., Annis, J. E., Garibov, M., Jarjour, J., Rawlings, D. J., et al. (2011). Tracking genome engineering outcome at individual DNA breakpoints. *Nature Methods*, *8*(8), 671–676.

Davis, K. M., Pattanayak, V., Thompson, D. B., Zuris, J. A., & Liu, D. R. (2015). Small molecule-triggered Cas9 protein with improved genome-editing specificity. *Nature Chemical Biology*, *11*(5), 316–318.

Dow, L. E., Fisher, J., O'Rourke, K. P., Muley, A., Kastenhuber, E. R., Livshits, G., et al. (2015). Inducible in vivo genome editing with CRISPR-Cas9. *Nature Biotechnology*, *33*(4), 390–394.

Gonzalez, F., Zhu, Z., Shi, Z. D., Lelli, K., Verma, N., Li, Q. V., et al. (2014). An iCRISPR platform for rapid, multiplexable, and inducible genome editing in human pluripotent stem cells. *Cell Stem Cell*, *15*(2), 215–226.

Liu, K. I., Ramli, M. N., Woo, C. W., Wang, Y., Zhao, T., Zhang, X., et al. (2016). A chemical-inducible CRISPR-Cas9 system for rapid control of genome editing. *Nature Chemical Biology*, *12*(11), 980–987.

Wright, A. V., Nunez, J. K., & Doudna, J. A. (2016). Biology and applications of CRISPR systems: Harnessing nature's toolbox for genome engineering. *Cell*, *164*(1–2), 29–44.

Zetsche, B., Volz, S. E., & Zhang, F. (2015). A split-Cas9 architecture for inducible genome editing and transcription modulation. *Nature Biotechnology*, *33*(2), 139–142.

CHAPTER FIVE

A guide for drug inducible transcriptional activation with HIT systems

Chen Zhao[a,*], Shixian Wei[b], Yu Wang[a,b,c,*]

[a]State Key Laboratory of Stem Cell and Reproductive Biology, Institute of Zoology, Chinese Academy of Sciences, Beijing, China
[b]University of Chinese Academy of Sciences, Beijing, China
[c]Institute for Stem Cell and Regeneration, Chinese Academy of Sciences, Beijing, China
*Corresponding authors: e-mail address: zhaochen@ioz.ac.cn; yuwang@post.harvard.edu

Contents

Abstract

Precise investigation and manipulation of gene function often require modulation in a controlled and dynamic manner. In this chapter, we describe the methods to apply HIT systems for drug inducible transcriptional activation or simultaneous activation and genome editing in human cells. Together with those for editing, which are described in another chapter, HIT systems herein provide a valuable toolbox toward many biological applications, especially when precision and dynamics are required for a functional perturbation.

Methods in Enzymology, Volume 621
ISSN 0076-6879
https://doi.org/10.1016/bs.mie.2019.02.032

1. Introduction

Precise dissection of dynamic biological processes often requires perturbation on genetic and transcriptional levels in a controlled inducible manner. Programmable DNA-binding proteins provide powerful tools for targeted genome modulations. The clustered, regularly interspaced, short palindromic repeats (CRISPR)/CRISPR-associated protein 9 (Cas9) and transcription activator-like effectors (TALEs), as representatives widely used for site-specific gene editing, have been repurposed for transcriptional activation in mammalian cells (Bultmann et al., 2012; Cheng et al., 2013; Gilbert et al., 2013; Maeder et al., 2013; Mali, Aach, et al., 2013; Perez-Pinera et al., 2013; Zhang et al., 2011). Estrogen receptor (ER), as a representative of nuclear hormone receptors, when coupled with Cre recombinase, is one of the most broadly used chemical inducible technologies in genetic research (Branda & Dymecki, 2004; Indra et al., 1999). Utilizing its ligand-dependent translocation from the cytoplasm to the nucleus, ER defines the access of Cre to the loxP sites in the engineered genome, thus controlling the loxP recombination in a ligand-dependent manner (Metzger, Clifford, Chiba, & Chambon, 1995). To further decrease background noise, ER bearing three mutations G400V/M543A/L544A, also known as ER^{T2}, was identified to be highly selective to the synthetic estrogen antagonist 4-OHT over β-estradiol, the endogenous ER ligand, a crucial property for minimal perturbation from endogenous signal (Feil, Wagner, Metzger, & Chambon, 1997). Encouraged by the broad utility of $CreER^{T2}$ in many areas of biomedical research by putting genome engineering under drug control (Branda & Dymecki, 2004; Indra et al., 1999), we devised and comprehensively optimized a series of novel drug inducible transcriptional activation systems, which we named Hybrid Inducible Technologies (HIT) based on coupling CRISPR or TALE systems with ER^{T2} (Lu et al., 2018; Zhao, Zhang, Zhao, et al., 2018). Furthermore, we granted extra functionality of simultaneous genome editing within one device we called HIT2 (hitting two birds with one stone) (Lu et al., 2018). In this chapter, we describe protocols for applying these HIT systems for transcriptional activation of endogenous genes, and simultaneous genome editing in a drug inducible manner.

1.1 Introduction to drug inducible transcriptional activation devices

Clustered regularly interspaced short palindromic repeats (CRISPR)/CRISPR-associated (Cas) systems enable bacteria and archaea's defense against mobile genetic elements such as viruses, transposable elements and conjugative plasmids (Deltcheva et al., 2011). Upon their RNA-guided DNA binding and endonuclease activity, CRISPR/Cas9, the most widely used variant as a biotechnological tool, can bind and cleave DNA in a sequence-specific manner. DNA double-strand breaks (DSB) produced by the Cas9-gRNA complex initiates the recruitment of DNA repair machinery to the cleaved site, resulting in targeted gene editing either through non-homologous DNA end-joining (NHEJ) or homologous recombination (HR) (Cong et al., 2013; Jinek et al., 2012; Mali, Yang, et al., 2013). The utility of the CRISPR/Cas9 systems has been expanded to other molecular functions, including transcription activation, repression, genomic DNA labeling, and epigenetic programming, by fusing with various effector domains (Dominguez, Lim, & Qi, 2016; Maeder et al., 2013; Mali, Esvelt, & Church, 2013). Several techniques have been developed using catalytically inactive Cas9 (dCas9) variant retaining the DNA-binding activity (Jinek et al., 2012; Mali, Aach, et al., 2013). CRISPR/Cas9-based transcriptional activation (CRISPRa) system is a well-characterized tool for systematic gene overexpression studies (Chavez et al., 2016) by recruiting activation domains (ADs) to the dCas9-gRNA complex. Several groups generated different systems to render dCas9 the ability to activate gene expression, including direct fusion of ADs to dCas9 (Chavez et al., 2015), SAM (Hsu, Lander, & Zhang, 2014), and SunTag system (Tanenbaum, Gilbert, Qi, Weissman, & Vale, 2014). Based on these designs, we further developed drug inducible transcriptional activation systems by engrafting ER^{T2} to either dCas9 or ADs, and optimized each strategy for tight and effective transcriptional activation of endogenous genes. Finally, we obtained three optimized drug inducible transcriptional activation devices, which we named HIT-dCas9 (Fig. 1A), HIT-SAM (Fig. 1B) and HIT-SunTag (Fig. 1C), respectively. Among all three systems, HIT-SunTag design displayed most efficient transcriptional activation upon drug treatment. Therefore, we describe the protocol for using the HIT-SunTag system.

Transcription activator-like effectors (TALEs) is another powerful tool for targeted gene editing and transcriptional programming (Huang et al., 2011; Joung & Sander, 2013; Miller et al., 2011). Thanks to the well-deciphered recognition code in the central repeat domain of TAL effecter (Boch et al., 2009; Moscou & Bogdanove, 2009), it is convenient to customize TALEs with sequence-specific DNA-binding capabilities. Fusing a TAL array that targets a promoter with a transcription activator domain enables activation of gene transcription specifically (Zhang et al., 2011). To provide more options in the toolbox (Ding et al., 2013; Gao et al., 2014),

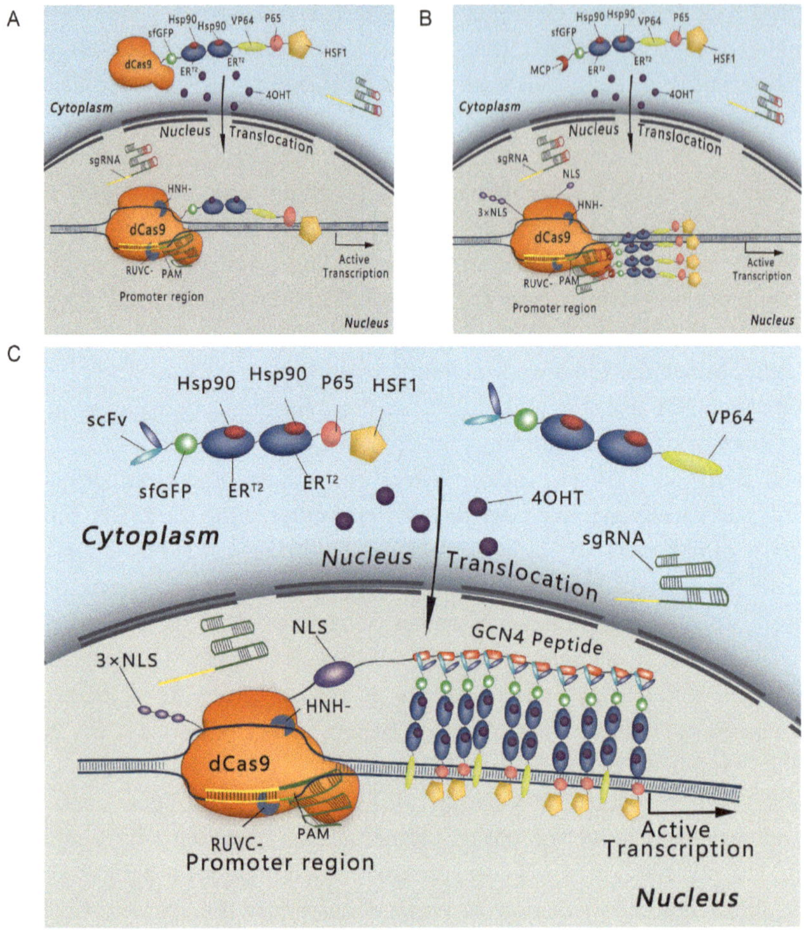

Fig. 1 Cartoons illuminating the mechanism of drug inducible transcriptional activation with HIT system based on CRISPR/Cas9. (A) HIT-dCas9 system; (B) HIT-SAM system; and (C) HIT-SunTag system.

we also developed and optimized a series of drug inducible TALEs by coupling with ER^{T2} for transcription activation. Similarly, the inducible TALE based on SunTag system, we named HIT-TALE-SunTag (Fig. 2), showed higher activation efficiency while keeping background activity low when compared with direct fusion TALEs. Therefore, we recommend to apply HIT-TALE-SunTag for drug inducible transcriptional activation when a TALE-based approach is preferred.

1.2 Introduction to simultaneous transcriptional activation and genome editing

Complex biological processes often involve functional regulation of multiple genes in multi-modes. It is of great use to develop techniques that can achieve multi-mode functional perturbations. It was reported previously that the nuclease activity of wild-type SpCas9 could be eliminated while its binding capability to target DNA retained when guided by a shortened sgRNA with less than 16 nucleotides (nt) (Dahlman et al., 2015; Kiani et al., 2015). Accordingly, we envisioned a drug inducible CRISPR/Cas9 device capable of both transcriptional activation and genome editing (Fig. 3) by combining the designs of HIT-Cas9 (Zhao, Zhao, Zhang, et al., 2018) and HIT-SunTag systems (Lu et al., 2018). By engrafting two tandem ER^{T2} to Cas9 and ADs separately, we observed significant 4-OHT induced activation via shortened sgRNA (14 nt) and editing activity via

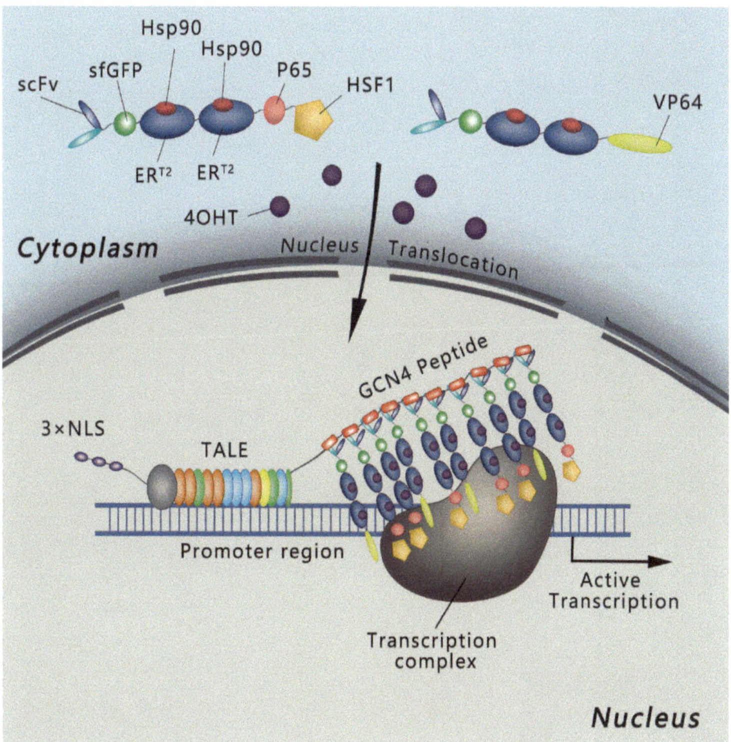

Fig. 2 Cartoon illuminating the mechanism of HIT-TALE-SunTag system.

normal sgRNA (20 nt) without significant background activity. This system was named HIT2, meaning hitting two birds with one stone, may provide a platform for multi-mode functional modulations in a precise and controlled manner.

2. Drug inducible transcriptional activation

Knowing that HIT systems can modulate the expression of a given gene in a drug inducible manner, a stable cell line expressing all effectors of HIT systems best suits consistent and controlled modulation. Nonetheless, drug inducible activation can also be accomplished with great performance upon transient transfection. The readers are recommended to refer to our

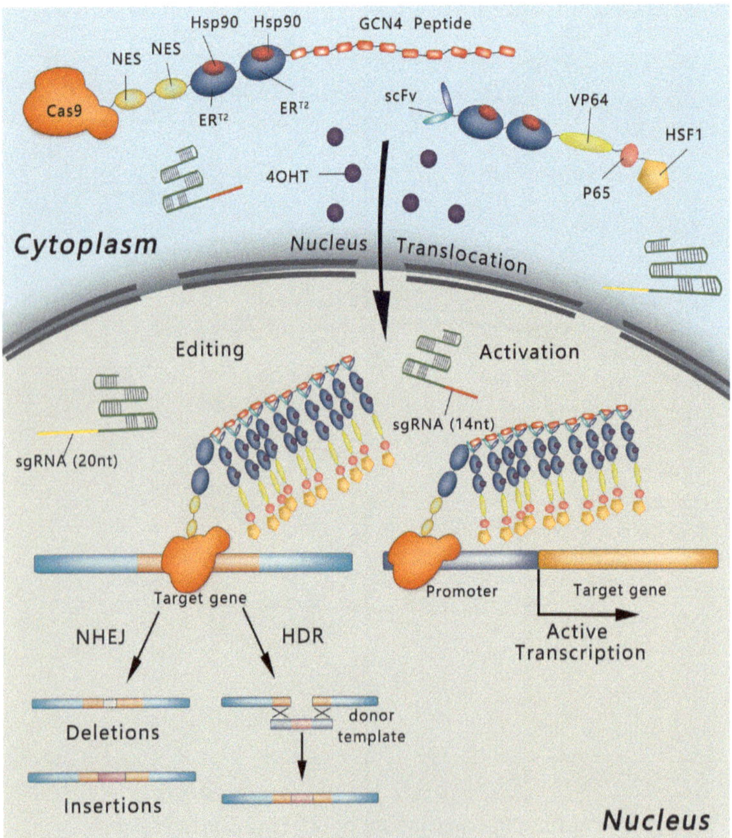

Fig. 3 Cartoon illuminating the mechanism of HIT2 system.

accompanying methods paper for drug inducible editing (Zhao, Zhao, Zhang, et al., 2018), for which similar protocol can be followed while modifying the readout. Among all developed drug inducible devices for transcriptional activation, SunTag-based systems (HIT-SunTag and HIT-TALE-SunTag) showed the highest activation efficiency while keeping background activity minimal. All plasmids of both two HIT systems are constructed as lentiviral vectors co-expressing antibiotic resistant genes. Stable cell lines can be established by lentiviral transduction and antibiotic selection. The transcriptional activation of endogenous genes can be achieved in a temporally controlled manner by the addition of 4-OHT in the culture medium of monoclonal cell lines.

2.1 Lentiviral packaging and the generation of monoclonal cell lines

Generally, both HIT-SunTag and HIT-TALE-SunTag systems consist of two parts, one expressing GCN4 fused dCas9 or TAL, and the other expressing scFv and ER^{T2} fused activation domains (ADs). The activation domains are usually split as VP64 and P65-HSF1 complex (V + PH) producing higher activation efficiency. However, more constructs might result in lower co-delivery efficiency to the same cell. Therefore, we suggest to use one scFv-ADs constructs containing VP64, P65 and HSF1 tandem fusion (VPH). The second-generation three plasmid system is used for lentiviral packaging in this protocol. Considering the wide application of embryonic stem cells (ESCs) in regenerative biology research, human H9 ES cell line is used as an example.

2.1.1 Equipment

- CO_2 incubator
- Biosafety cabinet (Thermo Fisher)
- Refrigerated centrifuge with horizontal rotor (Eppendorf)

2.1.2 Buffer and reagents

- Phosphate-buffered saline (PBS)—137 mM NaCl, 2.7 mM KCl, 10 mM Na_2HPO_4, and 2 mM KH_2PO_4.
- 0.25% Trypsin (Thermo Fisher)
  - Dulbecco's Modified Eagle Medium (Thermo Fisher)
  - Fetal Bovine Serum (FBS) (Thermo Fisher)
  - GlutaMAX (Thermo Fisher)

- Transfection reagent—polyethylenimine (PEI), linear, MW 25,000, 1 mg/mL
- Opti-MEM (Thermo Fisher)
- Essential 8 Medium (Thermo Fisher)

- Polybrene (hexadimethrine bromide, Sigma)—dissolve 10 mg of polybrene in 1 mL of ultrapure water, sterilize by passage through a 0.2 μm filter and store at −20 °C until use
  - Zeocin (10 mg/mL, Invivogen)
  - Geneticin (G418) (10 mg/mL, Invivogen)
  - Rock inhibitor (Y-27632, 10 mM, Selleck)
  - TryplE Express (Thermo Fisher)
- 4-Hydroxytamoxifen (4-OHT)—dissolve 10 mg of 4-OHT in 2.58 mL of DMSO to reach a stock concentration of 10 mM, then aliquot and store it at −20 °C.

2.1.3 Procedure

1. The components necessary for virus production include lentiviral transfer plasmid encoding effectors of HIT systems, packaging plasmid and envelope plasmid. Transfect all plasmids to the HEK293T cells at a molar ratio of 8:6:3 by PEI.
2. HEK293T cells need to be pre-seeded to a 10 cm cell culture dish to reach around 70% confluence on the day of transfection.
3. Dilute plasmids mixture and PEI (1 mg/mL) in equal volume of Opti-MEM medium, respectively. The amount of PEI applied for the transfection is calculated as the following: DNA (μg):PEI (μL) = 1:3.
4. After 5 min of incubation at room temperature, add diluted DNA to the transfection reagent solution and mix gently. Incubate the transfection reagent: DNA mixture for 15–20 min at room temperature to form transfection complex.
5. Transfer all transfection complex to HEK293T cells maintained in Dulbecco's Modified Eagle's Medium (DMEM) supplemented with 10% FBS, 2 mM GlutaMAX, and mix gently by rocking the cell culture dish back and forth. Change the medium with fresh complete growth medium after 4–6 h incubation at 37 °C in a CO_2 incubator.
6. Collect cell culture supernatant two times, 48 and 72 h after transfection, and filter it through a 0.45 μm filter to remove any remaining cell debris and large particles.

7. Centrifuge the filtrate in the Amicon Ultra-15 centrifugal filter device at 4000 rpm for 20–30 min at 4 °C to concentrate the virions.
8. Transfer 200 μL of viral solution to pre-seeded H9 cells in 6-well plate maintained in E8 medium supplemented with 10 μg/mL of polybrene. Incubate the cells at 37 °C in a CO_2 incubator for 6–8 h and then replace with fresh E8 medium. Change the medium with selective medium containing antibiotic and rock inhibitor 2 days after infection.
9. We use two antibiotics to select stable cell lines of HIT systems, zeocin and G418. The working concentration for H9 ES cells is 2.5 and 100 μg/mL, respectively. The duration of antibiotic selection is at least 2 weeks until the cells in the un-infected control well all die.
10. When antibiotic resistant cells expand sufficiently, prepare individual cell solution by treating H9 cells with TryplE for 2–3 min at 37 °C and homogenize the cells by gently pipetting up and down 7–10 times using a P1000 pipette.
11. Quantitate the cell concentration and transfer the cells to Matrigel coated 96-well plate or 10 cm cell culture dish at a concentration ranging from 10 to 50 cells/mL. Culture the cells in E8 medium supplemented with antibiotic and rock inhibitor in an incubator until the colonies appear.
12. Transfer survived monoclones to 12-well plate and maintain the cells in E8 medium without the rock inhibitor. Prepare replicated cell stocks in liquid nitrogen before screening them for the transgene expression and performance in transcriptional activation.

2.1.4 Notes

1. Plate the appropriate density of HEK293T cells to reach approximately 70% confluence on the day of transfection.
2. Concentrate the viral supernatant to obtain higher titer of virions, and this will reduce the adverse effect of viral supernatant on the stemness and proliferation of stem cells.
3. The volume of concentrated viral solution is normally 1 mL. Prepare 200 μL of aliquots and store at −80 °C until use. Do not use the virus stored for over 6 months.
4. Generate monoclonal lines expressing dCas9-NLS-GCN4 or NLS-TALE-GCN4 prior to introducing scFv-ER^{T2}/2ER^{T2}-VPH and sgRNA constructs into H9 cells. The supplement of rock inhibitor is necessary for the survival of monoclonal H9 ES cells in its initial stage.

2.2 Examination of drug inducible transcriptional activation

To achieve drug inducible transcriptional activation by HIT-SunTag system, an effective sgRNA construct is required. The efficiency of sgRNA for activation of endogenous gene should be examined first. Synergistic transcriptional activation may be accomplished using multiple sgRNAs targeting the same promoter. Similarly, the activation efficiency of HIT-TALE-SunTag system should also be determined before using it to generate the monoclonal cell lines. Quantitative real-time PCR (qPCR) or western blot can be used to examine activation efficiency.

2.2.1 Equipment

- CO_2 incubator
- NanoDrop microvolume spectrophotometer (Thermo Fisher)
- Digital dry bath incubator
- Refrigerated centrifuge
- Gel imaging system (Bio-Rad)
- Real-time qPCR system (Roche)

2.2.2 Buffer and reagents

- Puromycin (10 mg/mL, Invivogen)
- 4-Hydroxytamoxifen (4-OHT, 10 mM)
- TRIzol (Thermo Fisher)
- Chloroform
- RNA purification kit (Thermo Fisher)
- Reverse transcription kit (Promega)
- SYBR Green qPCR Master Mix, 2× (Yeasen)
- Ethidium bromide (10 mg/mL)

2.2.3 Procedure

1. Treat H9 HIT-SunTag or HIT-TALE-SunTag cell lines with 250 nM 4-OHT. Leave 1 well without the treatment of 4-OHT as drug free control. All cell samples will be collected for RNA extraction after 3 days of 4-OHT induction.
2. Aspirate the medium and rinse the cells with PBS. Add 0.5 mL of TRIzol for each sample in a 24-well plate and mix vigorously to homogenize the sample. Transfer the solution to a DNase/RNase free tube and add 0.1 mL of chloroform for each sample. Close the tube and vortex vigorously for 10–15 s followed by incubation at room temperature for 10 min. Centrifuge at 12,000 rpm for 15 min at 4 °C and transfer aqueous

phase into a new tube. Purify the RNA by the RNA purification kit according to the manufacturer's instruction. Measure the concentration of extracted total RNA by nanoDrop microvolume spectrophotometer.

3. Convert 1 μg of total RNA into first-strand cDNA using reverse transcription kit according to the manufacturer's instruction in a 20 μL of reaction system. Dilute 1 μL of cDNA solution in 7 μL of distilled water (dH_2O) for the following real-time PCR experiment.
4. Examine the mRNA expression level of activated gene by HIT-SunTag or HIT-TALE-SunTag system using SYBR green quantitative PCR protocol. Prepare a 10 μL of reaction mixture for each well including 1 μL of diluted cDNA template, 0.4 μL of each primer (10 μM), 3.2 μL of nuclease-free water and 5 μL of 2× SYBR Green qPCR Master Mix. Run the sample in real-time qPCR system using a normal two-step program (45 cycles) with a melt curve analysis: 1 cycle of pre-incubation at 95 °C for 1 min; 45 cycles of 95 °C for 10 s and 60 °C for 40 s; and finally dissociation curve analysis for assessing amplicon specificity (95 °C for 5 s, 65 °C for 1 min, then increasing to 97 °C at 0.11 °C/s ramp rate).
5. Analyze the real-time PCR results with the software to obtain Ct value for each sample. Calculate $2^{-\Delta Ct}$ and $2^{-\Delta\Delta Ct}$ to evaluate relative mRNA expression of activated gene. The formats for the calculation of ΔCt and ΔΔCt are as follows:

$$\Delta Ct = Ct_T - Ct_H$$

where T is target gene and H is housekeeping gene (e.g., GAPDH).

$$\Delta\Delta Ct = \Delta Ct_S - \Delta Ct_N$$

where S is tested sample and N is negative control.

2.2.4 Notes

1. Design multiple sgRNAs for the activation by HIT-SunTag system is helpful to improve the efficiency. Construct all sgRNA sequences in one lentiviral vector to enhance co-delivery into the same cell.
2. Avoid contamination of RNase when extracting the total RNA. Load 500 ng of each sample to 1% (w/v) agarose gel to examine the quality of extracted RNA samples.
3. Perform melting curve analysis and electrophoresis on agarose gel for products of qPCR to verify the specificity of the reaction. Run 5 μL of the amplicons in 2–3% (w/v) agarose gel stained with ethidium bromide and visualized by gel imaging system.

3. Simultaneous transcriptional activation and genome editing in a drug inducible manner

We further developed a drug inducible device, HIT2 system, for simultaneous genome editing and transcriptional activation. HIT2 consists of two major components, C2N2E-GCN4 and scFv-2E-VPH. The former can independently function as drug inducible genome editing device guided by a full length of sgRNA, while drug inducible transcriptional activation is achievable by the combination of both two constructs and a shortened sgRNA with less than 16 nucleotides (nt) for the most prevalent SpCas9. Here we describe the generation of cell lines in which two HIT2 constructs are stably integrated. Meanwhile, a transient transfection-based approach also works efficiently.

3.1 Lentiviral packaging and the generation of monoclonal cell lines

The packaging of lentivirus and the screening of monoclonal cell lines are the same as described in Section 2.1. Different from HIT-SunTag system, the length of sgRNA in HIT2 system designed for activation of endogenous is shorter, usually 14 nt for spCas9. The efficiency and specificity of 14 nt sgRNA must be qualified before being employed for functional activation. Besides, it is necessary to investigate whether the shortened sgRNA can initiate DNA double-strand breaks at the target site.

3.1.1 Equipment

See Section 2.1.1.

3.1.2 Buffer and reagents

See Section 2.1.2.

3.1.3 Procedure

1. As for the procedure of lentiviral packaging, infection of target cells and generation of monoclonal cell lines, please refer to Section 2.1.
2. sgRNA constructs can be introduced into the target cells by either transient transfection or lentiviral infection depending on the technological context. Enrich the target cells by drug selection if the delivery efficiency is low.

3. Induce the genome editing and transcriptional activation by treating the cells with 4-OHT for over 3 days and then collect the cells to extract genomic DNA and total RNA for functional examination.

3.1.4 Notes

1. It is recommended to produce monoclonal cell line of C2N2E-GCN4 first as it can be used as a tool for genome editing.
2. Construct editing and activation sgRNAs in one plasmid to enhance co-delivery.

3.2 Examination of simultaneous genome editing and transcriptional activation

After the introduction of sgRNAs into the monoclonal HIT2 cell line, simultaneous genome editing and transcription activation can be introduced by the addition of 4-OHT. Quantitative real-time PCR can be used to examine the relative mRNA expression level of target gene, while genome editing can be examined by paralleled surveyor and TIDE assays. The procedure for qPCR is the same as described in Section 2.2. Here we layout the protocol to determine the genome editing efficiency taking hAAVS1 site as an example.

3.2.1 Equipment

- Equipments for the examination of transcriptional activation, see Section 2.2.1.
- Centrifuge (Eppendorf)
- PCR system (Thermo Fisher)
- Vertical and horizontal electrophoresis system
- Digital dry bath incubator
- Gel imaging system (Bio-Rad)

3.2.2 Buffer and reagents

- Reagents for the examination of transcriptional activation, see Section 2.2.2.
- Phosphate-buffered saline (PBS)—137 mM NaCl, 2.7 mM KCl, 10 mM Na_2HPO_4, and 2 mM KH_2PO_4.
- 0.25% Trypsin (Thermo Fisher).
- PCR lysis buffer—10 m*M* Tris-Cl (pH 8.0), 2 mM EDTA (pH 8.0), 2.5% (v/v) Tween-20, 2.5% (v/v) Triton-X 100, 100 μg/mL of proteinase K.

- AccuPrime Taq DNA polymerase system (Thermo Fisher).
- 1 × TAE—40 mM Tris base, 20 mM acetic acid and 1 mM EDTA.
- 1% Agarose gel—dissolve 1 g agarose in 100 mL distilled water.
- DNA dye—10 mg/mL ethidium bromide (EB).
- 6 × loading buffer (NEB).
- SURVEYOR mutation detection kit (Idtdan Transgenomics).
- 10 × TBE—0.9 M Tris base, 0.9 M boric acid, 20 mM EDTA (PH 8.0).
- 20% Ammonium persulfate (APS)—dissolve 2 g of APS in 10 mL distilled water, then aliquot and store it at −20 °C until use.
- 3.5% polyacrylamide gel (stacking gel)—850 μL of 30% acrylamide-bisacrylamide (29:1), 750 μL of 10 × TBE, 25 μL of 20% APS, 5 μL of *N,N,N′,N′*-tetramethylethylenediamine (TEMED), 5.9 mL of distilled water.
- 15% polyacrylamide gel (separating gel)—9 mL of 30% acrylamide-bisacrylamide (29:1), 1.8 mL of 10 × TBE, 75 μL of 20% APS, 7.5 μL of *N,N,N′,N′*-tetramethylethylenediamine (TEMED), 7.2 mL of distilled water.

3.2.3 Procedure

1. Isolate the cells by trypsination using 0.25% trypsin after 3 days of 4-OHT treatment. Centrifuge at 1000 rpm for 5 min to collect the cells. Wash the cells with PBS and then perform centrifugation to gain cell pellets.
2. Resuspend the cells with lysis buffer at a concentration of 1000 cells/μL. Homogenize the samples by gently mixing with pipette for several times and perform the following lysis procedure: 50 °C for 60 min, 95 °C for 15 min. Store the cell lysate at −20 °C until use.
3. Apply 1 μL of lysate in a 25 μL PCR reaction system to amplify DNA fragment containing edited DNA sequence. Except the DNA template, the mixture includes 2.5 μL of 10 × AccuPrime™ PCR Buffer II, 500 nM of each primer and 0.1 μL of AccuPrime™ Taq DNA polymerase. Set PCR program as follows: 1 cycle of pre-denaturalized at 94 °C for 4 min, 35 cycles of 94 °C for 20 s, annealing at 64 °C for 20 s and extension at 68 °C for 35 s, finally 5 min at 68 °C for 1 cycle.
4. Run 5 μL of PCR products in 1% agarose gel to determine the robustness and specificity of PCR amplification.
5. For mutation detection by SURVEYOR nuclease, mix 5 μL of amplified DNA with 4.5 μL of 1 × AccuPrime™ PCR Buffer II and perform the annealing program to form heteroduplexes: denatured by heating to

95 °C for 5 min, and slowly reannealed from 95 to 85 °C at −2 °C/s ramp rate followed by −0.1 °C/s from 85 to 25 °C for rehybridization using PCR system. Add 0.5 μL of transgenomic's SURVEYOR nuclease into the reannealed products. Mix it gently and incubate at 42 °C for 20–30 min. Add 1 × loading buffer to stop the reaction and transfer all products into the well of polyacrylamide gel (PAGE). Run the gel at 100 V for 2 h and then stained by EB to visualize cleaved DNA band using a gel imaging system.

6. The remaining PCR products are used for Sanger sequencing examination. Analyze the sequencing images using Tracking of Indels by DEcomposition (TIDE) web tool (https://tide.nki.nl/) to calculate genome editing efficiency. The indel efficiency calculated from SURVEYOR and TIDE analysis is expected to be consistent.

3.2.4 Notes

1. Use unrelated sgRNAs as negative control when analyzing the efficiency of editing and activation. The examination for cells without the treatment of 4-OHT is also indispensable to determine background activity.
2. Collect sufficient amount of cells and split for genomic DNA and total RNA extraction. The cell confluency is usually over 80% after 4-OHT induction.
3. The quality of amplified product is a key factor affecting the accuracy of sequencing and SURVEYOR mutation detection. The primers with high T_m value often produce high purity of PCR amplicons without the interference of primer dimer. The optimal length of PCR products ranges from 400 to 700 bp. Do not digest reannealed DNA heteroduplex beyond 40 min to minimize non-specific cleavage by SURVEYOR nuclease.
4. The quality of sequence reads determines the reliability of TIDE analysis results. It is recommended to sequence both strands of PCR products to confirm the decomposition results, particularly when R^2, a parameter to assess the goodness of fit between the sequence trace intensities of the control and mutated DNA, is below 0.9.

4. Summary and conclusions

HIT drug inducible transcriptional activation systems are useful tools for dynamic and controlled transcriptional modulation. Moreover, HIT2 device expanded the function of HIT-SunTag system with simultaneous genome editing capability. Here we described methods to use these HIT systems in human cells through generation of stable cell lines via lentiviral infections. Alternative routes of delivery into cells might be favored in other circumstances and in vivo application into an organism would be a different situation. Practical parameters would need to be customized in these scenarios.

Acknowledgments

We are grateful for all the members of the Wang lab for helpful discussions and technical advices. We thank support from Strategic Priority Research Program of the Chinese Academy of Sciences (XDA16010500), the National Basic Research Program of China (2015CB964800), the National Natural Science Foundation of China (31571514 and 21701280), and State Key Laboratory of Stem Cell and Reproductive Biology.

References

Boch, J., Scholze, H., Schornack, S., Landgraf, A., Hahn, S., Kay, S., et al. (2009). Breaking the code of DNA binding specificity of TAL-type III effectors. *Science*, *326*(5959), 1509–1512.

Branda, C. S., & Dymecki, S. M. (2004). Talking about a revolution: The impact of site-specific recombinases on genetic analyses in mice. *Developmental Cell*, *6*(1), 7–28.

Bultmann, S., Morbitzer, R., Schmidt, C. S., Thanisch, K., Spada, F., Elsaesser, J., et al. (2012). Targeted transcriptional activation of silent oct4 pluripotency gene by combining designer TALEs and inhibition of epigenetic modifiers. *Nucleic Acids Research*, *40*(12), 5368–5377.

Chavez, A., Scheiman, J., Vora, S., Pruitt, B. W., Tuttle, M., E, P. R. I., et al. (2015). Highly efficient Cas9-mediated transcriptional programming. *Nature Methods*, *12*(4), 326–328.

Chavez, A., Tuttle, M., Pruitt, B. W., Ewen-Campen, B., Chari, R., Ter-Ovanesyan, D., et al. (2016). Comparison of Cas9 activators in multiple species. *Nature Methods*, *13*(7), 563–567.

Cheng, A. W., Wang, H., Yang, H., Shi, L., Katz, Y., Theunissen, T. W., et al. (2013). Multiplexed activation of endogenous genes by CRISPR-on, an RNA-guided transcriptional activator system. *Cell Research*, *23*(10), 1163–1171.

Cong, L., Ran, F. A., Cox, D., Lin, S., Barretto, R., Habib, N., et al. (2013). Multiplex genome engineering using CRISPR/Cas systems. *Science*, *339*(6121), 819–823.

Dahlman, J. E., Abudayyeh, O. O., Joung, J., Gootenberg, J. S., Zhang, F., & Konermann, S. (2015). Orthogonal gene knockout and activation with a catalytically active Cas9 nuclease. *Nature Biotechnology*, *33*(11), 1159–1161.

Deltcheva, E., Chylinski, K., Sharma, C. M., Gonzales, K., Chao, Y., Pirzada, Z. A., et al. (2011). CRISPR RNA maturation by trans-encoded small RNA and host factor RNase III. *Nature*, *471*(7340), 602–607.

Ding, Q., Regan, S. N., Xia, Y., Oostrom, L. A., Cowan, C. A., & Musunuru, K. (2013). Enhanced efficiency of human pluripotent stem cell genome editing through replacing TALENs with CRISPRs. *Cell Stem Cell*, *12*(4), 393–394.

Dominguez, A. A., Lim, W. A., & Qi, L. S. (2016). Beyond editing: Repurposing CRISPR-Cas9 for precision genome regulation and interrogation. *Nature Reviews. Molecular Cell Biology*, *17*(1), 5–15.

Feil, R., Wagner, J., Metzger, D., & Chambon, P. (1997). Regulation of Cre recombinase activity by mutated estrogen receptor ligand-binding domains. *Biochemical and Biophysical Research Communications*, *237*(3), 752–757.

Gao, X., Tsang, J. C., Gaba, F., Wu, D., Lu, L., & Liu, P. (2014). Comparison of TALE designer transcription factors and the CRISPR/dCas9 in regulation of gene expression by targeting enhancers. *Nucleic Acids Research*, *42*(20), e155.

Gilbert, L. A., Larson, M. H., Morsut, L., Liu, Z., Brar, G. A., Torres, S. E., et al. (2013). CRISPR-mediated modular RNA-guided regulation of transcription in eukaryotes. *Cell*, *154*(2), 442–451.

Hsu, P. D., Lander, E. S., & Zhang, F. (2014). Development and applications of CRISPR-Cas9 for genome engineering. *Cell*, *157*(6), 1262–1278.

Huang, P., Xiao, A., Zhou, M., Zhu, Z., Lin, S., & Zhang, B. (2011). Heritable gene targeting in zebrafish using customized TALENs. *Nature Biotechnology*, *29*(8), 699–700.

Indra, A. K., Warot, X., Brocard, J., Bornert, J. M., Xiao, J. H., Chambon, P., et al. (1999). Temporally-controlled site-specific mutagenesis in the basal layer of the epidermis: Comparison of the recombinase activity of the tamoxifen-inducible Cre-ER(T) and Cre-ER(T2) recombinases. *Nucleic Acids Research*, *27*(22), 4324–4327.

Jinek, M., Chylinski, K., Fonfara, I., Hauer, M., Doudna, J. A., & Charpentier, E. (2012). A programmable dual-RNA-guided DNA endonuclease in adaptive bacterial immunity. *Science*, *337*(6096), 816–821.

Joung, J. K., & Sander, J. D. (2013). TALENs: A widely applicable technology for targeted genome editing. *Nature Reviews. Molecular Cell Biology*, *14*(1), 49–55.

Kiani, S., Chavez, A., Tuttle, M., Hall, R. N., Chari, R., Ter-Ovanesyan, D., et al. (2015). Cas9 gRNA engineering for genome editing, activation and repression. *Nature Methods*, *12*(11), 1051–1054.

Lu, J., Zhao, C., Zhao, Y., Zhang, J., Zhang, Y., Chen, L., et al. (2018). Multimode drug inducible CRISPR/Cas9 devices for transcriptional activation and genome editing. *Nucleic Acids Research*, *46*(5), e25.

Maeder, M. L., Linder, S. J., Cascio, V. M., Fu, Y., Ho, Q. H., & Joung, J. K. (2013). CRISPR RNA-guided activation of endogenous human genes. *Nature Methods*, *10*(10), 977–979.

Mali, P., Aach, J., Stranges, P. B., Esvelt, K. M., Moosburner, M., Kosuri, S., et al. (2013). CAS9 transcriptional activators for target specificity screening and paired nickases for cooperative genome engineering. *Nature Biotechnology*, *31*(9), 833–838.

Mali, P., Esvelt, K. M., & Church, G. M. (2013). Cas9 as a versatile tool for engineering biology. *Nature Methods*, *10*(10), 957–963.

Mali, P., Yang, L., Esvelt, K. M., Aach, J., Guell, M., DiCarlo, J. E., et al. (2013). RNA-guided human genome engineering via Cas9. *Science*, *339*(6121), 823–826.

Metzger, D., Clifford, J., Chiba, H., & Chambon, P. (1995). Conditional site-specific recombination in mammalian cells using a ligand-dependent chimeric Cre recombinase. *Proceedings of the National Academy of Sciences of the United States of America*, *92*(15), 6991–6995.

Miller, J. C., Tan, S., Qiao, G., Barlow, K. A., Wang, J., Xia, D. F., et al. (2011). A TALE nuclease architecture for efficient genome editing. *Nature Biotechnology*, *29*(2), 143–148.

Moscou, M. J., & Bogdanove, A. J. (2009). A simple cipher governs DNA recognition by TAL effectors. *Science*, *326*(5959), 1501.

Perez-Pinera, P., Kocak, D. D., Vockley, C. M., Adler, A. F., Kabadi, A. M., Polstein, L. R., et al. (2013). RNA-guided gene activation by CRISPR-Cas9-based transcription factors. *Nature Methods*, *10*(10), 973–976.

Tanenbaum, M. E., Gilbert, L. A., Qi, L. S., Weissman, J. S., & Vale, R. D. (2014). A protein-tagging system for signal amplification in gene expression and fluorescence imaging. *Cell*, *159*(3), 635–646.

Zhang, F., Cong, L., Lodato, S., Kosuri, S., Church, G. M., & Arlotta, P. (2011). Efficient construction of sequence-specific TAL effectors for modulating mammalian transcription. *Nature Biotechnology*, *29*(2), 149–153.

Zhao, C., Zhang, Y., Zhao, Y., Ying, Y., Ai, R., Zhang, J., et al. (2018). Multiple chemical inducible Tal effectors for genome editing and transcription activation. *ACS Chemical Biology*, *13*(3), 609–617.

Zhao, C., Zhao, Y., Zhang, J., Lu, J., Chen, L., Zhang, Y., et al. (2018). HIT-Cas9: A CRISPR/Cas9 genome-editing device under tight and effective drug control. *Molecular Therapy—Nucleic Acids*, *13*, 208–219.

CHAPTER SIX

Direct cloning and heterologous expression of natural product biosynthetic gene clusters by transformation-associated recombination

Jia Jia Zhang[a], Kazuya Yamanaka[b], Xiaoyu Tang[a,*], Bradley S. Moore[a,c,*]

[a]Center for Marine Biotechnology and Biomedicine, Scripps Institution of Oceanography, University of California at San Diego, La Jolla, San Diego, CA, United States
[b]Department of Life Science and Technology, Kansai University, Osaka, Japan
[c]Skaggs School of Pharmacy and Pharmaceutical Sciences, University of California at San Diego, La Jolla, San Diego, CA, United States
*Corresponding authors: e-mail address: xtang@microbechembio.org; bsmoore@ucsd.edu

Contents

Methods in Enzymology, Volume 621
ISSN 0076-6879
https://doi.org/10.1016/bs.mie.2019.02.026

Abstract

Heterologous expression of natural product biosynthetic gene clusters (BGCs) is a robust approach not only to decipher biosynthetic logic behind natural product (NP) biosynthesis, but also to discover new chemicals from uncharacterized BGCs. This approach largely relies on techniques used for cloning large BGCs into suitable expression vectors. Recently, several whole-pathway direct cloning approaches, including full-length RecE-mediated recombination in *Escherichia coli*, Cas9-assisted *in vitro* assembly, and transformation-associated recombination (TAR) in *Saccharomyces cerevisiae*, have been developed to accelerate BGC isolation. In this chapter, we summarize a protocol for TAR cloning large NP BGCs, detailing the process of choosing TAR plasmids, designing pathway-specific TAR vectors, generating yeast spheroplasts, performing yeast transformation, and heterologously expressing BGCs in various host strains. We believe that the established platforms can accelerate the process of discovering new NPs, understanding NP biosynthetic logic, and engineering biosynthetic pathways.

1. Introduction

Genetic elements directly associated with the production of a natural product (NP) are generally clustered along a contiguous region of the microbial chromosome (Medema et al., 2015). These biosynthetic gene clusters (BGCs) code for proteins associated with NP biosynthesis, regulation, transport, and resistance. Recent advances in genome sequencing technologies and bioinformatics tools for NP BGC predication have led to the rapid accumulation of a large number of orphan or cryptic NP biosynthetic pathways in publicly available databases (Medema & Fischbach, 2015), providing unprecedented opportunities to (1) convert this genetic potential into chemical reality for the discovery of new molecules and (2) identify new biosynthetic capabilities and new enzymes associated with microbial metabolism. Genetic manipulation of native producing strains is a conventional strategy for exploring new BGCs and deciphering their biosynthetic mechanisms. However, this strategy relies heavily on the ability to genetically manipulate strains of interest by homologous recombination or transposon mutagenesis, which is often time consuming and not applicable to genera without established genetic approaches or for uncultured microbes. In contrast, expression of BGCs in a well-investigated and genetically amenable host is often more facile and broadly applicable (Kim, Moore, & Yoon, 2015; Luo, Enghiad, & Zhao, 2016). Heterologous expression requires (1) cloning of entire BGCs into suitable expression vectors and (2) accessibility of various heterologous hosts. Traditionally, BGC cloning involves the construction of large random libraries of genomic DNA fragments that are then screened for sequences of interest. This process is both inefficient and laborious, particularly given current accessibility of microbial genome sequences. Recently, several targeted approaches, including full-length RecE-mediated recombination (Fu et al., 2012) and *oriT*-directed capture (Kvitko, McMillan, & Schweizer, 2013) in *Escherichia coli*, Gibson assembly (Kvitko et al., 2013) or Golden-gate assembly (Liang, Liu, Low, Ang, & Zhao, 2017) based DNA ligation in a test tube, transformation-associated recombination (TAR) cloning and assembly in the yeast *Saccharomyces cerevisiae* (Kim et al., 2010; Yamanaka et al., 2014), have been developed for direct cloning of large BGCs. These methods have greatly accelerated the process of NP discovery and biosynthetic research.

TAR cloning, which leverages high rates of homologous recombination in *S. cerevisiae*, is arguably the most robust method for selective isolation of large chromosomal fragments. This method has enabled the direct cloning of complex chromosomal fragments up to 300 kb into yeast artificial chromosome (YAC) vectors, with positive clone rates up to 32% (Kouprina & Larionov, 2006). The highly recombinant nature of *S. cerevisiae* was first reported in the early 1980s (Orr-Weaver, Szostak, & Rothstein, 1981), but the use of TAR for selective isolation of large genomic regions from complex genomic DNA was not established until the late 1990s by a group of scientists at the National Institutes of Health (Kouprina et al., 1998; Larionov et al., 1996; Larionov, Kouprina, Solomon, Barrett, & Resnick, 1997). Because early applications of this method involved the cloning of mammalian DNA, which possess yeast origin-like replicons (*ARS* elements) at a frequency of approximately one per 20–30 kb, propagation of TAR-based vectors in yeast cells relied on the acquisition of an *ARS* element from the targeted chromosomal DNA. Later, a modified TAR system was established by integrating an *ARS* element into the TAR vector backbone (Noskov et al., 2003), enabling the isolation of the genomic regions that do not possess *ARS*-like sequences, such as centromere and telomere regions or bacterial genomic DNA. Co-introduction of the counter-selectable marker *URA3* alongside the *ARS* element helps to counteract high levels of vector re-circularization made possible by non-homologous end joining (NHEJ). For more details regarding the principles, development, and current applications of TAR cloning, please see other thorough review articles (Kouprina & Larionov, 2003, 2006, 2016).

The first application of TAR cloning for NP research was the assembly of large NP BGCs from soil-derived eDNA cosmid libraries (Kim et al., 2010). Later, our group developed a TAR-based genetic platform for direct cloning, genetic manipulation, and heterologous expression of large BGCs to identify new NPs and investigate their biosynthesis (Yamanaka et al., 2014). This platform was initially designed specifically for the study of large NP BGCs from Actinobacteria, an important source of bioactive NPs. Based on this platform, we successfully expressed a large nonribosomal peptide synthetase BGC from a marine actinomycete in the model expression host *Streptomyces coelicolor*, resulting in the discovery of a group of new lipopeptide antibiotics called the taromycins. Following this successful application, we have improved the approach by streamlining the process of capture vector construction and introducing the *URA3* counter-selectable marker to improve cloning efficiency (Tang et al., 2015). Furthermore, we have expanded the range of heterologous hosts compatible with our platform by constructing two new TAR vectors, which are compatible with Gram-positive *Bacillus subtilis* (Li et al., 2015) and various Gram-negative Proteobacteria (*i.e., Pseudomonas putida* and *Agrobacterium tumefaciens*) (Zhang, Tang, Zhang, Nguyen, & Moore, 2017). Here, we summarize a protocol detailing all elements of our TAR platform, including choosing TAR plasmids, designing pathway-specific TAR vectors, TAR cloning in yeast, and heterologous expression of large BGCs in various host strains (Fig. 1).

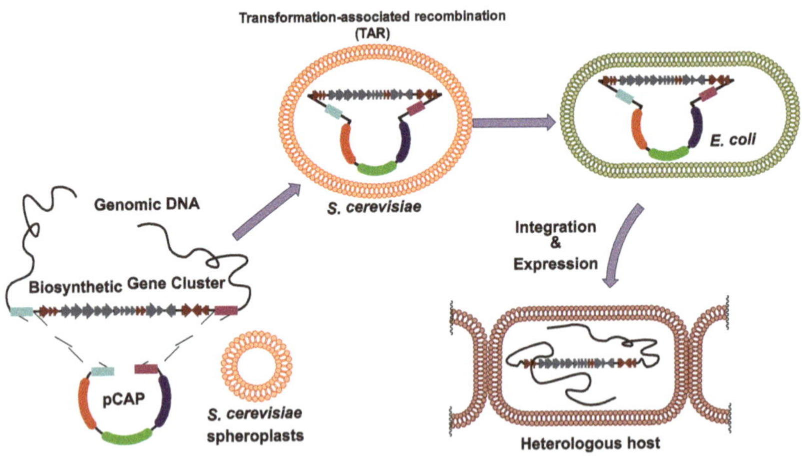

Fig. 1 Scheme for TAR cloning and heterologous expression of biosynthetic gene clusters using pCAP vectors.

2. Vector design

We have published three categories of TAR vectors for direct capture of large BGCs in yeast and heterologous expression in various host organisms. All vectors consist of elements for maintenance and selection in yeast, *E. coli*, and heterologous host organisms (Addgene plasmid numbers are provided in parenthesis).

2.1 pCAP01 (plasmid #59981)/pCAP03 (plasmid #69862)

pCAP01 is a yeast-*E. coli* shuttle vector that can be integrated into the genome of Actinobacterial hosts. It is composed of a low copy-number yeast origin of replication (*CEN6/ARSH4* from pRS314), a high copy-number *E. coli* origin of replication (pUC *ori* from SuperCos1), and a ΦC31 cassette for chromosomal integration in diverse *Streptomyces* species (*int-attP-oriT-aph* elements from pSET152) (Fig. 2A). For yeast selection, the *TRP1* auxotrophic marker is used for selection of transformants on tryptophan-deficient

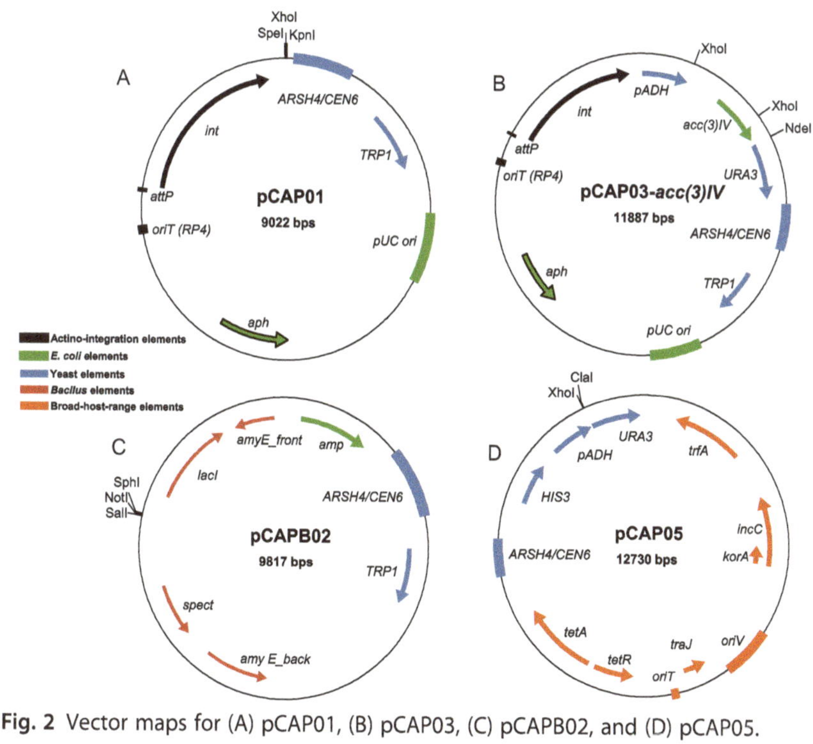

Fig. 2 Vector maps for (A) pCAP01, (B) pCAP03, (C) pCAPB02, and (D) pCAP05.

media. Additionally, the antibiotic resistance marker *neoR/kanR* (neomycin transferase) is used for kanamycin selection in both *E. coli* and Actinobacteria. This marker can also be selected for using neomycin for hosts that possess the apramycin resistance gene (*aac(3)IV*) or are otherwise kanamycin resistant. This vector is maintained as a single copy yeast artificial chromosome in *S. cerevisiae* to avoid multiple copies of recombinants in a single yeast colony. This feature also stabilizes large constructs, and we have experimentally confirmed that DNA inserts up to 80 kb can be maintained in yeast.

Because the pCAP01 vector backbone contains a yeast origin of replication, we observed high rates of self-recircularization of empty TAR vectors by NHEJ, greatly reducing the efficiency of successful BGC cloning. To counteract this, pCAP03 was constructed through the addition of the *ura3* gene under the control of the strong promoter of the *Schizosaccharomyces pombe* ADH1 gene (pADH1) into the *Spe*I and *Kpn*I restriction sites of pCAP01 (Fig. 2B). pADH1 can tolerate an insertion of up to 130 bp between the TATA box and the transcription initiation site and still act upon the downstream *ura3* gene. Thus, vector-recircularization can be effectively selected against in the presence of 5-fluoroorotic acid (5-FOA). Additionally, the *acc(3)IV* cassette was cloned into pCAP03 to obtain pCAP03-*acc(3)IV*, which can be efficiently digested with XhoI and NdeI and agarose gel purified to generate ready-to-use pCAP03 (RTU-pCAP03).

2.2 pCAPB02 (plasmid #65628)

In collaboration with Dr. Pei-Yuan Qian and his group at Hong Kong University of Science and Technology, we initially built a self-replicating plasmid (pCAPB01) from pCAP01 by replacing the *Streptomyces* integration elements with *Bacillus* elements from plasmid pBU4. However, instability was observed following transfer of pCAPB01-sert (containing the 38 kb surfactin BGC from *B. subtilis* 1779) into the host organism *B. subtilis* JH642, resulting in no production of surfactins in the host. To overcome this issue, we constructed the integration vector pCAPB02 (Fig. 2C), which consists of yeast elements from pCAP01, *E. coli* elements from pBR322, and a homologous recombination-based integration system for *B. subtilis* from pDR111. pCAPB02 can be selected for in *E. coli* and *B. subtilis* with ampicillin and spectinomycin, respectively. The BGC insertion site is flanked by the 5′ and 3′ gene fragments of the *B. subtilis amyE* gene, enabling double-crossover recombination of BGCs into the chromosome of various *B. subtilis* hosts, as *amyE* is conserved across many *B. subtilis* strains.

2.3 pCAP05 (plasmid #102920)

pCAP05 (Fig. 2D) was constructed by combining yeast elements for cloning with self-replicating Gram-negative broad-host-range elements for heterologous expression. Broad-host-range elements were amplified from plasmid pRK442(H), a broad-host-range vector of the RK2 family. The RK2 replicon and its derivatives, which belong to the incompatibility group P plasmids, have been shown to replicate at low copy number (<30 in *E. coli*) in a wide range of Gram-negative bacteria using the *oriV* origin of replication. The plasmid copy number and host range are controlled by the essential *trfA* gene. *HIS3* was chosen as the yeast selective marker to make pCAP05 orthogonal to pCAP01/03 and pCAPB02 vectors, which use *TRP1* for yeast selection. Additionally, the counter-selective system from pCAP03 was introduced into pCAP05, which allows for the use of short capture arm sequences and improves cloning efficiency. pCAP05 can be selected for in *E. coli* and bacterial hosts using tetracycline. Although we imagine that pCAP05 can be transferred to a broad range of Gram-negative host organisms, we have thus far only validated this vector system in two proteobacterial expression hosts beyond *E. coli*—*P. putida* KT2440 and *A. tumefaciens* LBA4404.

3. TAR cloning

3.1 Yeast strains

Highly transformable yeast strains *S. cerevisiae* VL6-48 (ATCC MYA-3666: *MATα, his3-D200, trp1-Δ1, ura3-52, lys2, ade2-101, met14, psi + cir°*) and *S. cerevisiae* VL6-48N (*MATα, his3-D200, trp1-Δ1, ura3-Δ1, lys2, ade2-101, met14, psi + cir°*) (Larionov et al., 1997) were chosen as hosts for TAR cloning experiments. *S. cerevisiae* VL6-48N is more suitable for counter selection using *ura3*, as *ura3* has been completely deleted in this strain. The yeast cells were grown in liquid or solid yeast extract-peptone-dextrose medium (YPD: 2% D-glucose, 1% yeast extract, and 2% peptone with or without agar), supplemented with 100 mg/L adenine. Adenine is not necessary for yeast growth but is recommended for propagating yeast for generating spheroplasts. Yeast cells turn red if adenine is not added to YPD agar plates.

3.2 Solutions

All solutions can be sterilized by autoclave unless otherwise specified.

1. 10× nitrogen bases (100 mL): 1.7 g of Yeast Nitrogen Base without amino acids and ammonium sulfate (Sigma Y1251); 1.9 g of Yeast

Synthetic Drop-out Medium Supplements without tryptophan or histidine; 5 g of ammonium sulfate. Dissolve above components in 100 mL warmed up Milli-Q water and filter through a 0.22 μm filter to make 10 × stock solution (store at 4 °C).

2. 100 × adenine (50 mL): 0.5 g adenine, 3.7 mL of 1 M HCl, and Milli-Q water up to 50 mL. The solution can be sterilized by filtration through a 0.22 μm filter to make 100 × stock solution.
3. Top selective agar (50 mL, for 6 transformations): sorbitol, 9.1 g, 1.1 M; dextrose, 1.1 g, 2.2% (w/v); agar, 1.0 g. Before using, microwave to melt and add 10 × nitrogen bases and 100 × adenine.
4. Bottom selective agar (500 mL): sorbitol, 91 g, 1.1 M; dextrose, 11 g, 2.2% (w/v); agar, 15 g. Before using, microwave and add 10 × nitrogen bases and 100 × adenine. For counter selection, add 100 × 5-fluoroorotic acid (5-FOA) stock solution (100 mg/mL in DMSO) before pouring plates.
5. Dextrose, 40% (w/v).
6. 1 M Sorbitol (500 mL): 91 g sorbitol in 500 mL Milli-Q water (store at 4 °C).
7. Milli-Q water, 200 mL.
8. HEPES buffer, 1 M and pH 7.5.
9. EDTA solution, 0.5 M, adjust the pH to 8.0 with NaOH.
10. 100 mM $CaCl_2$ solution: 1.1 g $CaCl_2$ in 100 mL Milli-Q water.
11. SPE (50 mL) solution: 500 μL HEPES buffer (8), 1 mL EDTA solution (9), 9.1 g sorbitol, Milli-Q water to 50 mL. Store up to 2 months at room temperature after autoclaving.
12. SOS medium (50 mL): 3.25 mL 100 mM $CaCl_2$ solution (10), 0.125 g yeast extract, 9.1 g sorbitol, 0.5 g peptone, Milli-Q water to 50 mL. Store up to 2 months at room temperature after autoclaving.
13. Tris buffer, 1 M and pH 7.5.
14. STC solution (50 mL): 500 μL 1 M tris buffer (13), 5 mL 100 mM $CaCl_2$ solution (10), 9.1 g sorbitol, Milli-Q water to 50 mL. Store up to 2 months at room temperature after autoclaving.
15. PEG8000 solution (10 mL): 100 μL 1 M tris buffer (13), 1 mL 100 m*M* $CaCl_2$ solution (10), 2.0 g PEG8000, Milli-Q water to 10 mL. Filter and store up to 4 weeks at room temperature.
16. Zymolyase 20T solution: 10 mg/mL zymolyase-20T (MP Biomedicals), 25% (w/v) glycerol, 50 mM Tris-HCl (pH 7.5). No sterilization required. Dispense into 500 μL aliquots and store up to 2 years at −20 °C.

3.3 Construction of pathway-specific capture vectors

There are currently two ways to construct pathway-specific capture vectors (Fig. 3). The first approach, which applies to pCAP01 and pCAPB02, requires (1) the amplification of two 1 kb homologous arms that flank the upstream and downstream regions of the BGC of interest, and (2) the insertion of the two 1 kb sequences into a TAR vector by traditional cloning methods (digestion and ligation) or Gibson Assembly (Fig. 3A). Constructs prepared in this manner do not utilize the pADH-*ura3* counter selection system, which we introduced to pCAP03 and pCAP05 to counteract high levels of plasmid recircularization in yeast due to NHEJ, resulting in capture rates below 2%. For second generation capture vectors pCAP03 and pCAP05, a 144 bp dsDNA fragment, which contains two 18 bp overlapping fragments with the vector backbone for Gibson Assembly, two 50 bp pathway-specific capture hooks for recombination, and an 8 bp PmeI blunt end restriction site for linearization, has been designed as shown in Fig. 3B. This approach simplifies the procedure for constructing pathway-specific cloning vectors. While use of the counter selection system significantly decreases the percentage of yeast transformants containing empty vectors, it also dramatically reduces the total number of yeast transformants and may necessitate preparation of additional transformation plates. Additionally, there should be no ATG start codons within the homology arm sequences, which could cause mis-translation of the *ura3* counter-selection gene. Finally, homology arm sequences, no matter their length, should be designed as closely as possible to restriction or fragmentation sites used for preparation of the genomic DNA (gDNA) (see below for more details).

3.4 Preparation of genomic DNA

gDNA must be appropriately fragmented before being transformed into yeast spheroplasts. Previous studies have demonstrated that homologous recombination combining two linear pieces of DNA into a yeast artificial chromosome is much more efficient if the homologous arm sequences are closer to the ends of the targeted DNA fragment (Noskov et al., 2003). Ideally, site specific double-stand breaks (DSBs) can be introduced directly upstream and downstream of BGCs of interest by restriction enzyme digestion, and homology arm sequences should be designed as close to these restriction sites as possible. However, this approach can be limited due to the lack of appropriate restriction sites directly beyond and not within BGCs of interest.

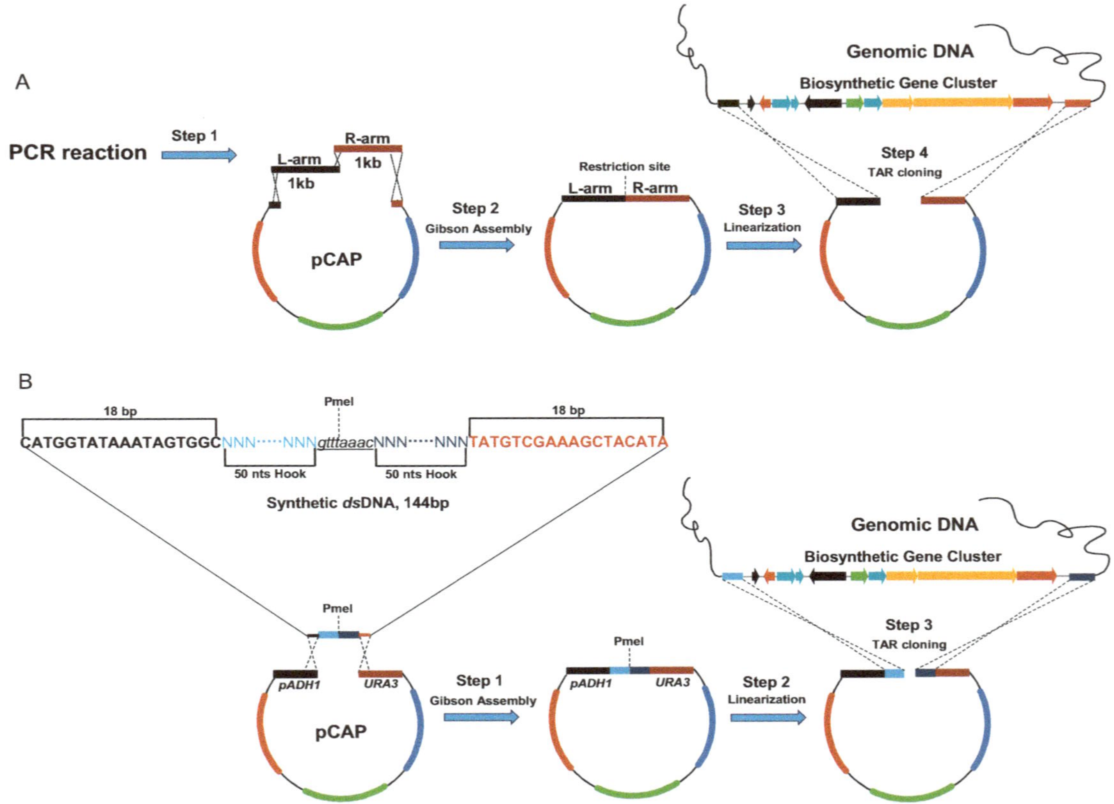

Fig. 3 Two methods for generating pathway-specific capture vectors involving (A) PCR amplification and assembly of two 1 kb homology arms or (B) assembly of synthetic ds-DNA fragment containing two 50 bp homology arms.

Alternatively, gDNA can be randomly fragmented by shearing with a needle or pipette tip. Recently, Dr. Natalay Kouprina and colleagues have successfully introduced DSBs to the ends of targeted gDNA fragments using the CRISPR-Cas9 endonuclease *in vitro*, resulting in high efficiency DNA capture, up to 32% for large fragments isolated from complex gDNA (Lee, Larionov, & Kouprina, 2015). However, *in vitro* CRISPR/Cas9-mediated DNA digestion necessitates several troubleshooting steps and thus may not be an attractive approach if commercial restriction enzymes can be used. The quality of the gDNA also greatly affects recombination efficiency. For isolation of gDNA from a single organism in pure culture, standard gDNA isolation procedures or commercial isolation kits are generally adequate, but gDNA should be checked by Nanodrop, Qubit, and agarose gel to confirm integrity. Finally, if the target of interest is low in quantity (such as from metagenomic DNA), PCR and/or genomic library preparation are encouraged for enrichment.

3.5 Preparation of yeast spheroplasts and transformation

The protocol for the preparation of yeast spheroplasts and transformation is outlined in the original TAR cloning protocol that was published elsewhere (Noskov et al., 2011). Although the authors of this protocol suggest using mid-log phase yeast cells (OD600 = 3.0–5.0) to generate spheroplasts, our protocol instead calls for early-log phase yeast cells (OD600 around 1.0) osmotically stabilized overnight in 1 M ice-cold sorbitol. Our slightly modified 3-day protocol is detailed below:

Day 1

1. Inoculate a large single colony of *S. cerevisiae* VL6-48 or *S. cerevisiae* VL6-48N from a YPD agar plate supplemented with adenine into 10 mL YPD liquid medium (in a 50 mL falcon tube) with adenine overnight at 30 °C with shaking at 220 rpm. Although colonies grown on YPD plates can be kept at 4 °C for up to a week, we suggest growing fresh cells on YPD plates directly from a glycerol stock.

Day 2

1. Inoculate 0.5–1 mL overnight culture into 50 mL YPD media containing adenine (in a 250 mL-flask). Grow for 5–8 h at 30 °C shaking at 220 rpm to OD600 of ~0.7–1.0. This will give you enough early-log phase cells for 10 transformations.
2. Put the flask on ice for 10 min and then centrifuge the cells in a 50 mL falcon tube for 3 min at 1800 × *g*, 4 °C. The cells must be well pelleted; if they are not, re-streak fresh yeast and start the protocol again.

3. Decant the supernatant and resuspend the pellets in 50 mL ice-cold sterile water by vortexing. Centrifuge for 3 min at 1800 × *g*, 4 °C. Again, the cells must be well pelleted.
4. Decant the water and resuspend the cells in 50 mL of 1 M ice-cold sorbitol by vortexing. Leave the cells at 4 °C overnight for osmotic stabilization.

Day 3

1. Invert the tube several times and then centrifuge for 3 min at 1800 × *g*, 4 °C. The cells must be well pelleted.
2. Remove the supernatant completely by standing the tube upside down on a paper towel and resuspend the cells in 20 mL SPE solution (room temperature) by vortexing. Add 40 μL of 2-mercaptoethanol (2-ME) and invert to mix. Then add 80 μL zymolyase-20T solution and invert to mix. The activity of the zymolyase enzyme may not be equivalent from brand to brand and even from batch to batch. Therefore, we suggest carefully monitoring the cells over the course of the incubation time.
3. Incubate the cell suspension in a 30 °C incubator, inverting the tube gently every 5 min or using a shaker set at slow speed <25 rpm.
4. After 20 min, sample zymolyase treated cells by diluting five times with 1 M sorbitol and 2% SDS. Mix 200 μL cell suspension with 800 μL 1 M sorbitol and another 200 μL of cells with 800 μL 2% SDS. Monitor the progress of spheroplast formation by taking absorbance measurements of the two cell suspensions at 600 nm. Spheroplasts are determined to be ready when the absorbance of the 1 M sorbitol cell suspension is 5- to 20-fold higher than the cell suspension diluted in 2% SDS (when 80–95% of cells have been converted to spheroplasts). This process often takes about 40 min. If the fold change is not great enough, keep incubating for an additional 10 min and measure the OD difference again. The total incubation time, however, should not be longer than 50 min, as the OD of cells suspended in 1 M sorbitol will decrease over time, indicating that the cells have been over-digested and lysed. The level of spheroplast formation can greatly affect transformation efficiency (insufficiently digested cells do not take up large DNA fragments). You can use a large construct (any pCAP captured gene cluster) as a control to measure spheroplast quality.
5. Add 1 M sorbitol (4 °C) up to 50 mL and gently invert to mix. Harvest the cells by centrifuging for 10 min at 600 × *g*, 4 °C.

6. Decant the supernatant and stand the tube upside down on a paper towel carefully (the cell pellet is not tightly packed) to remove the supernatant completely.
7. Gently resuspend the yeast spheroplasts in 20 mL of 1 M sorbitol (4 °C) by pipetting up and down with a 25-mL serological pipette. It generally takes 2 min to resuspend the spheroplasts completely (spheroplasts are extremely unstable, do not vortex to resuspend). Add ice-cold 1 M sorbitol to a total volume of 50 mL and invert to mix.
8. Harvest the cells by centrifuging for 10 min at 600 × *g*, 4 °C. Remove supernatant completely by standing the tube upside down on a paper towel.
9. Resuspend the spheroplasts in up to 2 mL STC solution (room temperature) by pipetting up and down with a 10-mL serological pipette (it takes about 2 min to resuspend the spheroplasts completely). The volume of STC can be adjusted depending on cell density and how many transformations will be made (200 μL per transformation).
10. Incubate the spheroplasts for 10 min at room temperature (22 °C).
11. Mix 200 μL spheroplast solution with no more than 40 μL DNA solution (0.2–0.5 μg linearized vector and 2–3 μg gDNA) in a 1.5 mL microcentrifuge tube gently (wide-bore tip is recommended).
12. Incubate cell mixture for 10 min at room temperature.
13. Add 0.8 mL 20% PEG8000 solution and mix by inverting the tube 10 times.
14. Incubate the tube for 20 min at room temperature.
15. Microcentrifuge the cells for 10 min at 700 × *g*, 4 °C.
16. Remove the supernatant with 1-mL pipette.
17. Add 800 μL SOS solution and resuspend the cells by pipetting with 1-mL wide-bore pipette tip.
18. Incubate the tube for 30–40 min at 30 °C.
19. Add a cell suspension into 8 mL melted top agar prepared in 15 mL tube (keep at 50–55 °C) and invert a few times to mix. Immediately pour cells onto selective bottom agar plate.
20. Incubate the plates for 4–7 days at 30 °C.

Use circular capture vector (~0.1 μg) as a control to determine the quality of the spheroplast cells. Cells harboring an empty pCAP vector can form visible colonies in ~3 days, but cells carrying a recombined vector generally grow much slower.

3.6 Screening of candidate colonies

3.6.1 Picking colonies

After 4–7 days incubation, yeast transformants form visible colonies on the selective agar plate. Normally, you should observe dozens to several hundred colonies on a regular selection plate from one transformation. We suggest patching at least 100 colonies for initially screening; however, more colonies would need to be collected if no positive hit could be detected from the first screening. For the experiments using the counter selection marker *ura3*, dozens of colonies or even less are commonly found on one selection plate with 5-FOA. In this case, you can pick all colonies from the plate to screen.

1. Pick colonies from two to three plates with sterile toothpicks and transfer them onto new selective agar plates.
2. Incubate new plates for 2 days at 30 °C.
3. Make three replica plates by using filter paper or sterile velvet.
4. Incubate replica plates for 1–2 days at 30 °C.

3.6.2 PCR screening

DNA isolated from yeast colonies is generally harder to use as a template for PCR (especially for colony PCR) in comparison with gDNA. Therefore, primers and PCR conditions must be optimized using gDNA (giving bold and clear PCR products). Multiplex PCR is not recommended for screening.

1. Scoop cells from one replica plate with loop or 200-μL pipette tip, and resuspend into 200 μL of solution 1 (10% sucrose, 50 mM Tris-HCl (pH 8.0), 10 m*M* EDTA) containing 2-ME (1/1000 dilution) and 0.2 mg/mL zymolyase 20T.
2. Incubate and shake for 2 h at 37 °C.

Option 1 (colony PCR, quick method)

1. Boil for 5 min (thermocycler can be used, 98 °C/5 min).
2. Use 0.2–0.5 μL as template for 50 μL scale PCR reaction.

Option 2 (plasmid isolation and PCR, reliable method)

1. Mix the suspension with 400 μL solution 2 (NaOH 0.2 M and 1% SDS (w/v)) by inversion. The solution should become clear.
2. Add 300 μL solution 3 (KAc, 3 M, pH 4.8). Mix well by inversion and centrifuge for 10 min at maximum speed.
3. Transfer clear supernatant into a fresh 1.5 mL microcentrifuge tube. Precipitate the DNA by addition of 1 volume isopropanol (IPA) and

centrifugation (maximum speed, 10 min). A white DNA pellet should be visible on the bottom of the tube.

4. Discard the supernatant. Dissolve the DNA pellet in 200 μL 10 mM Tris-HCl (pH 8.0).
5. Use 0.2–0.5 μL as a template for 50 μL scale PCR reactions.

3.6.3 DNA extraction from PCR positive yeast clones

1. Pick up PCR-confirmed clone and inoculate into 10 mL yeast selective liquid medium.
2. Incubate for 20–30 h at 30 °C with shaking at 220 rpm.
3. Resuspend cells in 200 μL of solution 1 (10% sucrose, 50 mM Tris-HCl (pH 8.0), 10 m*M* EDTA) containing 2-ME (1/500 dilution) and 0.5 mg/mL zymolyase.
4. Incubate and shake for 2 h at 37 °C.
5. Mix the suspension with 400 μL solution 2 (NaOH 0.2 M and 1% SDS (w/v)) by inversion. Solution should become clear.
6. Add 300 μL solution 3 (KAc, 3 M, pH 4.8) and centrifuge for 10 min at maximum speed.
7. Remove residual proteins by phenol-chloroform denaturation, and then precipitate and clean up DNA using IPA and 70% ethanol.
8. Dissolve DNA in 50 μL of 10 mM Tris-HCl (pH 8.0). Use 1–2 μL of the DNA solution for *E. coli* transformation.

Stability issues are often observed in *E. coli* when the size of inserts is larger than 50 kb. For propagation of large constructs, use of a *recA*$^-$ and *endA*$^-$ strain of *E. coli* such as Stbl4 (Invitrogen, Genotype: *mcr*A Δ(*mcr*BC-*hsd*RMS-*mrr*) *rec*A1 *end*A1 *gyr*A96 *gal*− *thi*-1 *sup*E44 λ− *rel*A1 Δ(*lac-pro*AB)/F′ *pro*AB+ *lac*IqZΔM15 *Tn*10 (Tet^R)) is recommended to avoid unwanted rearrangement of the cloned cluster. In *recA*$^+$ and/or *endA*$^+$ *E. coli* strains such as BW25113 (for λ-red recombination) and ET12567 (used for conjugal DNA transfer), stability issues tend to arise more frequently. Cultivating *E. coli* at lower temperature (30 °C) might help to overcome these stability issues.

4. Heterologous hosts

Following the successful capture of BGCs of interest, constructs can be transferred to suitable heterologous hosts that can be analyzed for the production of associated metabolites. pCAP01/pCAP03 can be introduced to Gram-positive actinobacterial host organisms (*Streptomyces* species and an engineered strain of *Salinispora tropica*) by *E. coli*-mediated conjugation. pCAPB02 can be introduced to Gram-positive *B. subtilis* strains by natural competence transformation. Finally, pCAP05 can be introduced to Gram-negative proteobacterial hosts (*P. putida* and *A. tumefaciens*) by electrical transformation.

4.1 Gram-positive Actinobacteria: *Streptomyces* and *Salinispora*

pCAP01/pCAP03 can be directly introduced into most *Streptomyces* hosts by conjugation, since these vectors possess proper conjugative elements such as *oriT* (RK2). Inclusion of the phage ΦC31 attachment site *attP* in these vectors enables the genomic integration of whole constructs into any host that possesses the chromosomal attachment site *attB*, for example many established *Streptomyces* hosts. We recently engineered the marine actinomycete *S. tropica* to serve as a pCAP01/pCAP03-compatible heterologous host by introducing the phage ΦC31 attachment site *attB* into the *S. tropica* CNB-440 genome. If a host carries a methyl-sensing restriction system (as is the case for *S. coelicolor* and *Streptomyces avermitilis*), it is necessary to utilize non-methylating *E. coli* strains such as *E. coli* ET12567 as the donor strain for conjugation. If the target host does not carry a methyl-sensing restriction system (as is the case for *Streptomyces lividans* and *S. tropica*), common *E. coli* strains such as DH5α can be used instead. Exconjugants can be selected for on MS agar plates using kanamycin or neomycin (50 μg/mL).

1. Prepare competent cells of *E. coli* ET12567 grown at 37 °C in LB containing chloramphenicol (25 μg/mL) to maintain selection for the *dam* mutation.
2. Transform competent cells with pCAP-BGC and select for the incoming plasmid using kanamycin (50 μg/mL) and chloramphenicol (25 μg/mL).
3. Inoculate an *E. coli* ET12567/pCAP-BGC colony and *E. coli* ET12567/pUB307 into 10 mL LB medium (containing 25 μg/mL chloramphenicol and 50 μg/mL kanamycin). Grow overnight at 37 °C.
4. Inoculate 250–500 μL overnight culture into 10 mL fresh LB plus antibiotics as above and grow for about 4 h at 37 °C to an OD600 of 0.4–0.6. Normally, *E. coli* ET12567/pUB307 should grow faster than *E. coli* ET/pCAP-BGC.5.
5. Wash the cells twice with 10 mL LB to remove antibiotics that might inhibit *Streptomyces* or *Salinispora* acceptor strains and resuspend *E. coli*

ET12567/pUB307 in 0.5–1 mL LB and *E. coli* ET12567/pCAP-BGC in 0.2–0.4 mL LB.

6. While washing the *E. coli* cells, add 10–20 μL *Streptomyces* spore stock to 200 μL 2 × YT broth (Thermo Fisher Scientific) or 10–20 μL *Salinispora* spore stock to 200 μL Marine Broth (BD Difco) for each conjugation. Heat shock at 50 °C for 10 min and then allow to cool at room temperature.
7. Mix 0.1 mL of each *E. coli* cell suspension and 0.2 mL heat-shocked spores.
8. Plate out *Streptomyces* and *E. coli* mixture on MS agar + 10 m*M* $MgCl_2$ (for *Streptomyces*) or *Salinispora* and *E. coli* mixture on A1 agar (10 g soluble starch, 4 g yeast extract, 2 g peptone, 16 g agar, 1 L natural or artificial seawater) and incubate at 30 °C for 16–20 h.
9. Overlay the plate with 1.5 mL water containing 0.5 mg nalidixic acid (20 μL of 25 mg/mL stock; selectively kills *E. coli*) and 2–4 mg kanamycin (40–80 μL of 50 mg/mL stock). Use a spreader to lightly distribute the antibiotic solution evenly. Continue incubation at 30 °C for 5–7 days.
10. Pick a single colony and streak it on a MS agar plate (*Streptomyces*) or A1 agar plate (*Salinispora*) containing nalidixic acid (25 μg/mL) and kanamycin (50 μg/mL) for single colony growth and removal of *E. coli* contamination.
11. Prepare a spore suspension for two to three independent colonies.

4.2 Gram-positive Firmicutes: *B. subtilis*

pCAPB02-based vectors can be readily transformed into *B. subtilis* hosts by natural competence transformation (Albano, Hahn, & Dubnau, 1987). The region between *amyE_front* and *amyE_back* (containing the BGC of interest and the spectinomycin resistance gene) can be integrated into the *amyE* gene (encoding an alpha-amylase conserved in *B. subtilis* strains) by double-crossover recombination at high frequency. Transformants can be selected for on LB agar plates using spectinomycin (100 μg/mL).

1. Inoculate a single colony of *B. subtilis* from a LB plate into 10 mL LB liquid medium (in a 50 mL falcon tube) overnight at 30 °C with shaking at 220 rpm.
2. Inoculate 100 μL overnight culture into 10 mL fresh LB and grow for about 3 h at 37 °C to an OD600 of ~1. In the meantime, make MD media (see below).

3. Inoculate 20 mL MD media with 1 mL of LB *B. subtilis* culture.
4. Grow for about 4 h at 37 °C at 220 rpm.
5. Add 50% glycerol to the competent cells to get a final concentration of 20% glycerol. Divide into 1 mL aliquots and freeze at 80 °C.
6. Transformation: Add DNA (5–10 μL plasmid) to 200 μL competent cells and incubate for 30 min at 37 °C at 220 rpm.
7. Plate out appropriate dilutions on plates containing the appropriate antibiotics.

4.2.1 Solutions

10× Phosphate citrate (1000 mL): 107 g K_2HPO_4, 60 g KH_2PO_4, 10 g sodium citrate·$2H_2O$

MD media (50 mL): 45.6 mL 10× phosphate citrate, 2.0 mL glucose (50%, w/v), 500 μL L-tryptophan (10 mg/mL), 250 μL ferric ammonium citrate (2.2 mg/mL), 1.25 mL potassium aspartate (100 mg/mL), 150 μL 1 M $MgSO_4$, and 250 μL phenylalanine (10 mg/mL).

4.3 Gram-negative Proteobacteria: *Pseudomonas* and *Agrobacterium*

pCAP05-based vectors are broad-host range vectors that are maintained as self-replicating plasmids in a broad range of Gram-negative bacteria. pCAP05 has been successfully transferred to two proteobacterial expression hosts beyond *E. coli*—*P. putida* KT2440 and *A. tumefaciens* LBA4404—using standard electrical transformation. pCAP05 can also be transferred by conjugation (see vector map details in Fig. 1D) and selected for using tetracycline.

1. For preparation of electrocompetent *P. putida* or *A. tumefaciens*, grow strains overnight in LB or YM medium (0.04% yeast extract, 1.0% mannitol, 1.7 mM NaCl, 08 mM $MgSO_4 \cdot 7H_2O$, 2.2 mM $K_2HPO_4 \cdot 3H_2O$), respectively, at 30 °C.
2. In the morning, inoculate 5–10 mL of *P. putida* or *A. tumefaciens* precultures into 50 mL of fresh LB or YM media, respectively. Grow at 30 °C for approximately 4 h or until cultures reach an OD600 of ~1.0–3.0 for *P. putida* or 0.5 for *A. tumefaciens*. *P. putida* will grow quickly; *A. tumefaciens* will grow more slowly.
3. Place cultures on ice for 10–15 min. Wash *P. putida* with ice-cold 10% glycerol twice and resuspend cells in a small volume of glycerol.

Wash *A. tumefaciens* with ice-cold 1 mM HEPES buffer (pH 7.5) twice and cells resuspend in a small volume of 1 mM HEPES-10% glycerol. Cells can be aliquoted and stored at −80 °C for future use.

4. Add 0.1–1 μg of pCAP05-BGC DNA to cells and electroporate at 2500 V in a 2-mm cuvette.
5. Recover *P. putida* for 2 h at 30 °C in LB medium before plating on LB agar supplemented with 15 μg/mL tetracycline. Recover *A. tumefaciens* for 3 h at 30 °C in YM medium before plating on YM agar supplemented with 5 μg/mL tetracycline.
6. For future heterologous expression experiments, both *P. putida* and *A. tumefaciens* can be grown on LB medium.

Because pCAP05 exists as a self-replicating plasmid in the host, we have observed occasional genetic instability in host organisms when carrying large and repetitive DNA sequences. Thus, we are currently developing and refining methods for facile and site-specific genomic integration of BGCs into Gram-negative hosts.

5. Conclusions

Heterologous expression of whole NP BGCs offers several advantages for discovery of new NPs and understanding their biosynthetic logic. First, successful expression of a genomic region containing a predicted BGC clearly defines all specialized genes that are essential for NP production. Second, genetic manipulation of cloned BGCs leveraging established *E. coli* or yeast recombineering strategies is comparatively faster, easier, and more broadly applicable than genetic manipulation of wild-type strains. Third, working in a sequenced model host organism provides the opportunity to rationally manipulate NP production by metabolic engineering. Thus, we believe our established platforms for cloning and heterologous expression provide a routine approach not only for the discovery of metabolites from "orphan" or "cryptic" BGCs but also to probe natural product biosynthetic logic and engineer biosynthetic pathways. We summarize per Table 1 a list of BGCs that have been successfully cloned and expressed using the pCAP vector systems. We believe that lessons learned through developing and optimizing these platforms will continue to be broadly useful to the natural products research community in future studies of microbial BGCs.

Table 1 List of selected NPs cloned and expressed using the pCAP vector platforms.

NP	BGC family	Native host	Size (kb)	Vector	Expression host	References
Taromycin	NRPS	*Saccharomonospora* sp. CNQ-490	67	pCAP01	*Streptomyces coelicolor* M1146	Yamanaka et al. (2014); Reynolds et al. (2018)
Marinopyrrole	PKS-NRPS	*Streptomyces* sp. CNQ418	30	pCAP01	*S. coelicolor* M512	Yamanaka et al. (2014)
Enterocin	Type II PKS	*Salinispora pacifica* CNT-150	18	pCAP01	*S. coelicolor* M1146	Bonet, Teufel, Crusemann, Ziemert, and Moore (2015)
Alterochromide	NRPS-FAS	*Pseudoalteromonas piscicida* JCM 20779	34	pCAP01 (capture) pETDuet-1 (expression)	*E. coli* BL21(DE3)	Ross, Gulland, Dorrestein, and Moore (2015)
Ammosamide	Alkaloid	*Streptomyces* sp. CNR-698	37.5	pCAP01	*S. coelicolor* M1146	Jordan and Moore (2016)
Cosmomycin	Type II PKS	*Streptomyces* sp. CNT-302	54	pCAP01	*S. coelicolor* M152	Larson, Crusemann, and Moore (2017)
Salinamide	PKS-NRPS	*Streptomyces* sp. CNB-091	48.2	pCAP01	*S. coelicolor* M1146	Ray, Yamanaka, and Moore (2016)
Jomthonic acid	PKS-NRPS	*S. caniferus* GUA-06-05-006A	35	pCAP01	*S. coelicolor* M1152 *Streptomyces albus* J1074	Garcia-Salcedo et al. (2018)
Chuangxinmycin	Alkaloid	*Actinoplanes tsinanensis* CPCC 200056	unknown	pCAP01	*S coelicolor* M1146	Shi et al. (2018)

Nataxazole	Alkaloid	*Streptomyces* sp. Tü 6176	unknown	pCAP01	*S. albus* JT46	Cano-Prieto et al. (2015)
Demethylchlortetracycline	Type II PKS	*S. aureofaciens* DM-1	43.5	pCAP01	*Streptomyces aureofaciens* DM-1	Wu, Huang, Min, and Hu (2017)
Precolibactins	PKS-NRPS	*Escherichia coli* CFT073	51	pCAP01	*Escherichia coli* DH10B	Li et al. (2016)
Thiolactomycin	PKS-NRPS	*Sa. pacifica* CNS863	22	pCAP03	*S. coelicolor* M1152 and *Salinispora tropica* CNB4401	Tang et al. (2015); Zhang, Moore, and Tang (2018)
Thiotetroamide	PKS-NRPS	*S. afghaniensis* NRRL 5621	33	pCAP03	*S. coelicolor* M1152	Tang et al. (2015)
Scleric acid	NRPS	*S. sclerotialus* NRRL ISP-5269	33	pCAP03	*S. coelicolor* M1152	Alberti et al. (2019)
Amicoumacin	PKS-NRPS	*Bacillus subtilis* 1779	47.4	pCAPB02	*Bacillus subtilis* JH642	Li et al. (2015)
Plipastatin	NRPS	*B. amyloliquefaciens* HYM12	39	pCAPB02	*B. subtilis* 1A751	Hu et al. (2018)
Violacein	Alkaloid	*P. luteoviolacea* 2ta16	9	pCAP05	*Pseudomonas putida* KT2440 *Agrobacterium tumefaciens* LBA4404	Zhang et al. (2017)

NRPS, nonribosomonal peptide synthetase; PKS, polyketide synthase; FAS, fatty acid synthase.

Acknowledgments

We kindly thank all past and current members of the Moore lab for providing their valuable experiences and discussion in developing the TAR-based platforms. We acknowledge Dr. Pei-Yuan Qian and his group for joint efforts in developing pCAPB plasmids. Finally, we thank Dr. Vladimir Larionov and Dr. Natalay Kouprina of the National Cancer Institute (NCI) at the National Institutes of Health for sharing their TAR expertise and yeast materials with our group. The research reported has been supported by a graduate fellowship from the National Science Foundation to J.J.Z. and National Institutes of Health grants F31-AI129299 to J.J.Z. and R01-GM085770 and R01-AI117712 to B.S.M.

References

Albano, M., Hahn, J., & Dubnau, D. (1987). Expression of competence genes in *Bacillus subtilis*. *Journal of Bacteriology*, *169*, 3110–3117.

Alberti, F., Leng, D. J., Wilkening, I., Song, L., Tosin, M., & Corre, C. (2019). Triggering the expression of a silent gene cluster from genetically intractable bacteria results in scleric acid discovery. *Chemical Science*, *10*, 453–463. https://doi.org/10.1039/C8SC03814G.

Bonet, B., Teufel, R., Crusemann, M., Ziemert, N., & Moore, B. S. (2015). Direct capture and heterologous expression of *Salinispora* natural product genes for the biosynthesis of enterocin. *Journal of Natural Products*, *78*, 539–542.

Cano-Prieto, C., Garcia-Salcedo, R., Sanchez-Hidalgo, M., Brana, A. F., Fiedler, H. P., Mendez, C., et al. (2015). Genome mining of *Streptomyces* sp. Tü 6176: Characterization of the nataxazole biosynthesis pathway. *Chembiochem*, *16*, 1461–1473.

Fu, J., Bian, X., Hu, S., Wang, H., Huang, F., Seibert, P. M., et al. (2012). Full-length RecE enhances linear-linear homologous recombination and facilitates direct cloning for bioprospecting. *Nature Biotechnology*, *30*, 440–446.

Garcia-Salcedo, R., Alvarez-Alvarez, R., Olano, C., Canedo, L., Brana, A. F., Mendez, C., et al. (2018). Characterization of the jomthonic acids biosynthesis pathway and isolation of novel analogues in *Streptomyces caniferus* GUA-06-05-006A. *Marine Drugs*, *16*, 259.

Hu, Y., Nan, F., Maina, S. W., Guo, J., Wu, S., & Xin, Z. (2018). Clone of plipastatin biosynthetic gene cluster by transformation-associated recombination technique and high efficient expression in model organism *Bacillus subtilis*. *Journal of Biotechnology*, *288*, 1–8.

Jordan, P. A., & Moore, B. S. (2016). Biosynthetic pathway connects cryptic ribosomally synthesized posttranslationally modified peptide genes with pyrroloquinoline alkaloids. *Cell Chemical Biology*, *23*, 1504–1514.

Kim, J. H., Feng, Z., Bauer, J. D., Kallifidas, D., Calle, P. Y., & Brady, S. F. (2010). Cloning large natural product gene clusters from the environment: Piecing environmental DNA gene clusters back together with TAR. *Biopolymers*, *93*, 833–844.

Kim, E., Moore, B. S., & Yoon, Y. J. (2015). Reinvigorating natural product combinatorial biosynthesis with synthetic biology. *Nature Chemical Biology*, *11*, 649–659.

Kouprina, N., Annab, L., Graves, J., Afshari, C., Barrett, J. C., Resnick, M. A., et al. (1998). Functional copies of a human gene can be directly isolated by transformation-associated recombination cloning with a small 3′ end target sequence. *Proceedings of the National Academy of Sciences of the United States of America*, *95*, 4469–4474.

Kouprina, N., & Larionov, V. (2003). Exploiting the yeast *Saccharomyces cerevisiae* for the study of the organization and evolution of complex genomes. *FEMS Microbiology Reviews*, *27*, 629–649.

Kouprina, N., & Larionov, V. (2006). TAR cloning: Insights into gene function, long-range haplotypes and genome structure and evolution. *Nature Reviews. Genetics*, 7, 805–812.

Kouprina, N., & Larionov, V. (2016). Transformation-associated recombination (TAR) cloning for genomics studies and synthetic biology. *Chromosoma, 125*, 621–632.

Kvitko, B. H., McMillan, I. A., & Schweizer, H. P. (2013). An improved method for *oriT*-directed cloning and functionalization of large bacterial genomic regions. *Applied and Environmental Microbiology, 79*, 4869–4878.

Larionov, V., Kouprina, N., Graves, J., Chen, X. N., Korenberg, J. R., & Resnick, M. A. (1996). Specific cloning of human DNA as yeast artificial chromosomes by transformation-associated recombination. *Proceedings of the National Academy of Sciences of the United States of America, 93*, 491–496.

Larionov, V., Kouprina, N., Solomon, G., Barrett, J. C., & Resnick, M. A. (1997). Direct isolation of human *BRCA2* gene by transformation-associated recombination in yeast. *Proceedings of the National Academy of Sciences of the United States of America, 94*, 7384–7387.

Larson, C. B., Crusemann, M., & Moore, B. S. (2017). PCR-independent method of transformation-associated recombination reveals the cosmomycin biosynthetic gene cluster in an ocean streptomycete. *Journal of Natural Products, 80*, 1200–1204.

Lee, N. C., Larionov, V., & Kouprina, N. (2015). Highly efficient CRISPR/Cas9-mediated TAR cloning of genes and chromosomal loci from complex genomes in yeast. *Nucleic Acids Research, 43*, e55.

Li, Z. R., Li, J., Gu, J. P., Lai, J. Y., Duggan, B. M., Zhang, W. P., et al. (2016). Divergent biosynthesis yields a cytotoxic aminomalonate-containing precolibactin. *Nature Chemical Biology, 12*, 773–775.

Li, Y., Li, Z., Yamanaka, K., Xu, Y., Zhang, W., Vlamakis, H., et al. (2015). Directed natural product biosynthesis gene cluster capture and expression in the model bacterium *Bacillus subtilis*. *Scientific Reports, 5*, 9383.

Liang, J., Liu, Z., Low, X. Z., Ang, E. L., & Zhao, H. (2017). Twin-primer non-enzymatic DNA assembly: An efficient and accurate multi-part DNA assembly method. *Nucleic Acids Research, 45*, e94.

Luo, Y., Enghiad, B., & Zhao, H. (2016). New tools for reconstruction and heterologous expression of natural product biosynthetic gene clusters. *Natural Product Reports, 33*, 174–182.

Medema, M. H., & Fischbach, M. A. (2015). Computational approaches to natural product discovery. *Nature Chemical Biology, 11*, 639–648.

Medema, M. H., Kottmann, R., Yilmaz, P., Cummings, M., Biggins, J. B., Blin, K., et al. (2015). Minimum information about a biosynthetic gene cluster. *Nature Chemical Biology, 11*, 625–631.

Noskov, V. N., Chuang, R. Y., Gibson, D. G., Leem, S. H., Larionov, V., & Kouprina, N. (2011). Isolation of circular yeast artificial chromosomes for synthetic biology and functional genomics studies. *Nature Protocols, 6*, 89–96.

Noskov, V. N., Kouprina, N., Leem, S. H., Ouspenski, I., Barrett, J. C., & Larionov, V. (2003). A general cloning system to selectively isolate any eukaryotic or prokaryotic genomic region in yeast. *BMC Genomics, 4*, 16.

Orr-Weaver, T. L., Szostak, J. W., & Rothstein, R. J. (1981). Yeast transformation: A model system for the study of recombination. *Proceedings of the National Academy of Sciences of the United States of America, 78*, 6354–6358.

Ray, L., Yamanaka, K., & Moore, B. S. (2016). A peptidyl-transesterifying type I thioesterase in salinamide biosynthesis. *Angewandte Chemie (International Ed. in English), 55*, 364–367.

Reynolds, K. A., Luhavaya, H., Li, J., Dahesh, S., Nizet, V., Yamanaka, K., et al. (2018). Isolation and structure elucidation of lipopeptide antibiotic taromycin B from the activated taromycin biosynthetic gene cluster. *Journal of Antibiotics (Tokyo), 71*, 333–338.

Ross, A. C., Gulland, L. E., Dorrestein, P. C., & Moore, B. S. (2015). Targeted capture and heterologous expression of the *Pseudoalteromonas* alterochromide gene cluster in *Escherichia coli* represents a promising natural product exploratory platform. *ACS Synthetic Biology*, *4*, 414–420.

Shi, Y., Jiang, Z., Li, X., Zuo, L., Lei, X., Yu, L., et al. (2018). Biosynthesis of antibiotic chuangxinmycin from *Actinoplanes tsinanensis*. *Acta Pharmaceutica Sinica B*, *8*, 283–294.

Tang, X., Li, J., Millan-Aguinaga, N., Zhang, J. J., O'Neill, E. C., Ugalde, J. A., et al. (2015). Identification of thiotetronic acid antibiotic biosynthetic pathways by target-directed genome mining. *ACS Chemical Biology*, *10*, 2841–2849.

Wu, N., Huang, H., Min, T., & Hu, H. (2017). TAR cloning and integrated overexpression of 6-demethylchlortetracycline biosynthetic gene cluster in *Streptomyces aureofaciens*. *Acta Biochimica et Biophysica Sinica Shanghai*, *49*, 1129–1134.

Yamanaka, K., Reynolds, K. A., Kersten, R. D., Ryan, K. S., Gonzalez, D. J., Nizet, V., et al. (2014). Direct cloning and refactoring of a silent lipopeptide biosynthetic gene cluster yields the antibiotic taromycin A. *Proceedings of the National Academy of Sciences of the United States of America*, *111*, 1957–1962.

Zhang, J. J., Moore, B. S., & Tang, X. (2018). Engineering *Salinispora tropica* for heterologous expression of natural product biosynthetic gene clusters. *Applied Microbiology and Biotechnology*, *102*, 8437–8446.

Zhang, J. J., Tang, X., Zhang, M., Nguyen, D., & Moore, B. S. (2017). Broad-host-range expression reveals native and host regulatory elements that influence heterologous antibiotic production in Gram-negative bacteria. *MBio*, *8*, e01291–17.

CHAPTER SEVEN

Salt-sensitive intein for large-scale polypeptide production

Yi-Zong Lee[a,b], Shih-Che Sue[b,c,*]
[a]Instrument Center, National Tsing Hua University, Hsinchu, Taiwan
[b]Institute of Bioinformatics and Structural Biology, National Tsing Hua University, Hsinchu, Taiwan
[c]Department of Life Science, National Tsing Hua University, Hsinchu, Taiwan
*Corresponding author: e-mail address: scsue@life.nthu.edu.tw

Contents

Methods in Enzymology, Volume 621
ISSN 0076-6879
https://doi.org/10.1016/bs.mie.2019.02.024

Abstract

In this chapter, we propose to use salt-sensitive intein as a fusion protein to promote polypeptide expression; the removal of intein from the target sequence requires no enzyme, only a buffer change. The method will be particularly helpful for large-scale polypeptide preparations. Intein is an enzyme that can perform N- and C-terminal self-cleavage. Upon introduction of a mutation to eliminate the N-terminal cleavage activity, the C-terminal cleavage function can still be preserved. This feature was used to develop intein as a fusion protein through conjugation with a given sequence to promote protein expression in a biosynthesis system. Fused intein could later be separated from the target sequence through a C-terminal self-cleavage reaction. Here, a type of salt-sensitive intein is characterized in which ionic strength becomes an effector to control the self-cleavage activity. Low salt concentrations favor the cleavage reaction. Thus, using salt-sensitive intein as a fusion protein simply requires a buffer change to activate the self-cleavage mechanism, which makes it an enzyme-free process. This process has many advantages, including low cost, no extra residue remaining after cleavage, feasibility for preparing proteins starting from a non-Met codon and a special benefit for producing isotope-labeled peptides.

1. Introduction

A methodology for producing polypeptides can generally be addressed through two major pathways, chemical synthesis and biosynthesis (Bedard & Biron, 2018; Hojo, 2014). Chemical synthesis occurs when amino acids are chemically linked via peptide bonds using several repeating cycles to allow for the assembly of a polypeptide chain. The most common method is known as solid-phase peptide synthesis (SPPS) (Bedard & Biron, 2018; Hojo, 2014). The advantage of chemical synthesis is the efficiency and flexibility in preparing different polypeptides at a laboratory scale. It also provides the benefit of incorporating non-natural amino acids. The challenge of chemical synthesis comes from the gradually reduced yield and purity as the length of the polypeptide increases. For long polypeptide cases (>25 aa), the recovery rate and purity will be reduced (Bedard & Biron, 2018; Hojo, 2014). Therefore, for industrial usage, the concern of cost will favor to use biological systems to alternatively produce long polypeptides (Bray, 2003).

Biosynthesis methods use a biological system to express the target polypeptide. Depending on the various strategies, the final products can be purified after expression. There is a common strategy that conjugates a fusion tag at the beginning of the polypeptide to enhance target polypeptide expression and purification (Paraskevopoulou & Falcone, 2018;

Rosano & Ceccarelli, 2014; Zhao, Xu, Xing, & Lin, 2016). The fusion tag can be any well-known protein domain or specific sequence that has great expression yield and stability. After purification, the tag can be removed by enzyme digestion to obtain the target polypeptide (Rosano & Ceccarelli, 2014). Biosynthesis has the advantages of high yield and low cost, especially for long polypeptides (Paraskevopoulou & Falcone, 2018). However, the use of enzyme digestion creates uncertainty during purification. Enzyme digestion is expensive and represents another cost during production. Meanwhile, depending enzyme-cutting site, residual residues may remain after digestion. The additional residues could affect target polypeptide activity or stability (Waugh, 2011). In some cases, enzyme digestion is not fully functional, therefore reducing the final product yield. Thus, choosing a proper fusion tag and enzyme cutting site becomes an important issue for polypeptide production. The conditions affect expression, stability and purification.

In this chapter, we will propose a method using intein as a fusion protein. Intein is a protein containing self-cleavage activity. For certain inteins, the cleavage activity can be regulated by ionic strength. Changing the salt concentration to induce intein self-cleavage can be developed into an enzyme-free method for polypeptide preparation. This method can be generally used for polypeptide biosynthesis.

2. Intein

2.1 Background of mechanism

Intein is a protein domain that can spontaneously splice its flanking N- and C-terminal domains to become a mature protein and excise itself from a sequence (Fig. 1A). Similar to introns at the mRNA level, the protein is an intron-like protein and is therefore named intein (Anraku, Mizutani, & Satow, 2005). Protein splicing is a post-translational process that includes peptide bond cleavage and backbone conjugation activities. The reaction requires no cofactor or ATP hydrolysis (Shah & Muir, 2014).

There are over 600 inteins identified from difference species with various lengths from 130 to 800 residues. The protein family can be grouped into two types, wild-type and split intein. Wild-type intein is a "single" chain containing a full-length intein domain that performs *cis*-splicing (Fig. 1A) (Vila-Perello & Muir, 2010). Split intein contains "two" chains, as a backbone disconnection is present in the intein sequence to separate it into two fragments. Split intein requires the combination of its N- and C-terminal

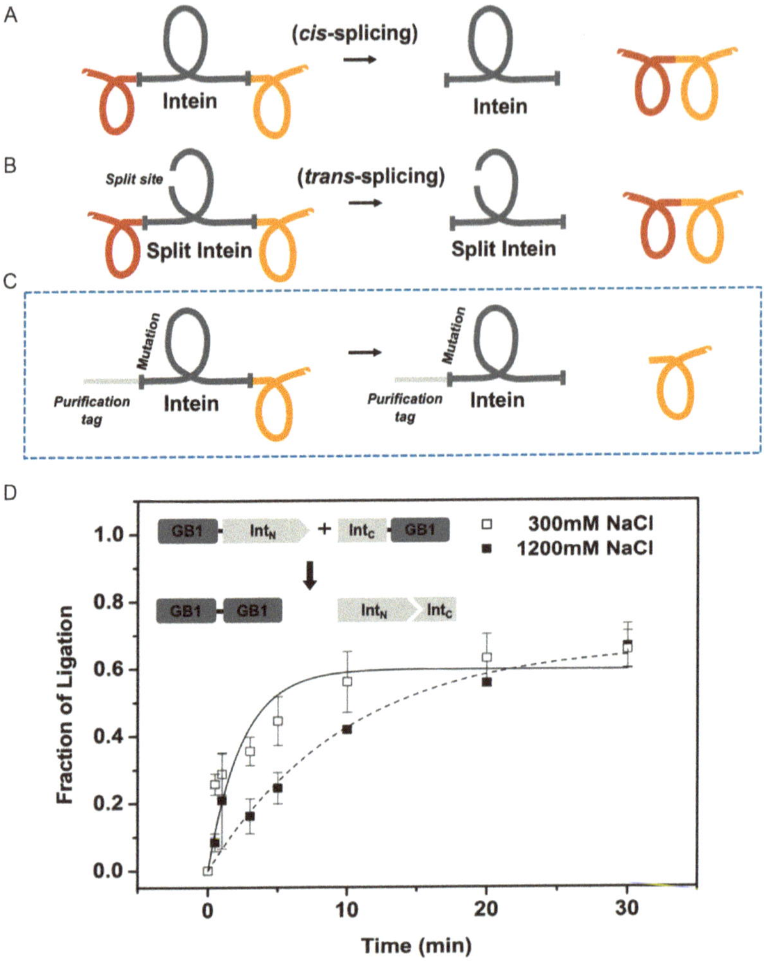

Fig. 1 Schematic of the intein reaction. (A) The single-chain *cis*-splicing and (B) two-chain *trans*-splicing reactions. Two N- and C-terminal flanking sequences are ligated through the activity of (A) intein or (B) split intein. (C) The mutated intein is used as a fusion protein. After mutating the first residue in intein, only the C-terminal self-cleavage activity remained. After C-terminal cleavage, intein is separated from the target polypeptide. (D) Time course for protein ligation of GB1 and GB1 with split *Npu* intein. The in vitro *trans*-splicing assay was performed in the presence of 0.3 *M* and 1.2 *M* NaCl. The kinetics curves were analyzed from SDS-PAGE after the reaction. The curves are fit by a pseudo-first-order reaction.

chains to form a complete intein fold to recover its splicing activity. The two-chain intein performs *trans*-splicing (Fig. 1B). Split intein has been widely applied in bioengineering in the past two decades, such as in the case of protein semi-synthesis. The advantage comes from the fact that the N- and C-terminal chains can be separately prepared through different methods.

Overall, active intein molecules execute three reactions, including N-terminal cleavage, N to C ligation and C-terminal cleavage (Wood & Camarero, 2014). The three activities can be individually eliminated by mutations (Mathys et al., 1999). For example, the first residue (usually Cys but can be Ser or Thr in some types) is mainly responsible for N-terminal cleavage. Mutation of the first residue removes the nucleophilic center and abolishes the N-terminal cleavage site, which subsequently prevents the splicing reaction. However, even without the N-terminal cleavage function, the last residue (usually Asn) can independently perform side chain cyclization (succinimide formation) to break the peptide bond between intein and the C-terminal flanking sequence, causing a C-terminal cleavage (Cheriyan, Chan, & Perler, 2014). If the last residue is mutated to disable C-terminal cleavage, intein can still perform N-terminal cleavage (Eryilmaz, Shah, Muir, & Cowburn, 2014; Mathys et al., 1999; Shah & Muir, 2014).

Since the N- and C-terminal cleavage reactions can be performed independently, if we can prepare a functionally deficient intein that has no N-terminal cleavage activity but preserves its C-terminal cleavage activity, the mutant will have potential to act as a fusion tag that can perform self-cleavage to remove itself from a target sequence (Fig. 1C).

2.2 Salt-sensitive intein

The *Nostoc punctiforme (Npu)* DnaE intein from cyanobacteria has been characterized as one of the most active inteins. *Npu* intein is used for many applications because of its compact size (137 residues) and extraordinary protein splicing activity (Lee, Lee, Lin, Chen, & Sue, 2014; Lee, Lo, & Sue, 2017; Lee, Su, Lo, Lyu, & Sue, 2012; Zettler, Schutz, & Mootz, 2009). *Npu* intein is very sensitive to the presence of ion and ionic additives (Shah, Eryilmaz, Cowburn, & Muir, 2013; Shah, Vila-Perello, & Muir, 2011). This was revealed by testing split *Npu* intein function under low and high salt conditions (Fig. 1D). We performed a protein *trans*-splicing assay (Zettler et al., 2009). Two GB1 domains are conjugated to the termini of the two split *Npu* intein chains. The ligation yield was quantified from the amount of GB1-GB1 product. We employed a pseudo-first-order equation to obtain the *trans*-splicing rates. The results showed that the *trans*-splicing protein rate is $9.0 \times 10^{-3}\,s^{-1}$ in 300 m*M* NaCl and $1.5 \times 10^{-3}\,s^{-1}$ in 1.2 *M* NaCl (Fig. 1D). High NaCl concentrations reduced the protein *trans*-splicing reaction. We used a similar experiment to examine ionic additives, such as ionic detergents containing sodium dodecyl sulfate (SDS) and cetyltrimethylammonium bromide (CTAB), which eliminated the splicing activity, while non-ionic detergents had very mild effects. *Npu* intein resisted the denaturation induced by urea but cannot resist the effects induced by guanidine hydrochloride. Considering the chemical nature of denaturants, guanidine hydrochloride contains charged group and urea has no charge. Thus, the increase in the ionic strength influences intein activity. Salt also reduced the assemblage affinity of the two intein chains. The affinity (K_D) measured by ITC under 300 mM NaCl conditions was 10^5-fold stronger than that measured under the 1.2 M NaCl conditions. We also noticed that *Npu* intein started to lose its structural stability when the NaCl concentration was higher than 300 mM.

To understand the structural basis, two *Npu* intein X-ray structures were compared. We determined the structure of split *Npu* intein (PDB code 4QFQ) and compared it with the "single-chain" *Npu* intein structure (PDB code 4KL5) (Aranko, Oeemig, Kajander, & Iwai, 2013). The two structures contain nearly the same structural patterns. Positively and negatively charged residues formed stable salt-bridges that are distributed on the assembled interface, and total six ionic clusters were involved in the interaction (Fig. 2A). Most acidic residues were distributed on the N-terminal chain while basic residues were on the C-terminal chain. The charge segregation is significantly important for assembly and maintaining the *Npu* intein fold. Therefore, shielding the electrostatic interactions would reduce the structural integrity, explaining the loss of protein splicing activity in the presence of high concentrations of salt or ionic additives.

2.3 Using ionic-strength to regulate intein cleavage

The charge-charge interaction facilitates the association of structural elements. The presence of high salt concentrations destabilizes the *Npu* intein fold. Ionic strength could manipulate the activity of the *Npu* intein. Generally, *Npu* intein is less active when salt concentrations are high. By increasing the ionic strength, we can slow down the cleavage and splicing reactions. The reactions can be reactivated by reducing the salt concentration. Thus, ionic strength becomes a regulator.

This "charge segregation" feature in the *Npu* intein has been noticed in other inteins. Some sequences have been selected and compared (Fig. 2B). The negative and positive charges distributed on the N-terminal and C-terminal portions, respectively, provide specificity for intramolecular interactions. We expect that splicing reactions in these inteins sharing similar salt sensitivity features could be manipulated by ionic strength. Notably, the

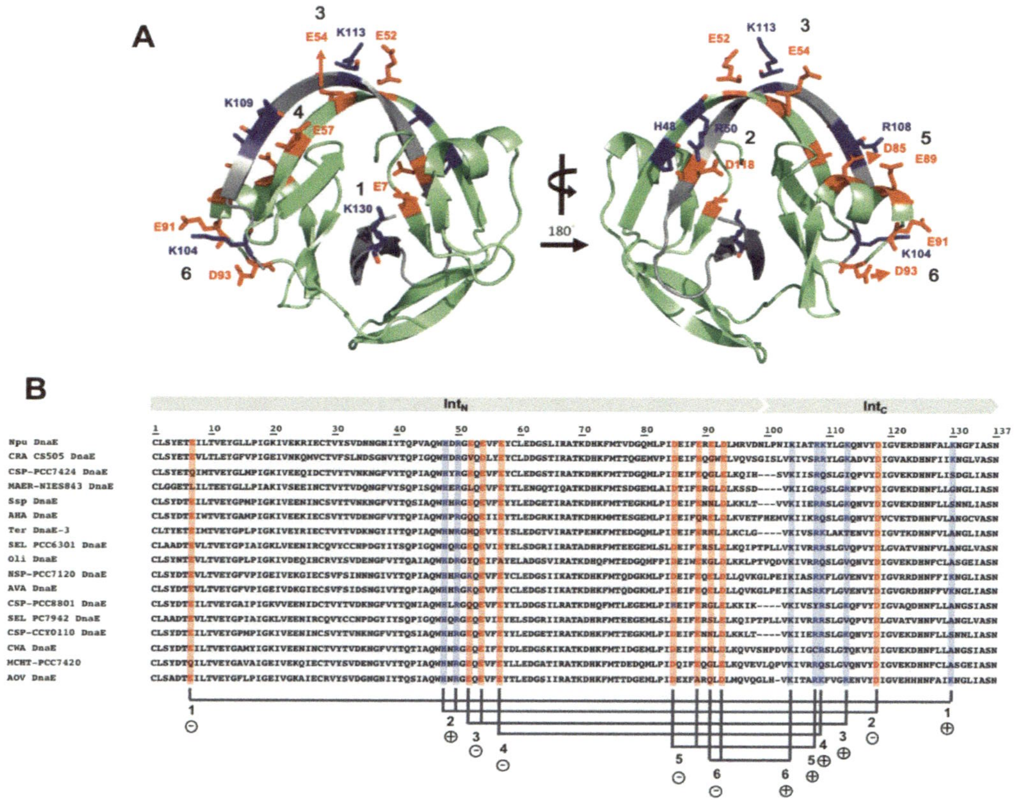

Fig. 2 Salt-sensitive DnaE intein. (A) Salt-bridge distribution in the *Npu* intein N- (*green*) and C-terminal chains (*gray*). Positive residues are depicted as blue sticks and negative residues as red sticks. Charge clusters 1–6 are labeled. (B) Sequence alignment of salt-sensitive DnaE inteins. The sequence numbering refers to the *Npu* intein sequence. The six charge clusters are connected by lines.

removal of salt to accelerate intein activity could be opposite in other types of inteins. HutMCM intein has been shown to have protein *trans*-splicing activity in the presence of high concentrations of salt (Ciragan, Aranko, Tascon, & Iwai, 2016). The structural model of HutMCM intein reported no salt-bridges on the assembled interface. The presence of salt alternatively stabilized the HutMCM intein structure.

2.4 Strategy for polypeptide production

The self-cleavage function of intein provides a solution for breaking the peptide bond without an enzyme. We devised a general strategy using intein in our polypeptide preparation (Fig. 3). To remove the N-terminal cleavage and protein splicing functions, the first residue of intein was mutated.

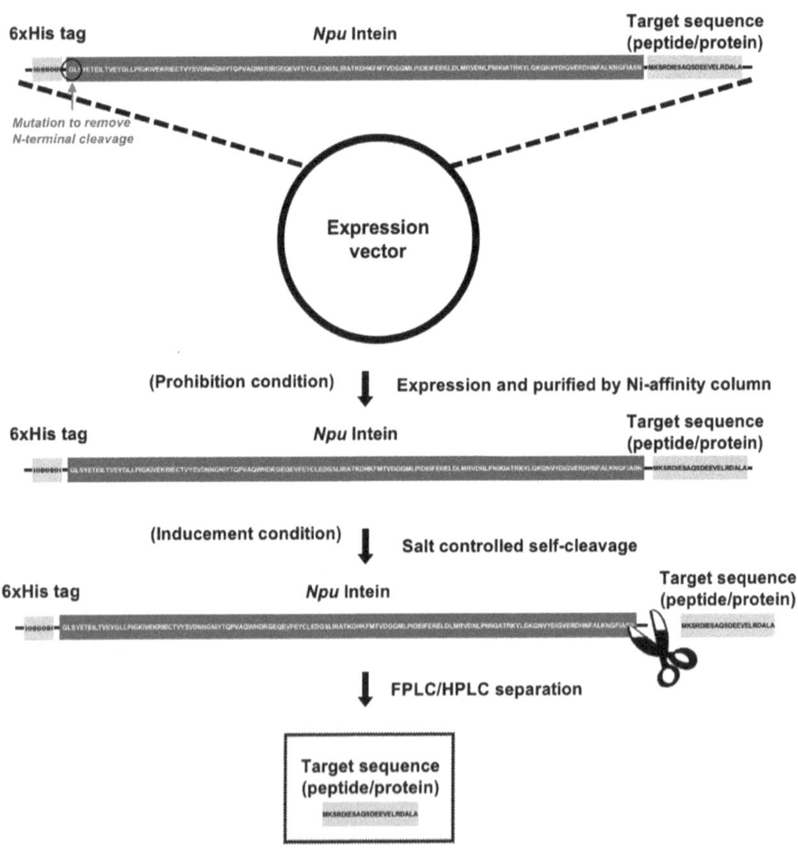

Fig. 3 Flowchart for how to incorporate the intein C-terminal self-cleavage reaction when preparing a target polypeptide.

Thus, intein becomes a fusion domain with spontaneous C-terminal cleavage that is fused at the N-terminal end of target sequence. An affinity tag, such as 6xHis (hexa-Histidine) tag, could be incorporated at the intein N-terminus to promote purification. After expression, the target polypeptide with mutated intein can be purified through the affinity tag. Subsequently, we could recover the C-terminal self-cleavage activity by reducing the ionic strength. With a proper reaction period, intein will be separated from the target polypeptide. After another purification step to remove the intein fraction, we can obtain the pure target polypeptide.

2.5 Conditions to control intein activity

Although we noticed the ionic strength dependency of *Npu* intein, the strategy would be feasible if we can precisely control the C-terminal cleavage activity. Together with the salt concentration, we performed self-cleavage at different pH values and temperatures (Fig. 4). We found that the cleavage reaction favors alkaline pH conditions. The activity decreased when pH is reduced (Fig. 4A–C). We detected very less activity under acidic conditions at pH 5 (Fig. 4C), while the activity was dramatically increased when the pH reached 10 (Fig. 4A). The correlation between peptide production yield and

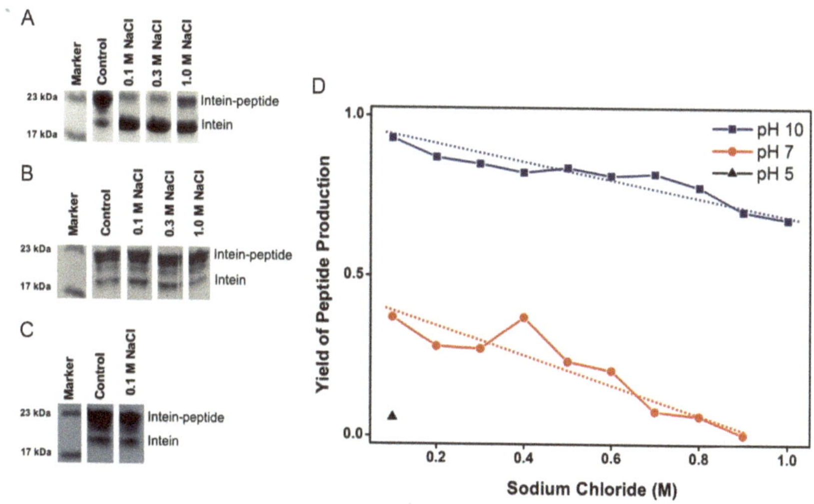

Fig. 4 *Npu* intein C-cleavage analysis under different NaCl concentrations. SDS-PAGEs of (A) pH 10.0, (B) pH 7.0 and (C) pH 5.0. Intact intein-peptide and cleaved intein fractions are indicated. All reactions were performed at 37 °C for 6 h. (D) Yield of peptide production under different NaCl concentrations and pH conditions. The intensities of the bands of cleaved intein are measured to report the yield of peptide production.

pH condition is summarized in Fig. 4D. By regulating the temperature, we showed that self-cleavage has higher activity at 37 °C than 10 °C. Low temperature can be used to slow down the reaction.

Taken together, we propose a "prohibition condition" to inhibit the C-terminal cleavage and an "induction condition" to active it. The "prohibition condition" includes a condition with preferably >300 m*M* NaCl. Considering high intein activity occurred under basic pH conditions, the "prohibition condition" could further include a neutral or acidic pH condition, preferably lower than pH 7. To perform the C-terminal self-cleavage, the reaction is activated under "an induction condition," including an alkaline pH condition, preferably pH 7–11, and a low ionic strength environment, preferably <300 m*M* NaCl. The temperature could be another feature for controlling the reaction, which favors activity at 37 °C, whereas 10 °C slows down the reaction.

3. Polypeptide preparation using salt-sensitive intein

3.1 Equipment

Shaking incubator with temperature control
Centrifuge
UV/Vis spectrophotometer
Polyacrylamide gel electrophoresis apparatus
Micro-pipettor
Homogenizer
High Performance Liquid Chromatography (HPLC)
PCR machine
Lyophilizer
MALDI-TOF mass spectrometer
Open Column
C18 reversed phase column
Dialysis bag and the related dialysis kit
Centricon or Amicon (1 kDa)

3.2 Materials

KOD plus polymerase (or other high-fidelity DNA polymerase)
BL21(DE3) competent *Escherichia coli*
Luria-Bertani (LB) growth medium with antibiotic

IMAC resin or equivalents
Sodium phosphate monobasic and dibasic (NaH_2PO_4 and Na_2HPO_4)
Sodium chloride (NaCl)
Nickel sulfate hexahydrate ($NiSO_4 \cdot 6H_2O$)
Acetonitrile (ACN)
Trifluoroacetic acid (TFA)

3.3 Solutions and buffers

Nickel buffer

Component	Final concentration (m*M*)
Nickel sulfate	400

Lysis buffer

Component	Final concentration (m*M*)
Sodium phosphate	50
Sodium chloride	>300

pH 7.0.

Wash buffer

Component	Final concentration (m*M*)
Sodium phosphate	50
Sodium chloride	>300
Imidazole	15

pH 7.0.

Elution buffer

Component	Final concentration (m*M*)
Sodium phosphate	50
Sodium chloride	>300
Imidazole	400

pH 7.0.

Dialysis buffer

Component	Final concentration (m*M*)
Sodium phosphate	50
Sodium chloride	100

pH 9.5.

HPLC buffer A

Component	Final concentration (%)
99.9% Trifluoroacetic acid (TFA)	0.1

HPLC buffer B

Component	Final concentration (%)
99.9% Trifluoroacetic acid (TFA)	0.1
99.9% Acetonitrile (ACN)	99.9

4. Protocol

4.1 Expression plasmid

The first step is to prepare the plasmid harboring the intein sequence. We designed a plasmid with a 6xHis tag placed before the intein sequence for later purification. The *Npu* intein DNA sequence had been codon-optimized for *E. coli* expression, and the DNA sequence was subsequently chemically synthesized (GENOMICS, TW) and cloned into the pUC57 vector. The sequence was further amplified by KOD polymerase and ligated to an expression vector, pET6H, modified from pET-11b with an inserted hexa-histidine sequence. Then, the first residue of intein was mutated, for example, from Cys to Gly or Ala, to remove the intein splicing activity (Fig. 3).

4.2 Plasmid design and construction

The target polypeptide DNA sequence is fused at the 3′-end of the intein sequence. The target DNA sequence is then inserted into the intein vector using a standard mutagenesis procedure. Depending on the length of the DNA sequence, the target DNA sequence could be divided into several pieces and inserted several times. The final plasmid is validated by DNA sequencing.

4.3 Protein expression

The constructed plasmid is transformed into the *E. coli* BL21 (DE3) strain for protein expression. A traditional heat shock method is applicable. The *E. coli* BL21 (DE3) strain is cultured on an LB agar plate with 0.1 mg/mL ampicillin overnight at 37 °C. Randomly select a single clone, transfer it into 10 mL of LB medium with 0.1 mg/mL ampicillin, and culture it overnight at 37 °C. The *E. coli* culture is further transferred into 500 mL of LB medium. When the culture reaches an OD_{600} of 0.6, add the inducer IPTG to the medium to final concentration of 1 m*M*, at which point culture growth continues for 24 h at 16 °C. Harvest the cells by centrifugation (4 °C, 6000 × *g*, and 30 min). The culture medium is discarded and the cell pellet is frozen for further purification.

4.4 Ni column preparation

Suitable amounts of IMAC resin (GE Healthcare) are packed into an open column (Bio-Rad) and three column volumes of ddH_2O are added to wash the resin. Nickel buffer is further transferred into the column to couple the nickel ions to the IMAC resin. After overnight coupling, the Nickel buffer is removed with ddH_2O and the column is ready to use.

4.5 Ni-column purification

Resuspend the cells in 50 mL Lysis buffer and lyse with a high-pressure homogenizer with 15,000 psi. The lysed cells are centrifuged at 15,000 × *g* for 30 min. The supernatant is further used for nickel-column purification. The supernatant is mixed with nickel resin. The target protein bound to the nickel resin, is washed with Wash buffer, and then eluted with Elution buffer. SDS PAGE is used to check the purification. To prevent self-cleavage from occurring, these steps are performed at low temperature, high ionic strength (>300 m*M* NaCl) and neutral pH. In some cases, you can notice small fractions of cleaved intein, indicating that mild self-cleavage occurs during protein expression. However, intact protein product is still the major fraction.

4.6 Self-cleavage reaction

After purification from the nickel column, use a dialysis bag to exchange the buffer. Dialysis is performed against 50–100-fold Dialysis buffer. Dialysis is performed two times at 4 °C, with each dialysis cycle taking 8 h. The dialysis process aims to remove imidazole and reduce the NaCl concentration to 100 m*M*. Meanwhile, the pH value is increased to 9–10. Intein self-cleavage starts during dialysis. The choice of the dialysis bag with the proper pore size depends on the molecular weight of the target polypeptide. A molecular weight cut-off (MWCO) of 1 kDa is a proper choice to retain most polypeptides. SDS PAGE is used to check the reaction. It usually takes <12 h to finish the self-cleavage reaction.

4.7 HPLC purification

Use Amicon or Centricon (1 kDa) to concentrate the product after dialysis. The target polypeptide is purified by HPLC using a C18 reverse-phase column. An ACN gradient (HPLC buffers A and B) is used to separate intein and the polypeptide. High concentrations of intein sometimes aggregate; thus, filter the concentrated sample before HPLC injection. The polypeptide is eluted at 20–40% ACN depending on the hydrophobicity of individual samples, while intein is eluted at 45–50% ACN. If HPLC is not available, a Ni column could be used to separate intein from the target polypeptide. The final product should be validated using MALDI-TOF mass spectrometry.

5. Extension of application

As mentioned, this method can be used to express different polypeptides without enzyme or chemical requirements. We have employed this strategy to successfully prepare several polypeptides. We will further discuss two unique cases here.

5.1 Disulfide bond-rich chemokine

We successfully applied this method to express chemokine platelet factor 4 (PF4) (Chen et al., 2017; Quemener et al., 2016), a small protein (8 kDa) that contains 70 residues and two intramolecular disulfide bonds. Native PF4 is initially translated into a 101-residue protein with an additional signal peptide region (31 residues) incorporated at its N-terminus. When the signal peptide is cleaved, PF4 becomes functionally active. Mature PF4 is stored in platelet α-granules and has the first codon Glu, not Met.

Conventional expression in bacteria usually start from Met. This creates difficulty when preparing the protein. The previous method fused thioredoxin as a fusion protein, which was later cleaved by enterokinase, a specific protease that cleaves after Lys at the cleavage site Asp-Asp-Asp-Asp-Lys. However, the usage of enterokinase increased the cost for preparation. Meanwhile, enterokinase activities vary from batch to batch. The cleavage efficiency was not consistent. The current protocol overcomes the disadvantages associated with the previous production method. It requires no enzyme and shows high self-cleavage activity when proper induction conditions were used. Here, we used 50 m*M* sodium phosphate, pH 10.0 and 150 m*M* NaCl as the buffer to trigger the self-cleavage. The current method shows the ability to express a disulfide bond-rich protein as well as proteins that start with non-Met codons.

5.2 NMR isotope-labeled peptide

The current method would greatly benefit in the preparation of isotope-labeled peptides. Nuclear magnetic resonance (NMR) requires ^{13}C and/or ^{15}N-isotope-labeled samples to perform multi-dimensional NMR experiments. The ^{13}C and ^{15}N nuclei provide information for characterizing protein structure. Together with ^{1}H, we could use triple-resonance experiments to detect ^{1}H, ^{13}C and ^{15}N nuclei. The resonances derived from ^{1}H, ^{13}C and ^{15}N nuclei will be valuable for realizing backbone and side-chain structures, dynamics and even calculating protein structure. Meanwhile, ^{13}C and ^{15}N resonances give you wider distribution in spectra, bringing advantages especially for disordered peptides where backbone amide ^{1}H signals collapse in the narrow spectral range from 8.0 to 8.5 ppm.

In protein NMR, it is common to use isotope-labeled samples to permit the assignments of the backbone and side chains. Isotope-labeled proteins can be routinely produced from *E. coli*, which is grown on minimal medium supplemented with $^{15}NH_4Cl$ and ^{13}C-glucose. However, for polypeptide preparation, it is rather difficult to directly express peptides from *E. coli*. Rapid degradation and difficulty in purification hamper the idea of direct expression. Target polypeptides conjugated with a fusion domain have become a common strategy. In an alternative way, we could chemically synthesize isotope-labeled peptides using isotope-labeled amino acids as materials. However, the cost is not always affordable. Using the current method to prepare isotope-labeled polypeptides by fusing with intein domain could by a feasible way to obtain the isotope-labeled target polypeptides at a decent cost.

We present a case for preparing the ^{13}C- and ^{15}N-labeled magnetite-interacting peptide, Mms6 (Fig. 5) (Nudelman et al., 2018). The peptide

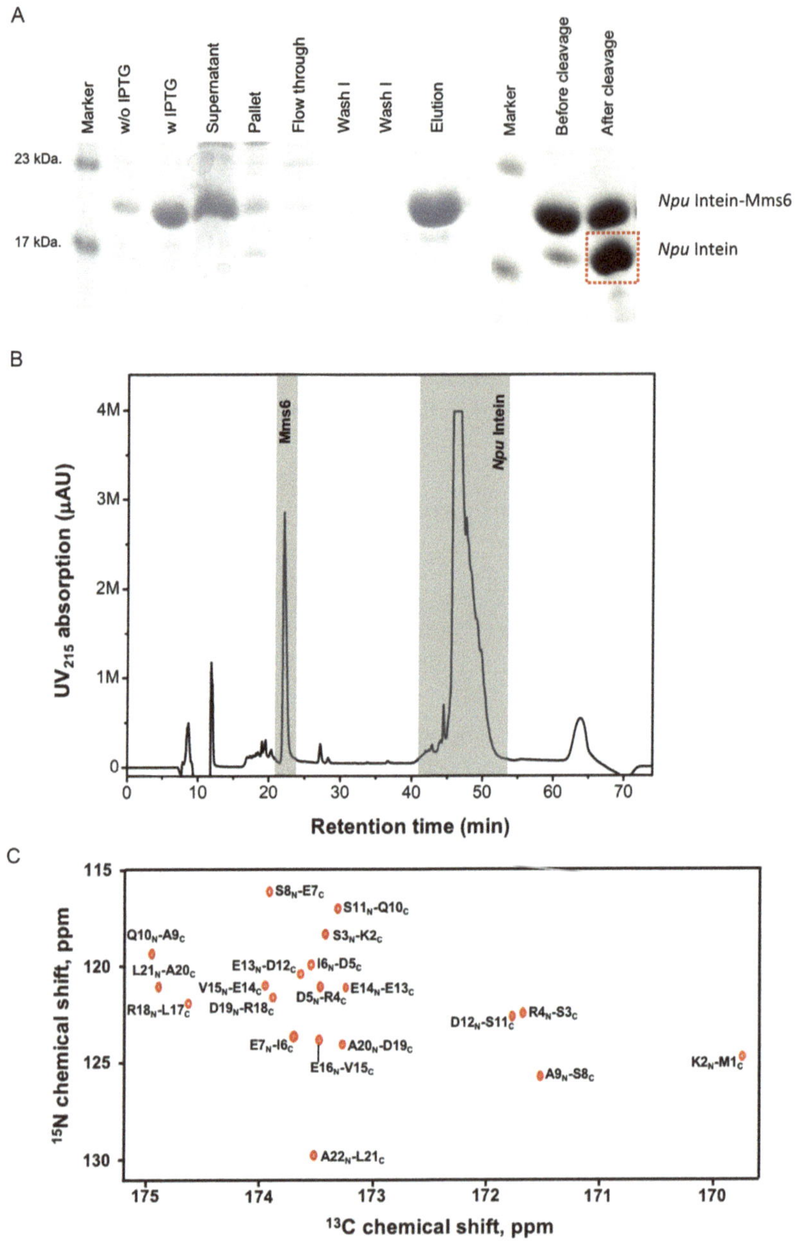

Fig. 5 Preparation of an isotope-labeled polypeptide. (A) SDS-PAGE of the expression and purification of ^{13}C- and ^{15}N-labeled intein-Mms6. Intein-Mms6 expression was induced by IPTG and the fusion protein was purified using a Ni-column. After elution, the protein is exchanged into an induction buffer to activate intein self-cleavage. >50% of the protein is cleaved after 2 h of incubation at 37 °C (boxed). (B) The HPLC profile of the Mms6 polypeptide and *Npu* intein. A C18 reverse-phase column and ACN linear gradient were used for purification. (C) The NCO heteronuclear NMR spectra for Mms6. The correlations between CO and N are depicted for each residue.

has been known to specifically interact with Fe^{2+} and Ni^{2+}. We prepared an appropriate expression system in which the Mms6 N-terminus was linked to *Npu* intein and an affinity 6xHis tag was conjugated with the N-terminal end of intein. The protein product was induced by IPTG. After cell lysis, the over-expressed product was distributed in both the supernatant and pellet (Fig. 5A). We purified the supernatant portion using a Ni-affinity column. The proteins were eluted with Elution buffer (Fig. 5A). By exchanging into buffer conditions consisting of 50 m*M* sodium phosphate, pH 10.0 and 150 m*M* NaCl, Mms6 peptide was cleaved from the encoded intein. The peptide and intein were separated by HPLC using a C18 column (Fig. 5B). The purified Mms6 was then dissolved in NMR buffer (25 m*M* MES and 10% D_2O, pH 5.6). We demonstrated the NCO spectrum with the correlations between CO and N (Fig. 5C) and the heteronuclear NMR spectra provided extraordinary resolution (Bermel et al., 2005).

5.3 Residue type limitations

C-terminal cleavage is theoretically based on the mechanism that the Asn (the last residue in intein) side chain NH_2 breaks the peptide bond between intein and the target sequence. Thus, the first residue of the target sequence is not critical. C-terminal cleavage shows tolerance for most residue types. However, there is still an exception, which cleavage might not be favored when the first residue of the target sequence is Cys, Ser or Thr (Zettler et al., 2009). These residues with hydroxyl or sulfhydryl side chains are not recommended in the junction. The hydroxyl or sulfhydryl groups could alternatively provide a nucleophilic center for splicing to compensate for the role of the first residue in intein that is already mutated from Cys to other residues in our method. Cys, Ser or Thr as the first residue in the target sequence might provide unpredicted splicing reactions; for example, the N-terminal 6xHis tag could be ligated to the target polypeptide when the intein domain is excised from the sequence.

6. Discussion

Chemical synthesis methods are used to synthesize polypeptides. However, when considering large-scale preparation, chemical synthesis is not feasible when a sequence is longer than 25 residues. Longer sequences are usually accompanied by reduced yield and increased cost. When people use biosynthesis methods to prepare polypeptides, fusion tags are helpful to promote expression and purification. However, the issues of enzyme digestion and residual residues will affect the entire strategy; for example, residual residues left in the final product usually influence function and activity. Therefore, salt-sensitive intein provides a potential option for preparing polypeptides with a very low cost. Together with bioreactor usage, individuals can easily achieve large scale preparation. In our experience, we obtained approximately 10 mg of final polypeptide product from 1 L of bacterial culture. When using a bioreactor, the final product reached >0.5 g per liter when the final OD_{600} of the culture reached 180.

Several intein-mediated purifications have been developed to promote target polypeptide preparation (Chong et al., 1997, 1998; Muir, Sondhi, & Cole, 1998), and these methods have been successfully applied in many systems. We compared our method with others here. The first compared method used split intein to process self-cleavage. In this case, it is necessary to prepare two individual fragments, an intein N-terminal chain with an affinity tag and an intein C-terminal chain conjugated with the target polypeptide. Upon mixing the two fragments, the two chains assemble and form an entire intein fold. The target polypeptide is subsequently cleaved from the intein C-terminal chain and the affinity tag is used to separate the intein portion. This method has more complicated sample preparation. Meanwhile, the individual intein chains only represent a partial fold of intein and are less stable than the full-length intein. The intein C-terminal chain, which is conjugated with the target polypeptide, demonstrates protein aggregation and precipitation. Another popular case for intein-mediated purification is through conjugation with an affinity chitin-binding tag. This method is IMPACT (Muir et al., 1998). IMPACT links mature Mxe intein and a chitin-binding domain as a fusion tag that is subsequently fused with a target polypeptide. Chitin is used for purification. IMPACT uses dithiothreitol (DTT) as a regulator to trigger the self-cleavage process. However, in some cases, DTT is not favorable because of the presence of disulfide bonds. The presence of DTT will affect target polypeptide stability and structure. The use of DTT also increases the need to remove the chemical in the final preparation.

7. Summary

We employed the salt-sensitive feature of *Npu* intein to develop a self-cleaving fusion tag. By removing the intein N-terminal cleavage activity and maintaining its C-terminal cleavage activity, we develop a system in which a target polypeptide is fused with the novel self-cleaving intein for bacterial expression. The self-cleavage can be controlled by buffer conditions in that it is inhibited under high ionic strength, acidic and low temperature conditions and induced under low ionic strength, alkaline and high temperature conditions. Therefore, salt (i.e., NaCl) and ionic strength can be used to control the cleavage reaction, and no additional enzyme is required for the process. This method provides an easy and efficient way to produce polypeptides. The advantages of this method include low cost, enzyme-free, no extra residual residue left in the final product, and feasibility for preparing proteins that start with non-Met codons. This method also has a special benefit for producing isotope-labeled peptide for NMR experiments. This method could widely benefit different polypeptide preparations.

Acknowledgments

We thank National Tsing-Hua University (NTHU), Taiwan for their financial support and the NMR facility at the NTHU instrument center for performing the NMR measurements.

References

Anraku, Y., Mizutani, R., & Satow, Y. (2005). Protein splicing: Its discovery and structural insight into novel chemical mechanisms. *IUBMB Life*, *57*(8), 563–574.

Aranko, A. S., Oeemig, J. S., Kajander, T., & Iwai, H. (2013). Intermolecular domain swapping induces intein-mediated protein alternative splicing. *Nature Chemical Biology*, *9*(10), 616–622.

Bedard, F., & Biron, E. (2018). Recent progress in the chemical synthesis of class II and S-glycosylated bacteriocins. *Frontiers in Microbiology*, *9*, 1048.

Bermel, W., Bertini, I., Duma, L., Felli, I. C., Emsley, L., Pierattelli, R., et al. (2005). Complete assignment of heteronuclear protein resonances by protonless NMR spectroscopy. *Angewandte Chemie (International ed. in English)*, *44*(20), 3089–3092.

Bray, B. L. (2003). Large-scale manufacture of peptide therapeutics by chemical synthesis. *Nature Reviews. Drug Discovery*, *2*(7), 587–593.

Chen, Y. P., Wu, H. L., Boye, K., Pan, C. Y., Chen, Y. C., Pujol, N., et al. (2017). Oligomerization state of CXCL4 chemokines regulates G protein-coupled receptor activation. *ACS Chemical Biology*, *12*(11), 2767–2778.

Cheriyan, M., Chan, S. H., & Perler, F. (2014). Traceless splicing enabled by substrate-induced activation of the *Nostoc punctiforme* Npu DnaE intein after mutation of a catalytic cysteine to serine. *Journal of Molecular Biology*, *426*(24), 4018–4029.

Chong, S. R., Mersha, F. B., Comb, D. G., Scott, M. E., Landry, D., Vence, L. M., et al. (1997). Single-column purification of free recombinant proteins using a self-cleavable affinity tag derived from a protein splicing element. *Gene*, *192*(2), 271–281.

Chong, S. R., Montello, G. E., Zhang, A. H., Cantor, E. J., Liao, W., Xu, M. Q., et al. (1998). Utilizing the C-terminal cleavage activity of a protein splicing element to purify recombinant proteins in a single chromatographic step. *Nucleic Acids Research*, *26*(22), 5109–5115.

Ciragan, A., Aranko, A. S., Tascon, I., & Iwai, H. (2016). Salt-inducible protein splicing in cis and trans by inteins from extremely halophilic archaea as a novel protein-engineering tool. *Journal of Molecular Biology*, *428*(23), 4573–4588.

Eryilmaz, E., Shah, N. H., Muir, T. W., & Cowburn, D. (2014). Structural and dynamical features of inteins and implications on protein splicing. vol 289, pg 14506, 2014 *Journal of Biological Chemistry*, *289*(28), 19278.
Hojo, H. (2014). Recent progress in the chemical synthesis of proteins. *Current Opinion in Structural Biology*, *26*, 16–23.
Lee, Y. Z., Lee, Y. T., Lin, Y. J., Chen, Y. J., & Sue, S. C. (2014). A streamlined method for preparing split intein for NMR study. *Protein Expression and Purification*, *99*, 106–112.
Lee, Y. Z., Lo, W. C., & Sue, S. C. (2017). Computational prediction of new intein split sites. *Methods in Molecular Biology*, *1495*, 259–268.
Lee, Y. T., Su, T. H., Lo, W. C., Lyu, P. C., & Sue, S. C. (2012). Circular permutation prediction reveals a viable backbone disconnection for split proteins: An approach in identifying a new functional split intein. *PLoS One*, 7(8), e43820.
Mathys, S., Evans, T. C., Chute, I. C., Wu, H., Chong, S. R., Benner, J., et al. (1999). Characterization of a self-splicing mini-intein and its conversion into autocatalytic N- and C-terminal cleavage elements: Facile production of protein building blocks for protein ligation. *Gene*, *231*(1–2), 1–13.
Muir, T. W., Sondhi, D., & Cole, P. A. (1998). Expressed protein ligation: A general method for protein engineering. *Proceedings of the National Academy of Sciences of the United States of America*, *95*(12), 6705–6710.
Nudelman, H., Lee, Y. Z., Hung, Y. L., Kolusheva, S., Upcher, A., Chen, Y. C., et al. (2018). Understanding the biomineralization role of magnetite-interacting components (MICs) from magnetotactic bacteria. *Frontiers in Microbiology*, *9*, 2480.
Paraskevopoulou, V., & Falcone, F. H. (2018). Polyionic tags as enhancers of protein solubility in recombinant protein expression. *Microorganisms*, *6*(2), 47.
Quemener, C., Baud, J., Boye, K., Dubrac, A., Billottet, C., Soulet, F., et al. (2016). Dual roles for CXCL4 chemokines and CXCR3 in angiogenesis and invasion of pancreatic cancer. *Cancer Research*, *76*(22), 6507–6519.
Rosano, G. L., & Ceccarelli, E. A. (2014). Recombinant protein expression in *Escherichia coli*: Advances and challenges. *Frontiers in Microbiology*, *5*, 172.
Shah, N. H., Eryilmaz, E., Cowburn, D., & Muir, T. W. (2013). Naturally split inteins assemble through a "capture and collapse" mechanism. *Journal of the American Chemical Society*, *135*(49), 18673–18681.
Shah, N. H., & Muir, T. W. (2014). Inteins: Nature's gift to protein chemists. *Chemical Science*, *5*(2), 446–461.
Shah, N. H., Vila-Perello, M., & Muir, T. W. (2011). Kinetic control of one-pot trans-splicing reactions by using a wild-type and designed split intein. *Angewandte Chemie (International ed. in English)*, *50*(29), 6511–6515.
Vila-Perello, M., & Muir, T. W. (2010). Biological applications of protein splicing. *Cell*, *143*(2), 191–200.
Waugh, D. S. (2011). An overview of enzymatic reagents for the removal of affinity tags. *Protein Expression and Purification*, *80*(2), 283–293.
Wood, D. W., & Camarero, J. A. (2014). Intein applications: From protein purification and labeling to metabolic control methods. *The Journal of Biological Chemistry*, *289*(21), 14512–14519.
Zettler, J., Schutz, V., & Mootz, H. D. (2009). The naturally split Npu DnaE intein exhibits an extraordinarily high rate in the protein trans-splicing reaction. *FEBS Letters*, *583*(5), 909–914.
Zhao, Q., Xu, W. H., Xing, L., & Lin, Z. L. (2016). Recombinant production of medium-to large-sized peptides in *Escherichia coli* using a cleavable self-aggregating tag. *Microbial Cell Factories*, *15*, 136.

CHAPTER EIGHT

Methods for the recombinant expression of active tyrosine kinase domains: Guidelines and pitfalls

M. Escarlet Díaz Galicia, Abdullah Aldehaiman, SeungBeom Hong, Stefan T. Arold*, Raik Grünberg*

Division of Biological and Environmental Sciences and Engineering (BESE), Computational Bioscience Research Center (CBRC), King Abdullah University of Science and Technology (KAUST), Thuwal, Saudi Arabia

*Corresponding authors: e-mail address: stefan.arold@kaust.edu.sa; raik.grunberg@kaust.edu.sa

Contents

Abstract

Protein tyrosine kinases (PTKs) are key signaling molecules and important drug targets. Although the efficient recombinant production of active PTKs is important for both pharmaceutical industry and academic research, most PTKs are still obtained from conventional, expensive and time-consuming insect-cell based expression. Host toxicity, kinase inactivity, insolubility and heterogeneity are among the reasons thought to preclude PTK expression in *Escherichia coli*. Herein we review these presumed roadblocks and their possible solutions for bacterial expression of PTKs, and give an overview on kinase activity assays. Finally, we report our experiences and observations with the kinases Src, Lyn and FAK as examples to illustrate implementation, effects and pitfalls of *E. coli* expression and *in vitro* assaying of PTKs.

Methods in Enzymology, Volume 621
ISSN 0076-6879
https://doi.org/10.1016/bs.mie.2019.02.027

1. Introduction

Kinase-mediated phosphorylation of serine, threonine or tyrosine side chains rapidly and reversibly alters protein function, conformation and interaction partners. In particular, this mechanism is widely used in metazoans where a large part of the proteome experiences some type of phosphorylation (Sharma et al., 2014). Within the eukaryotic protein kinase family, protein tyrosine kinases (PTKs) appear to be the most recent addition to the cell's signaling repertoire. (Herein, we distinguish between "PTKs," which can have additional non-kinase domains on the same polypeptide chain, and "kinase domains" (KDs), which refer to the catalytic domain of any type of kinase). In fact, tyrosine phosphorylation was suggested to have opened a new biochemical "communication channel" which may have been a prerequisite for the development of multicellularity (Hunter, 2009; Lim & Pawson, 2010). This view is supported by the role of eukaryotic PTKs as important regulators of processes closely linked to multicellular organization in addition to proliferation, hormone response and immune reactions (Hunter, 2009). In contrast to serine or threonine phosphorylation, human cells tightly downregulate phosphorylation on tyrosines. In the absence of specific signaling events, most tyrosine phosphorylation sites remain "switched off" (Sharma et al., 2014). This is also reflected in the human genome, where protein tyrosine phosphatases (PTPs) outnumber PTKs 107–90 (Alonso et al., 2004). By contrast, the 428 serine/threonine kinases are countered by only 28 serine/threonine phosphatases (Sacco, Perfetto, Castagnoli, & Cesareni, 2012).

Multiple human health threats such as cancer or diabetes, as well as immune, neurological, infectious and inflammatory diseases have been associated with the dysregulation of protein kinases (Bhullar et al., 2018; Patterson, Nibbs, McInnes, & Siebert, 2014). The targeted inhibition of PTKs is therefore of high therapeutic and pharmaceutical interest, which has spurred the quest to understand the structure, specificity and control of PTKs. In 2018 alone, seven thousand publications report on kinase inhibitors (Pub Med), in addition to undisclosed research within the pharmaceutical industry. To date, the US Food and Drug Administration (FDA) has approved 37 kinase inhibitors to treat cancer and 150 similar drugs are currently in clinical trials (Bhullar et al., 2018). Imatinib and Dasatinib are the best-known examples and have advanced the treatment of myeloid leukemia (for reviews, see Agafonov, Wilson, Otten, Buosi, & Kern, 2014; Fountas, Diamantopoulos, & Tsatsoulis, 2015; Wu, Nielsen, & Clausen, 2016).

Aside from the therapeutic interest in KDs, their modularity, compatibility with other signaling systems and their capacity to function as "protein interaction computation nodes" make them a promising building block for the engineering of synthetic protein systems. In native proteins, KDs are often found fused to other domains, which can direct the kinase to specific targets or regulate its enzymatic activity. In fact, the very concept of a modular protein domain arose from the study of the oncogenic Src kinase, which remains a model PTK to this day (Pawson, 1995). The artificial rewiring of the PTK's domain composition can redirect signaling flows and could pave the way to designed protein circuits with applications *in vitro* or in living cells (Gordley, Bugaj, & Lim, 2016; Grünberg & Serrano, 2010). Moreover, recently described "split kinase" systems may offer an engineering shortcut through the complex regulation of kinase activity (Camacho-Soto, Castillo-Montoya, Tye, & Ghosh, 2014; Diaz et al., 2017; Ma, Nagamune, & Kawahara, 2014).

Fundamental research, drug design, synthetic biology and biotechnology all depend heavily on the availability of pure, soluble and active PTKs. The fastest and cheapest expression system is *E. coli*. However, while many serine/threonine kinases can be expressed easily in bacteria, PTKs are considered more challenging expression targets. This difficulty is somewhat surprising as PTKs and serine/threonine kinases are structurally highly similar (Fig. 1A). Consequently, most PTKs are currently expressed from insect or mammalian cells. However, these methods are expensive and slow,

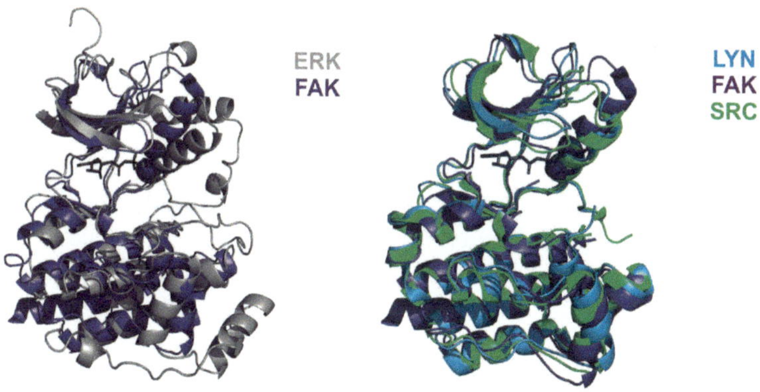

Fig. 1 Structural comparison of the catalytic domains of ERK, Src, Lyn and FAK. (A) Backbone superposition of serine kinase ERK2 (gray) and tyrosine kinase FAK (red), root mean square difference (RMSD) = 1.07 Å. (B) Superposition of tyrosine kinases Lyn (blue), FAK (purple), and Src (green) catalytic domains, RMSD = 1.70 Å. In both figures, bound ADP is shown as stick model.

hampering the rapid generation of mutants and new constructs as required for the trial-and-error workflow of bioengineering projects and mechanistic studies.

Herein, we briefly review PTK expression protocols from insect and mammalian cells. Given the marked advantages of *E. coli*-based protein expression methods, we will discuss the reasons that impede production of active PTKs in *E. coli*, and review possible solutions. Additionally, we will use our own experiences with the PTKs FAK, Src and Lyn (Fig. 1B) to highlight some of the pitfalls and oddities associated with the expression of PTKs. Finally, given that the ability to measure and quantify kinase activity is instrumental to evaluating the success of recombinant kinase production, we also review kinase activity assays.

2. Expression of PTKs in eukaryotic systems

In initial studies on the topic, PTKs appeared difficult to express in *E. coli* and were generally cytotoxic to yeast. In particular for structural studies, requiring large amounts of pure protein, insect cell expression quickly emerged as the method of choice. The baculovirus-insect cell system has long been a strong workhorse for the expression of eukaryotic recombinant proteins that are not well-expressed, folded and/or processed in *E. coli* (Smith, Summers, & Fraser, 1983; van Oers, Pijlman, & Vlak, 2015). In the most common workflow, the target gene(s) is first cloned between Tn7 transposon sites into a bacterial donor plasmid (i.e., pFASTBAC). This puts them under the control of either the polyhedrin or the P10 promoters which are derived from the *Autographa californica* multiple nucleopolyhedrovirus (AcMNPB). The expression cassette is then transferred by transposon-mediated insertion into a Bacmid shuttle vector that can be propagated in both *E. coli* and insect cells and encodes all the remaining components of the virus. Actual virus is produced after transfection of this Bacmid into *Spodoptera frugiperda* (fall armyworm) insect cells and then used to infect a larger production culture of the same cell type (Luckow, Lee, Barry, & Olins, 1993; Pijlman, van Schijndel, & Vlak, 2003). With this system, many eukaryotic kinases have been successfully expressed and purified to high yields of active protein. We have deposited a detailed example protocol for insect-cell production of multidomain constructs of the PTKs Src and FAK along with the resulting kinase activities at https://github.com/strubelab/kinaseexpress. Nevertheless, insect cell expression requires a minimum of 3–4 weeks from an initial cloning of the target gene into the bacterial vector to the actual protein expression.

Furthermore, when compared to the *E. coli* expression system, growth media and the additional infrastructure required, the maintenance of insect cell cultures is relatively expensive.

Nonetheless, insect cell expression still compares very favorably to traditional methods for the large-scale expression from mammalian cell lines. Although production from mammalian cells allows native post-translational modifications and chaperon-assisted folding of human proteins (Bruel, Cha, Reeves, Getmanova, & Khorana, 2000; Nettleship, Assenberg, Diprose, Rahman-Huq, & Owens, 2010), the time and resources involved are often only invested in an industrial setting. For example, more than half of biopharmaceutical (therapeutic) proteins are currently produced from mammalian cell culture (Khan, 2013). In an academic context, mammalian production of PTKs seems mostly limited to small-scale expression from transiently transfected cells (Deng et al., 2014). Nevertheless, recent lentiviral transfection protocols (Elegheert et al., 2018) or baculovirus-mediated gene transfer into mammalian cells (BacMam) (Goehring et al., 2014; Hofmann et al., 1995) obviates the generation of stable cell lines and shortens the mammalian expression protocol to a time frame that is comparable with insect cell expression. This method is thus also becoming attractive for the large-scale expression of PTKs.

3. PTK expression from *E. coli*

About 60% of all recombinant proteins mentioned in the literature (Correa & Oppezzo, 2015) and close to 90% of proteins with known structure (Nettleship et al., 2010) have been produced in *E. coli*. Indeed, where applicable, *E. coli* typically provides the easiest, fastest and cheapest access to larger quantities of recombinant protein. But not all proteins can be produced in this system. Eukaryotic proteins in particular are often subject to degradation or aggregation in inclusion bodies (Correa & Oppezzo, 2015; Rosano & Ceccarelli, 2014). In a direct comparison of 62 full-length human kinases, all but one showed soluble expression from insect cells compared to only a 54% overall success rate for expression in *E. coli*. Among the six major groups of kinases tested, PTKs performed the worst with less than 30% success (Chambers, Austen, Fulghum, & Kim, 2004). However, this result may have been biased by the use of sequences with the original mammalian codon usage in an expression vector that was optimized for insect-cell expression. By contrast, thanks also to a targeted structural genomics effort, the PDB now contains many human kinase structures, including also PTKs, that were produced from *E. coli*-expressed proteins (SGC, 2013). Starting from this record of successful expression in *E. coli*, Albanese et al. assembled a panel of 96 KD expression constructs, spanning 8 kinase families. Small-scale expression together with a phosphatase (but without any further optimization) succeeded for 10 out of the 17 tyrosine KDs tested, which compares well with the overall success rate of 56% (Albanese et al., 2018).

Also the earliest attempts of purifying mammalian c-Src KD from *E. coli* did not fail outright but reported low yields of soluble protein (Osusky, Taylor, & Shalloway, 1995; Saya et al., 1993). Meanwhile, expression attempts in yeast indicated a general cytotoxicity of PTKs in this system, although, in the case of c-Src, this cytotoxicity could be overcome by the co-expression of human PTP-PEST as an antagonizing phosphatase (Weijland et al., 1996). In *E. coli*, the picture was less clear. Sufficient expression and activity were, for example, shown for v-Abl (Garcia et al., 1993), for Csk (Bougeret, Rothhut, Jullien, Fischer, & Benarous, 1993) (each fused to GST) and later also claimed for the expression of c-Src, c-Abl, and c-Met KDs (Wang, Marimuthu, et al., 2006). In the meantime, co-expression of an antagonizing phosphatase was reported to improve kinase yields also in the *E. coli* expression system (Seeliger et al., 2005; Wang, Marimuthu, et al., 2006). Two different phosphatases, YopH and PTP1B, were introduced for this purpose: YopH is an unspecific tyrosine phosphatase derived from the bacterial pathogen *Yersinia* (Bliska, Guan, Dixon, & Falkow, 1991) and was employed to produce recombinant c-Src and c-Abl KDs (Seeliger et al., 2005). PTP1B had earlier been identified as the primary human phosphatase targeting and activating c-Src (Bjorge, Pang, & Fujita, 2000) and other PTKs (Fan, Lin, Lucito, & Tonks, 2013) and was used to improve yield and homogeneity of recombinant c-Src, c-Abl and c-Met KDs (Wang, Marimuthu, et al., 2006). Phosphatase co-expression was then reported in subsequent studies using either PTP1B (Cui & Sun, 2018; Tu, Wang, Cai, Zhou, & Zhang, 2014; Wang, Ayrapetov, Lin, & Sun, 2006) or YopH (Albanese et al., 2018; Filippakopoulos et al., 2008; Wilson et al., 2015). With one dissenting opinion (Wang, Marimuthu, et al., 2006), there seems to be a consensus that the active Src KD is cytotoxic to *E. coli* and that its recombinant expression benefits from phosphatase co-expression. This conclusion was reinforced by Kemble et al. who showed that different mutants of the Src KD could be expressed in *E. coli* as long as their enzymatic activity remained below a certain threshold (Kemble, Wang, & Sun, 2006). However, the view that PTKs are *generally* cytotoxic to *E. coli* is not backed up by the literature record. Indeed, in contrast to earlier assumptions, *E. coli* has its own tyrosine phosphorylation and dephosphorylation system (Vincent et al., 1999). Hence, cytotoxicity of PTKs might be related to both their particular intrinsic activity and substrate specificity as well as their level of overexpression.

Aside from potential cytotoxicity, bacterially expressed PTKs seem particularly prone to misfolding and accumulation in inclusion bodies (Chambers et al., 2004; Piserchio, Cowburn, & Ghose, 2012; Seeliger et al., 2005). Purification of active Src KD from such inclusion bodies was reported (Gong et al., 2006; Marin et al., 2010), but this approach might not be generally applicable. Fusion to solubility-enhancing tags (Kimple, Brill, & Pasker, 2013) such as GST (Bougeret et al., 1993; Garcia et al., 1993; Gong et al., 2006) or MBP (Piserchio et al., 2012) may increase the soluble fraction. Nevertheless, Piserchio et al. warn that large amounts of soluble MBP-Src fusion product were, in fact, soluble aggregates of still unfolded protein, a problem that they addressed through the co-over-expression of chaperons.

This overexpression of bacterial chaperons has earlier been touted as a general strategy for improved protein production in *E. coli* (Thomas, Ayling, & Baneyx, 1997) and was also used by Kemble et al. for the expression of an inactivated c-Src KD (Kemble et al., 2006). In a more systematic comparison, Haacke et al. subjected a set of 10 KDs (including 4 non-receptor PTKs, 4 receptor PTKs, and 2 Ser/Thr kinases) to expression with and without different mixtures of bacterial chaperons (Haacke, Fendrich, Ramage, & Geiser, 2009). Overexpression of chaperons generally reduced the amount of insoluble protein (inclusion bodies), but only increased the yield of affinity-captured soluble protein in about half of the tested KDs. c-Src and Hck KDs (which were also co-expressed with PTP1B) did not benefit from chaperon co-expression. However, also in most of the remaining cases, the formation of soluble aggregates [as assessed by size exclusion chromatography (SEC)] negated the advantage of bacterial chaperon co-expression which, therefore, does not seem to be of general help for PTK expression.

The efficiency of chaperon co-expression might also depend on the choice of chaperones used. Most studies use bacterial chaperones (i.e., GroEL, GroES, DnaK, DnaJ, others) (Haacke et al., 2009; Kemble et al., 2006; Piserchio et al., 2012). However, in their native eukaryotic cell, the folding of many PTKs depends on HSP90 and the kinase-specific co-chaperon CDC37. The two chaperons form a specific complex that preserves the client kinase in an inactive state and protects buried tyrosines from phosphorylation

(Taipale et al., 2012; Verba et al., 2016). Co-expression of CDC37 alone seems to be beneficial for the expression of active and soluble c-Src KD in *E. coli* (Piserchio et al., 2012). Further studies are needed to clarify the utility of human chaperon co-expression for PTK production in this host.

Beyond the lack of chaperons, misfolding could also be caused by the kinase itself. Shrestha, Hamilton, O'Neill, Knapp, and Elkins (2012) analyzed phosphorylation patterns of several Ser/Thr kinases produced in *E. coli* and suggested that the KDs expressed first "hyper-phosphorylate" subsequently produced kinase molecules while these are still in process of translation and elongation (Shrestha et al., 2012). Non-native phosphorylation of normally buried residues would likely impair folding and solubility and can also affect the functioning of affinity tags (Du et al., 2005). The prevention of such "hyper-phosphorylation" provides an additional rational for the co-expression of phosphatases during PTK production from *E. coli*. Hyper-autophosphorylation may, in fact, be a more severe problem for tyrosine kinases than for Ser/Thr kinases: Whereas Ser/Thr phosphorylation (and de-phosphorylation) is relatively wide-spread in both eukaryotic and bacterial cells, tyrosine phosphorylation is efficiently and globally suppressed in eukaryotic cells (Sharma et al., 2014). Overexpression in *E. coli* therefore exposes PTKs to their own uncontrolled high activity, which they have not evolved to cope with.

In summary, although PTKs and their isolated KDs can be difficult targets for expression in *E. coli*, the use of antagonistic phosphatases, optimized domain boundaries and, possibly, also chaperones or solubility tags often enables bacterial production. The proper validation of the expression system requires kinase activity assays, which we will discuss in the next section.

4. *In-vitro* kinase activity assays

The importance of tyrosine kinases as pharmaceutical targets (Breen & Soellner, 2015) has spurred the development of *in vitro* activity assays. Principles, advantages and disadvantages of such assays have been previously reviewed (Glickman, 2004; Hawes & Biesen, 1999; Lorenz, 2011) and we here only give a brief overview and update.

Traditional methods measure tyrosine phosphorylation through the combination of radiolabeling with $AT^{32}P$ or $AT^{33}P$, immunoprecipitation and western blotting. This strategy has been used to study tyrosine kinase activity for more than two decades with several protocol improvements made during these years (Beeler, LaRochelle, Chedid, Tronick, & Aaronson, 1994; Karra, Stippec, & Cobb, 2017; Saya et al., 1993). In a recent example of this approach, Marholz et al. enabled assessing the catalytic activity of orphan PTKs (i.e., PTKs for which the substrates are unknown) by the identification of a set of "universal" substrate sequences (Marholz, Zeringo, Lou, Turk, & Parker, 2018).

Radiolabeling, however, has drawbacks in terms of safety, risks to the environment and the logistics and cost of radioactive transport and waste disposal (Glickman, 2004). A common alternative is the immunoprecipitation with antibodies that either nonspecifically pull-down phosphorylated proteins from cell lysates, or specifically recognize certain phosphorylated target peptides (Amanchy et al., 2009; Hawes & Biesen, 1999). This method reliably identifies natural kinase substrates in cell samples. However, despite efforts to streamline the collection of quantitative temporal data (Bankston, Ku, & Feng, 2017; Mazet et al., 2015), this method is still limited in throughput and less amenable to accurate quantification.

Measurement kits from several commercial providers promise to address the shortcomings of traditional methods. Many use antibody detection of phosphorylated substrates followed by fluorescence or luminescence readouts (Altıntop et al., 2018; Gopinathan et al., 2018). *PamChip*, for example, employs a multiplex ELISA-like strategy in which a library of substrates is attached to a membrane and, following the kinase reaction, is exposed to a fluorescently labeled antibody (Dussaq, Anderson, Willey, & Almeida, 2016). This approach was used to measure kinase activity in cell lysates (Gopinathan et al., 2018), and as a diagnostic approach for the analysis of kinase inhibitor therapies in patient-derived cell lysates (Arni et al., 2017; Labots et al., 2016). By contrast, the *LanthaScreen*, another commercially available assay, operates in solution (Carazo & Pávek, 2015). *LanthaScreen* assays measure the fluorescence energy transfer (FRET) from a phosphotyrosine antibody labeled with a donor fluorophore to the acceptor fluorescent dye that is conjugated to a substrate peptide. The method uses the exceptionally long fluorescence life time of lanthanide ions (europium or terbium), which allows to delay the measurement of donor and acceptor emission until background fluorescence from other potentially interfering compounds has decayed (Carlson, Robers, Vogel, & Machleidt, 2009; Sumaoka, Akiba, & Komiyama, 2016).

There are also several assays that do not rely on expensive and fragile phosphopeptide antibodies. The *ADPGlo* assay measures kinase activity indirectly, through the generation of ADP. At the end of the chosen kinase reaction time, any remaining ATP is hydrolyzed to AMP. Kinase-generated ADP is converted back to ATP which is then driving a luciferase reaction and quantified by luminescence. This method has, for example, been employed for the characterization of recombinant ribosomal S6 kinase 2 (RSK2) and Tousled-like kinase 1B (Bhoir, Shaik, Thiruvenkatam, & Kirubakaran, 2018; Utepbergenov et al., 2016). In addition, there exist several variants of non-antibody FRET assays (Lawrence & Wang, 2007; Liu et al., 2010; Newman & Josiah, 2004). *Z'Lyte*, for example, quantifies the activity of various kinases with kinase-specific peptides that are labeled with both donor and acceptor fluorophores. After the kinase reaction, only non-phosphorylated peptides are cleaved by a protease, thus separating donor and acceptor fluorophores (Fig. 2). The proportion of uncleaved (phosphorylated) versus cleaved (unphosphorylated) peptide is then determined from the ratiometric FRET signal (Ma, Deacon, & Horiuchi, 2008; Rodems et al., 2002).

Thus, depending on her specific needs, a researcher can now choose from a variety of commercial or non-commercial kinase assays to assess catalytic activity of recombinant PTKs.

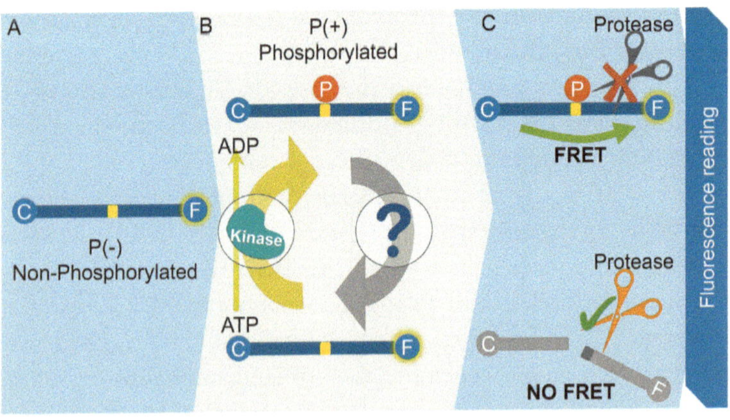

Fig. 2 Mechanism of Z'LYTE kinase activity assay. (A) Specific substrate peptides are labeled with FRET donor (coumarin) and acceptor (fluorescein) fluorophores. (B) Phosphorylation reaction in the presence of ATP. (C) A protease cleaves only non-phosphorylated substrate peptide which leads to an increase in donor and decrease in acceptor fluorescence signal.

5. Practical guidelines and hidden pitfalls

In the following, we will share our experiences with bacterial PTK expression and activity assays, illustrated by experiments with the KDs of three PTKs [Lyn (residues 211–512), FAK (residues 374–708 taken from an isoform used by (Camacho-Soto et al., 2014) and Src (residues 234–536)]. Complete sequences and detailed experimental protocols are available at https://github.com/strubelab/kinaseexpress.

5.1 Expression vectors

We codon-optimized PTK KD sequences, considering both *E. coli* codon and codon-pair usage, optimal GC content, enrichment of hidden stop codons and avoidance of sequence repetitions (Chin, Chung, & Lee, 2014). The synthesized genes were recombined into a synthetic protein expression vector (pJExpress411, originally from DNA 2.0, CA) which we had previously modified to include a RBS secondary structure insulation cassette from Mutalik et al. (2013). In our hands, the combination of codon optimization with 5′ mRNA secondary structure suppression very often gives a marked increase in protein expression yields. However, it did not clearly improve PTK KD yields over values reported in the literature (i.e., 22 μg/mL of Src by Albanese et al., 2018 or 5–15 mg/L of Src by Seeliger et al., 2005). In all constructs, the catalytic KD was fused to yellow fluorescent mCitrine (Griesbeck, Baird, Campbell, Zacharias, & Tsien, 2001) followed by a modified TwinStrep purification tag (Schmidt et al., 2013) (Fig. 3). The YopH catalytic domain (residues 164–468) was expressed from a low copy vector with Spectinomycin resistance (Addgene #79749) (Albanese et al., 2018), whereas the human PTP1B (fused to GST) was expressed from a high copy plasmid with ampicillin resistance (Addgene #8602).

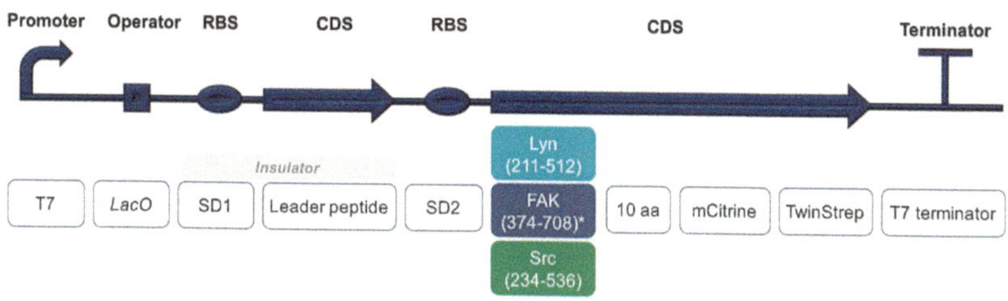

Fig. 3 Construct design for kinases expressed in *E. coli*. *: residues 374–708 correspond to residues 2–336 of genbank AAH28733.1.

5.2 Expression screen

We performed a small-scale (4 ml cultures) expression screen that varied incubation temperature and induction strength for three colonies of each kinase construct. Each KD was co-expressed (i) with YopH, (ii) with PTP1B or (iii) expressed alone. After lysis, approximate expression yields were quantified by fluorescence and normalized to cell density (Fig. 4). Quantification through fluorescent tags is fast and convenient as it does not require purification or SDS page gels. In our examples, Src KD showed approximately half the fluorescence of Lyn KD under most small-scale conditions and this result was reproduced in the large-scale affinity purification (Fig. 4; Table 1A). Yet, this detection method can be deceiving: based on fluorescence, we observed up to 10 times higher expression levels for the

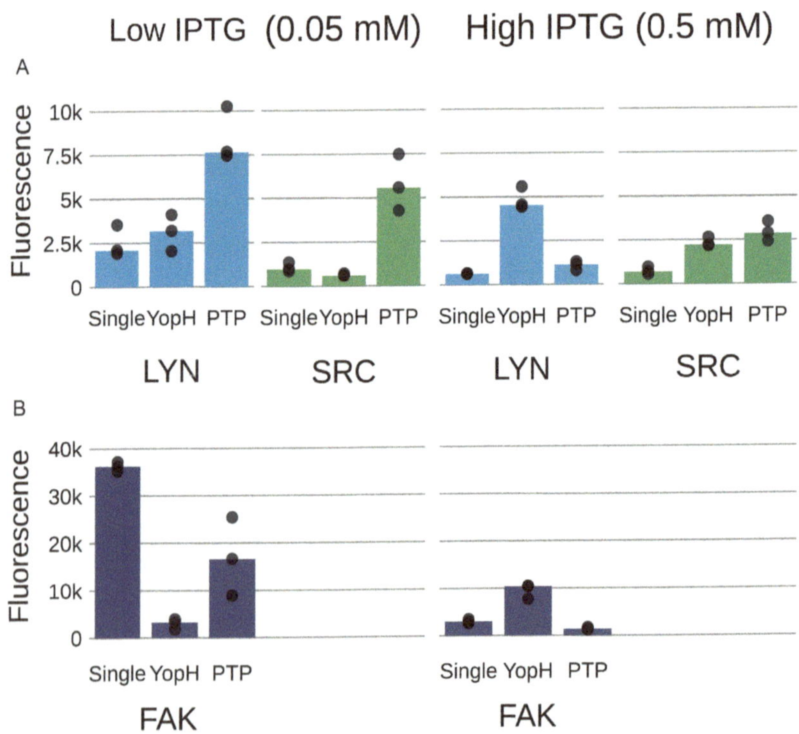

Fig. 4 Small-scale expression in *Escherichia coli.* Protein expression was evaluated by fluorescence from the fused mCitrine tag. Bars indicate median of three measurements (shown as dots), each starting from a different bacterial colony. "Single": expression of KD alone, without phosphatase. Fluorescence data were normalized to cell density. (A) Low induction (IPTG 0.05 mM) and PTP1B co-transformation was the overall best condition. (B) Expression of FAK KD-mCitrine showed high apparent expression which stemmed from a truncation product lacking the KD.

Table 1 Protein yield after large-scale expression and purification.

Kinase sample	Yield (mg)
(A) After Strep Trap affinity purification	
Single Lyn	6.8
Single Src	0.4
PTP1B Lyn	3.7
PTP1B Src	6.2
(B) After size exclusion purification	
Single Lyn	0.1
Single Src	N.D.
PTP1B Lyn	0.1
PTP1B Src	1.4

Yield is presented per liter of culture. "Single": expression of KD alone, without phosphatase.

FAK KD than for Src. However, SDS-PAGE analysis (Supplementary Material in the online version at https://doi.org/10.1016/bs.mie.2019.02.027) revealed that the fluorescence stemmed from truncation products containing mainly the C-terminal mCitrine.

5.3 Influence of phosphatase co-expression

In agreement with previous reports, we were able to express Src KD (Albanese et al., 2018; Seeliger et al., 2005) and Lyn KD (Chin et al., 1998; Li et al., 2016) with and without phosphatase co-expression (Fig. 4A). At high induction levels, YopH co-expression often performed better, but the overall best condition for both constructs was PTP1B co-expression at a low (0.05 mM) IPTG concentration and low (20 °C) temperature. These conditions (for comparison, with and without PTP1B co-expression) were then used for the purification of Src and Lyn KDs from 1 L cultures.

In line with previous reports (Albanese et al., 2018; Cui & Sun, 2018; Seeliger et al., 2005), phosphatase co-expression improved Src KD yield from 0.4 to 6.2 mg/L after affinity purification (Table 1A). By contrast, expression of Lyn KD alone performed two-fold better (6.8 mg/L) than its co-expression with PTP1B (3.7 mg/L). Phosphatase co-expression thus improved the primary yield of Src KD but not the yield of the highly similar Lyn KD. Although overall yields differed, all samples showed similar specific kinase activities after this initial stage of purification, regardless whether

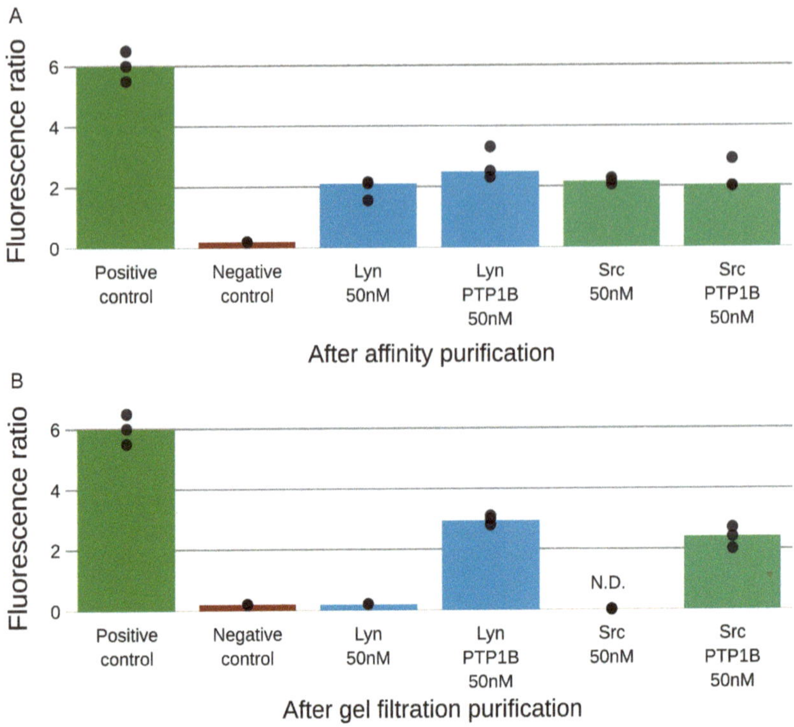

Fig. 5 Kinase activity of *E. coli* expressed proteins. Activity was measured with the Z'LYTE kinase assay. A high fluorescence ratio indicates high kinase activity. Positive control: signal obtained from the fully phosphorylated substrate peptide. (A) Kinase activity of proteins (expressed alone or co-expressed with PTP1B) after affinity purification. (B) Kinase activity after an additional SEC. N.D.: Owing to low yield, Src KD expressed alone was not further purified by SEC. Bars give the median of three measurements (represented by dots).

expressed alone or with a phosphatase (Fig. 5A). However, after SEC, only samples from the PTP1B co-expression retained kinase activity (Fig. 5B). Importantly, SEC reduced the final yield of Lyn KD and Src KD by 95% or 80%, respectively (Table 1B), indicating that most of the affinity-purified fraction had formed soluble aggregates. This effect has been observed by others (Haacke et al., 2009; Piserchio et al., 2012), and illustrates the problem of correlating soluble protein quantity with active monomeric KD. In our case, the affinity purification yield was highest for Lyn KD but a milligram quantity of high quality and active protein could, eventually, only be obtained for Src KD. Although the primary yield of Lyn KD did not benefit from the presence of PTP1B, monomeric and active Lyn KD or Src KD could only be obtained from cultures co-expressing the phosphatase. This may indicate that the phosphatase is not only protecting the host from cytotoxic effects (for which there is no clear evidence in the case of Lyn) but can also protect the kinase from uncontrolled (auto)phosphorylation and misfolding or aggregation (Shrestha et al., 2012).

5.4 PTP1B co-expression and kinase assays

We noticed that phosphopeptide-positive controls from the commercial kinase assay were significantly dephosphorylated when exposed to protein samples originating from phosphatase co-expression cultures. This phosphatase activity could not be removed by filtration over a GST affinity column and, given that phosphatases have similar sizes as KDs, even persisted after SEC (Fig. 6B). If undetected, phosphatase contamination would of course

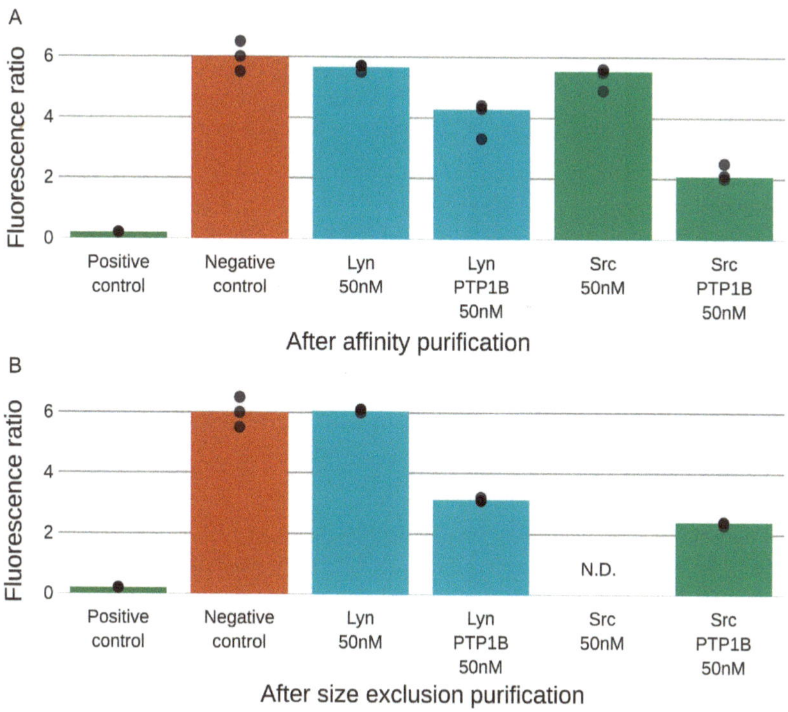

Fig. 6 Phosphatase activity of *E. coli* expressed proteins. High values indicate low phosphatase activity. The negative control (fully phosphorylated peptide) defines the fluorescence ratio without phosphatase activity. PTP1B co-expressed samples show phosphatase activity. Bars give the median of three measurements (represented by dots).

lead to a marked underestimation of kinase activity. It may or may not also be the reason for our observation that kinase assays worked best at lower kinase concentrations. We haven't explored this effect in detail but advise to test several concentrations when establishing the assay.

6. Conclusion

Protein tyrosine kinases are of high scientific and therapeutic interest as anti-cancer drug targets. They, moreover, hold promise as regulatory building blocks in protein-based synthetic biology applications. Both traditional structural biology and such bioengineering efforts critically depend on rapid access to larger quantities of pure protein. Insect cell expression remains the most predictable and reliable option for the expression of PTK catalytic domains but is prohibitively slow and expensive for higher throughput experiments or repeated design-build-test cycles.

Expression of PTKs from *E. coli* has been reported early on with varying success. Although the comparison is complicated by differences in sequence boundaries and fusion tags, which can have a marked impact (Albanese et al., 2018; Elkins, 2014; Wang, Marimuthu, et al., 2006), the literature record indicates two connected reasons for the difficulties associated with bacterial expression of many PTKs: (i) cytotoxicity of the unchecked tyrosine phosphorylation within a non-adapted organism, and (ii) the production of soluble or insoluble PTK aggregates due to misfolding. Misfolding may be caused by a lack of eukaryotic chaperons, by non-optimal domain boundaries, or it can result from the PTK's own unchecked (co-translational) autophosphorylation. Conveniently, phosphatase co-expression can simultaneously address both host cytotoxicity and autophosphorylation-induced misfolding, but may result in downstream phosphatase contamination that affects kinase activity.

Another common pitfall is the prevalence of truncation products or soluble aggregates, which may be both active and inactive. This effect complicates the interpretation of small-scale expression screening and one-step affinity purifications. Size-exclusion chromatography and kinase activity assays are highly recommended before embarking on further experiments. It is also worth exploring different solubility tags, protein boundaries or sequences from different species, which can all have a big impact on product quantity and quality, as even closely related kinase domains can behave very differently. Fast and affordable gene synthesis, codon optimization, modern cloning techniques and robotics now enable the rapid testing of different constructs and conditions, thus greatly facilitating successful *E. coli* production of PTKs.

Acknowledgments

This research was supported by the King Abdullah University of Science and Technology (KAUST) through the baseline fund and the Award No. URF/1/2602-01 and FCC/1/1976-21 from the Office of Sponsored Research (OSR). Supplemental data, detailed methods and complete sequences are available on http://github.com/strubelab/kinaseexpress.

References

Agafonov, R. V., Wilson, C., Otten, R., Buosi, V., & Kern, D. (2014). Energetic dissection of Gleevec's selectivity toward human tyrosine kinases. *Nature Structural & Molecular Biology*, *21*(10), 848–853. https://doi.org/10.1038/nsmb.2891.

Albanese, S. K., Parton, D. L., Işık, M., Rodríguez-Laureano, L., Hanson, S. M., Behr, J. M., et al. (2018). An open library of human kinase domain constructs for automated bacterial expression. *Biochemistry*, *57*(31), 4675–4689. https://doi.org/10.1021/acs.biochem.7b01081.

Alonso, A., Sasin, J., Bottini, N., Friedberg, I., Friedberg, I., Osterman, A., et al. (2004). Protein tyrosine phosphatases in the human genome. *Cell*, *117*(6), 699–711. https://doi.org/10.1016/j.cell.2004.05.018.

Altıntop, M. D., Sever, B., Akalın Çiftçi, G., Turan-Zitouni, G., Kaplancıklı, Z. A., & Özdemir, A. (2018). Design, synthesis, in vitro and in silico evaluation of a new series of oxadiazole-based anticancer agents as potential Akt and FAK inhibitors. *European Journal of Medicinal Chemistry*, *155*, 905–924. https://doi.org/10.1016/j.ejmech.2018.06.049.

Amanchy, R., Zhong, J., Hong, R., Kim, J. H., Gucek, M., Cole, R. N., et al. (2009). Identification of c-Src tyrosine kinase substrates in platelet-derived growth factor receptor signaling. *Molecular Oncology*, *3*(5–6), 439–450. https://doi.org/10.1016/j.molonc.2009.07.001.

Arni, S., Le, T. H. N., de Wijn, R., Garcia-Villegas, R., Dankers, M., Weder, W., et al. (2017). Ex vivo multiplex profiling of protein tyrosine kinase activities in early stages of human lung adenocarcinoma. *Oncotarget*, *8*(40), 68599–68613. https://doi.org/10.18632/oncotarget.19803.

Bankston, A. N., Ku, L., & Feng, Y. (2017). Active Cdk5 immunoprecipitation and kinase assay. *Bio-Protocol*, 7(13), e2363. https://doi.org/10.21769/BioProtoc.2363.

Beeler, J. F., LaRochelle, W. J., Chedid, M., Tronick, S. R., & Aaronson, S. A. (1994). Prokaryotic expression cloning of a novel human tyrosine kinase. *Molecular and Cellular Biology*, *14*(2), 982–988.

Bhoir, S., Shaik, A., Thiruvenkatam, V., & Kirubakaran, S. (2018). High yield bacterial expression, purification and characterisation of bioactive human tousled-like kinase 1B involved in cancer. *Scientific Reports*, *8*(1), 4796. https://doi.org/10.1038/s41598-018-22744-5.

Bhullar, K. S., Lagarón, N. O., McGowan, E. M., Parmar, I., Jha, A., Hubbard, B. P., et al. (2018). Kinase-targeted cancer therapies: Progress, challenges and future directions. *Molecular Cancer*, *17*(1), 48. https://doi.org/10.1186/s12943-018-0804-2.

Bjorge, J. D., Pang, A., & Fujita, D. J. (2000). Identification of protein-tyrosine phosphatase 1B as the major tyrosine phosphatase activity capable of dephosphorylating and activating c-Src in several human breast cancer cell lines. *The Journal of Biological Chemistry*, *275*(52), 41439–41446. https://doi.org/10.1074/jbc.M004852200.

Bliska, J. B., Guan, K. L., Dixon, J. E., & Falkow, S. (1991). Tyrosine phosphate hydrolysis of host proteins by an essential Yersinia virulence determinant. *Proceedings of the National Academy of Sciences of the United States of America*, *88*(4), 1187–1191.

Bougeret, C., Rothhut, B., Jullien, P., Fischer, S., & Benarous, R. (1993). Recombinant Csk expressed in Escherichia coli is autophosphorylated on tyrosine residue(s). *Oncogene*, *8*(5), 1241–1247.

Breen, M. E., & Soellner, M. B. (2015). Small molecule substrate phosphorylation site inhibitors of protein kinases: Approaches and challenges. *ACS Chemical Biology, 10*(1), 175–189. https://doi.org/10.1021/cb5008376.

Bruel, C., Cha, K., Reeves, P. J., Getmanova, E., & Khorana, H. G. (2000). Rhodopsin kinase: Expression in mammalian cells and a two-step purification. *Proceedings of the National Academy of Sciences of the United States of America, 97*(7), 3004–3009.

Camacho-Soto, K., Castillo-Montoya, J., Tye, B., & Ghosh, I. (2014). Ligand-gated split-kinases. *Journal of the American Chemical Society, 136*(10), 3995–4002. https://doi.org/10.1021/ja4130803.

Carazo, A., & Pávek, P. (2015). The use of the LanthaScreen TR-FRET CAR coactivator assay in the characterization of constitutive androstane receptor (CAR) inverse agonists. *Sensors (Basel, Switzerland), 15*(4), 9265–9276. https://doi.org/10.3390/s150409265.

Carlson, C. B., Robers, M. B., Vogel, K. W., & Machleidt, T. (2009). Development of LanthaScreen cellular assays for key components within the PI3K/AKT/mTOR pathway. *Journal of Biomolecular Screening, 14*(2), 121–132. https://doi.org/10.1177/1087057108328132.

Chambers, S. P., Austen, D. A., Fulghum, J. R., & Kim, W. M. (2004). High-throughput screening for soluble recombinant expressed kinases in Escherichia coli and insect cells. *Protein Expression and Purification, 36*(1), 40–47. https://doi.org/10.1016/j.pep.2004.03.003.

Chin, H., Arai, A., Wakao, H., Kamiyama, R., Miyasaka, N., & Miura, O. (1998). Lyn physically associates with the erythropoietin receptor and may play a role in activation of the Stat5 pathway. *Blood, 91*(10), 3734–3745.

Chin, J. X., Chung, B. K.-S., & Lee, D.-Y. (2014). Codon optimization OnLine (COOL): A web-based multi-objective optimization platform for synthetic gene design. *Bioinformatics, 30*(15), 2210–2212. https://doi.org/10.1093/bioinformatics/btu192.

Correa, A., & Oppezzo, P. (2015). Overcoming the solubility problem in E. coli: Available approaches for recombinant protein production. In E. García-Fruitós (Ed.), *Insoluble proteins: Methods and protocols* (pp. 27–44). New York, NY: Springer New York. https://doi.org/10.1007/978-1-4939-2205-5_2.

Cui, Y., & Sun, G. (2018). Structural versatility that serves the function of the HRD motif in the catalytic loop of protein tyrosine kinase, Src. *Protein Science, 28*, 533–542. https://doi.org/10.1002/pro.3554.

Deng, Y., Alicea-Velázquez, N. L., Bannwarth, L., Lehtonen, S. I., Boggon, T. J., Cheng, H.-C., et al. (2014). Global analysis of human nonreceptor tyrosine kinase specificity using high-density peptide microarrays. *Journal of Proteome Research, 13*(10), 4339–4346. https://doi.org/10.1021/pr500503q.

Diaz, J. E., Morgan, C. W., Minogue, C. E., Hebert, A. S., Coon, J. J., & Wells, J. A. (2017). A Split-Abl kinase for direct activation in cells. *Cell Chemical Biology, 24*, 1250–1258. https://doi.org/10.1016/j.chembiol.2017.08.007.

Du, P., Loulakis, P., Luo, C., Mistry, A., Simons, S. P., LeMotte, P. K., et al. (2005). Phosphorylation of serine residues in histidine-tag sequences attached to recombinant protein kinases: A cause of heterogeneity in mass and complications in function. *Protein Expression and Purification, 44*(2), 121–129. https://doi.org/10.1016/j.pep.2005.04.018.

Dussaq, A., Anderson, J. C., Willey, C. D., & Almeida, J. S. (2016). Mechanistic parameterization of the Kinomic signal in peptide arrays. *Journal of Proteomics & Bioinformatics, 9*(5), 151–157. https://doi.org/10.4172/jpb.1000401.

Elegheert, J., Behiels, E., Bishop, B., Scott, S., Woolley, R. E., Griffiths, S. C., et al. (2018). Lentiviral transduction of mammalian cells for fast, scalable and high-level production of soluble and membrane proteins. *Nature Protocols, 13*(12), 2991–3017. https://doi.org/10.1038/s41596-018-0075-9.

Elkins, J. M. (2014). Case study—Structural genomics and human protein kinases. In W. F. Anderson (Ed.), *Structural genomics and drug discovery: Methods and protocols* (pp. 325–336). New York, NY: Springer New York. https://doi.org/10.1007/978-1-4939-0354-2_24.

Fan, G., Lin, G., Lucito, R., & Tonks, N. K. (2013). Protein-tyrosine phosphatase 1B antagonized signaling by insulin-like growth factor-1 receptor and kinase BRK/PTK6 in ovarian cancer cells. *The Journal of Biological Chemistry*, *288*(34), 24923–24934. https://doi.org/10.1074/jbc.M113.482737.

Filippakopoulos, P., Kofler, M., Hantschel, O., Gish, G. D., Grebien, F., Salah, E., et al. (2008). Structural coupling of SH2-kinase domains links Fes and Abl substrate recognition and kinase activation. *Cell*, *134*(5), 793–803. https://doi.org/10.1016/j.cell.2008.07.047.

Fountas, A., Diamantopoulos, L.-N., & Tsatsoulis, A. (2015). Tyrosine kinase inhibitors and diabetes: A novel treatment paradigm? *Trends in Endocrinology and Metabolism: TEM*, *26*(11), 643–656. https://doi.org/10.1016/j.tem.2015.09.003.

Garcia, P., Shoelson, S. E., George, S. T., Hinds, D. A., Goldberg, A. R., & Miller, W. T. (1993). Phosphorylation of synthetic peptides containing Tyr-Met-X-Met motifs by nonreceptor tyrosine kinases in vitro. *The Journal of Biological Chemistry*, *268*(33), 25146–25151.

Glickman, J. F., Sittampalam, G. S., Coussens, N. P., Brimacombe, K., Grossman, A., Arkin, M., et al. (2004). *Assay development for protein kinase enzymes. In Assay guidance manual* Bethesda, MD: Eli Lilly & Company and the National Center for Advancing Translational Sciences. (pp. 1–24). Retrieved from http://www.ncbi.nlm.nih.gov/books/NBK91991/.

Goehring, A., Lee, C.-H., Wang, K. H., Michel, J. C., Claxton, D. P., Baconguis, I., et al. (2014). Screening and large-scale expression of membrane proteins in mammalian cells for structural studies. *Nature Protocols*, *9*(11), 2574–2585. https://doi.org/10.1038/nprot.2014.173.

Gong, X., Ji, J., Xie, J., Zhou, Y., Zhang, J., & Zhong, W. (2006). Expression, purification, and bioactivity of GST-fused v-Src from a bacterial expression system. *Journal of Zhejiang University Science B*, 7(1), 13–19. https://doi.org/10.1631/jzus.2006.B0013.

Gopinathan, U., Redalen, K. R., Trøseid, A.-M., Kierulf, P., Brandtzaeg, P., Ree, A. H., et al. (2018). Large-scale reduction of tyrosine kinase activities in human monocytes stimulated in vitro with N. Meningitidis. *PLoS One*, *13*(1) e0181912. https://doi.org/10.1371/journal.pone.0181912.

Gordley, R. M., Bugaj, L. J., & Lim, W. A. (2016). Modular engineering of cellular signaling proteins and networks. *Current Opinion in Structural Biology*, *39*, 106–114. https://doi.org/10.1016/j.sbi.2016.06.012.

Griesbeck, O., Baird, G. S., Campbell, R. E., Zacharias, D. A., & Tsien, R. Y. (2001). Reducing the environmental sensitivity of yellow fluorescent protein. Mechanism and applications. *The Journal of Biological Chemistry*, *276*(31), 29188–29194. https://doi.org/10.1074/jbc.M102815200.

Grünberg, R., & Serrano, L. (2010). Strategies for protein synthetic biology. *Nucleic Acids Research*, *38*(8), 2663–2675. https://doi.org/10.1093/nar/gkq139.

Haacke, A., Fendrich, G., Ramage, P., & Geiser, M. (2009). Chaperone over-expression in Escherichia coli: Apparent increased yields of soluble recombinant protein kinases are due mainly to soluble aggregates. *Protein Expression and Purification*, *64*(2), 185–193. https://doi.org/10.1016/j.pep.2008.10.022.

Hawes, B. E., & Biesen, T. v. (1999). Protein tyrosine kinase activity assays. *Current Protocols in Pharmacology*, *5*(1), 3.5.1–3.5.18. https://doi.org/10.1002/0471141755.ph0305s05.

Hofmann, C., Sandig, V., Jennings, G., Rudolph, M., Schlag, P., & Strauss, M. (1995). Efficient gene transfer into human hepatocytes by baculovirus vectors. *Proceedings of the National Academy of Sciences of the United States of America*, *92*(22), 10099–10103.

Hunter, T. (2009). Tyrosine phosphorylation: Thirty years and counting. *Current Opinion in Cell Biology*, *21*(2), 140–146. https://doi.org/10.1016/j.ceb.2009.01.028.

Karra, A. S., Stippec, S., & Cobb, M. H. (2017). Assaying protein kinase activity with radiolabeled ATP. *Journal of Visualized Experiments: JoVE*, (123), e55504. https://doi.org/10.3791/55504.

Kemble, D. J., Wang, Y.-H., & Sun, G. (2006). Bacterial expression and characterization of catalytic loop mutants of SRC protein tyrosine kinase. *Biochemistry*, *45*(49), 14749–14754. https://doi.org/10.1021/bi061664.

Khan, K. H. (2013). Gene expression in mammalian cells and its applications. *Advanced Pharmaceutical Bulletin*, *3*(2), 257–263. https://doi.org/10.5681/apb.2013.042.

Kimple, M. E., Brill, A. L., & Pasker, R. L. (2013). Overview of affinity tags for protein purification. *Current Protocols in Protein science*, *73*, Unit-9.9, 9.9.1–9.9.23. https://doi.org/10.1002/0471140864.ps0909s73.

Labots, M., Gotink, K. J., Dekker, H., Azijli, K., van der Mijn, J. C., Huijts, C. M., et al. (2016). Evaluation of a tyrosine kinase peptide microarray for tyrosine kinase inhibitor therapy selection in cancer. *Experimental & Molecular Medicine*, *48*(12), e279. https://doi.org/10.1038/emm.2016.114.

Lawrence, D. S., & Wang, Q. (2007). Seeing is believing: Peptide-based fluorescent sensors of protein tyrosine kinase activity. *Chembiochem: A European Journal of Chemical Biology*, *8*(4), 373–378. https://doi.org/10.1002/cbic.200600473.

Li, X., He, S., Zhou, X., Ye, Y., Tan, S., Zhang, S., et al. (2016). Lyn delivers bacteria to lysosomes for eradication through TLR2-initiated autophagy related phagocytosis. *PLoS Pathogens*, *12*(1), e1005363. https://doi.org/10.1371/journal.ppat.1005363.

Lim, W. A., & Pawson, T. (2010). Phosphotyrosine signaling: Evolving a new cellular communication system. *Cell*, *142*(5), 661–667. https://doi.org/10.1016/j.cell.2010.08.023.

Liu, M., Poulose, S., Schuman, E., Zaitsev, A. D., Dobson, B., Auerbach, K., et al. (2010). Development of a mechanism-based high-throughput screen assay for leucine-rich repeat kinase 2--discovery of LRRK2 inhibitors. *Analytical Biochemistry*, *404*(2), 186–192. https://doi.org/10.1016/j.ab.2010.05.033.

Lorenz, U. (2011). Protein tyrosine phosphatase assays. *Current Protocols in Immunology*, *93*, *Chapter*, Unit-11.7, 11.7.1–11.7.12. https://doi.org/10.1002/0471142735.im1107s93.

Luckow, V. A., Lee, S. C., Barry, G. F., & Olins, P. O. (1993). Efficient generation of infectious recombinant baculoviruses by site-specific transposon-mediated insertion of foreign genes into a baculovirus genome propagated in Escherichia coli. *Journal of Virology*, *67*(8), 4566–4579.

Ma, H., Deacon, S., & Horiuchi, K. (2008). The challenge of selecting protein kinase assays for lead discovery optimization. *Expert Opinion on Drug Discovery*, *3*(6), 607–621. https://doi.org/10.1517/17460441.3.6.607.

Ma, Y., Nagamune, T., & Kawahara, M. (2014). Split focal adhesion kinase for probing protein–protein interactions. *Biochemical Engineering Journal*, *90*, 272–278. https://doi.org/10.1016/j.bej.2014.06.022.

Marholz, L. J., Zeringo, N. A., Lou, H. J., Turk, B. E., & Parker, L. L. (2018). In silico design and in vitro characterization of universal tyrosine kinase peptide substrates. *Biochemistry*, *57*(12), 1847–1851. https://doi.org/10.1021/acs.biochem.8b00044.

Marin, V., Groveman, B. R., Qiao, H., Xu, J., Ali, M. K., Fang, X.-Q., et al. (2010). Characterization of neuronal SRC kinase purified from a bacterial expression system. *Protein Expression and Purification*, 74(2), 289–297. https://doi.org/10.1016/j.pep.2010.06.004.

Mazet, F., Dunster, J. L., Jones, C. I., Vaiyapuri, S., Tindall, M. J., Fry, M. J., et al. (2015). A high-density immunoblotting methodology for quantification of total protein levels and phosphorylation modifications. *Scientific Reports*, *5*, 16995. https://doi.org/10.1038/srep16995.

Mutalik, V. K., Guimaraes, J. C., Cambray, G., Lam, C., Christoffersen, M. J., Mai, Q.-A., et al. (2013). Precise and reliable gene expression via standard transcription and translation initiation elements. *Nature Methods*, *10*(4), 354–360. https://doi.org/10.1038/nmeth.2404.

Nettleship, J. E., Assenberg, R., Diprose, J. M., Rahman-Huq, N., & Owens, R. J. (2010). Recent advances in the production of proteins in insect and mammalian cells for structural biology. *Journal of Structural Biology*, *172*(1), 55–65. https://doi.org/10.1016/j.jsb.2010.02.006.

Newman, M., & Josiah, S. (2004). Utilization of fluorescence polarization and time resolved fluorescence resonance energy transfer assay formats for SAR studies: Src kinase as a model system. *Journal of Biomolecular Screening*, *9*(6), 525–532. https://doi.org/10.1177/1087057104264597.

Osusky, M., Taylor, S. J., & Shalloway, D. (1995). Autophosphorylation of purified c-Src at its primary negative regulation site. *The Journal of Biological Chemistry*, *270*(43), 25729–25732. https://doi.org/10.1074/jbc.270.43.25729.

Patterson, H., Nibbs, R., McInnes, I., & Siebert, S. (2014). Protein kinase inhibitors in the treatment of inflammatory and autoimmune diseases. *Clinical and Experimental Immunology*, *176*(1), 1–10. https://doi.org/10.1111/cei.12248.

Pawson, T. (1995). Protein modules and signalling networks. *Nature*, *373*(6515), 573–580. https://doi.org/10.1038/373573a0.

Pijlman, G. P., van Schijndel, J. E., & Vlak, J. M. (2003). Spontaneous excision of BAC vector sequences from bacmid-derived baculovirus expression vectors upon passage in insect cells. *The Journal of General Virology*, *84*(Pt. 10), 2669–2678. https://doi.org/10.1099/vir.0.19438-0.

Piserchio, A., Cowburn, D., & Ghose, R. (2012). Expression and purification of Src-family kinases for solution NMR studies. In *Vol. 831. Methods in molecular biology* (pp. 111–131). Clifton, NJ: Humana press. https://doi.org/10.1007/978-1-61779-480-3_7.

Rodems, S. M., Hamman, B. D., Lin, C., Zhao, J., Shah, S., Heidary, D., et al. (2002). A FRET-based assay platform for ultra-high density drug screening of protein kinases and phosphatases. *Assay and Drug Development Technologies*, *1*(1), 9–19. https://doi.org/10.1089/154065802761001266.

Rosano, G. L., & Ceccarelli, E. A. (2014). Recombinant protein expression in Escherichia coli: Advances and challenges. *Frontiers in Microbiology*, *5*, 172. https://doi.org/10.3389/fmicb.2014.00172.

Sacco, F., Perfetto, L., Castagnoli, L., & Cesareni, G. (2012). The human phosphatase interactome: An intricate family portrait. *FEBS Letters*, *586*(17), 2732–2739. https://doi.org/10.1016/j.febslet.2012.05.008.

Saya, H., Lee, P. S., Nishi, T., Izawa, I., Nakajima, M., Gallick, G. E., et al. (1993). Bacterial expression of an active tyrosine kinase from a protein a/truncated c-src fusion protein. *FEBS Letters*, *327*(2), 224–230.

Schmidt, T. G. M., Batz, L., Bonet, L., Carl, U., Holzapfel, G., Kiem, K., et al. (2013). Development of the twin-strep-tag® and its application for purification of recombinant proteins from cell culture supernatants. *Protein Expression and Purification*, *92*(1), 54–61. https://doi.org/10.1016/j.pep.2013.08.021.

Seeliger, M. A., Young, M., Henderson, M. N., Pellicena, P., King, D. S., Falick, A. M., et al. (2005). High yield bacterial expression of active c-Abl and c-Src tyrosine kinases. *Protein Science: A Publication of the Protein Society*, *14*(12), 3135–3139. https://doi.org/10.1110/ps.051750905.

SGC. (2018). *Progress in protein kinase structural biology*. 2013, November 5. Retrieved November 28, from. https://www.thesgc.org/scientists/resources/kinases.

Sharma, K., D'Souza, R. C. J., Tyanova, S., Schaab, C., Wiśniewski, J. R., Cox, J., et al. (2014). Ultradeep human phosphoproteome reveals a distinct regulatory nature of Tyr and Ser/Thr-based signaling. *Cell Reports*, *8*(5), 1583–1594. https://doi.org/10.1016/j.celrep.2014.07.036.

Shrestha, A., Hamilton, G., O'Neill, E., Knapp, S., & Elkins, J. M. (2012). Analysis of conditions affecting auto-phosphorylation of human kinases during expression in bacteria. *Protein Expression and Purification, 81*(1), 136–143. https://doi.org/10.1016/j.pep.2011.09.012.

Smith, G. E., Summers, M. D., & Fraser, M. J. (1983). Production of human beta interferon in insect cells infected with a baculovirus expression vector. *Molecular and Cellular Biology, 3*(12), 2156–2165.

Sumaoka, J., Akiba, H., & Komiyama, M. (2016). Selective sensing of tyrosine phosphorylation in peptides using terbium(III) complexes. *International Journal of Analytical Chemistry, 2016*, 3216523. https://doi.org/10.1155/2016/3216523.

Taipale, M., Krykbaeva, I., Koeva, M., Kayatekin, C., Westover, K. D., Karras, G. I., et al. (2012). Quantitative analysis of Hsp90-client interactions reveals principles of substrate recognition. *Cell, 150*(5), 987–1001. https://doi.org/10.1016/j.cell.2012.06.047.

Thomas, J. G., Ayling, A., & Baneyx, F. (1997). Molecular chaperones, folding catalysts, and the recovery of active recombinant proteins from E. coli. *Applied Biochemistry and Biotechnology, 66*(3), 197–238. https://doi.org/10.1007/BF02785589.

Tu, P. N., Wang, Y., Cai, M., Zhou, X., & Zhang, Y. (2014). Co-expression of protein tyrosine kinases EGFR-2 and PDGFRβ with protein tyrosine phosphatase 1B in Pichia pastoris. *Journal of Microbiology and Biotechnology, 24*(2), 152–159.

Utepbergenov, D., Hennig, P. M., Derewenda, U., Artamonov, M. V., Somlyo, A. V., & Derewenda, Z. S. (2016). Bacterial expression, purification and in vitro phosphorylation of full-length ribosomal S6 kinase 2 (RSK2). *PLoS One, 11*(10), e0164343. https://doi.org/10.1371/journal.pone.0164343.

van Oers, M. M., Pijlman, G. P., & Vlak, J. M. (2015). Thirty years of baculovirus-insect cell protein expression: From dark horse to mainstream technology. *The Journal of General Virology, 96*(Pt. 1), 6–23. https://doi.org/10.1099/vir.0.067108-0.

Verba, K. A., Wang, R. Y.-R., Arakawa, A., Liu, Y., Shirouzu, M., Yokoyama, S., et al. (2016). Atomic structure of Hsp90-Cdc37-Cdk4 reveals that Hsp90 traps and stabilizes an unfolded kinase. *Science (New York, N.Y.), 352*(6293), 1542–1547. https://doi.org/10.1126/science.aaf5023.

Vincent, C., Doublet, P., Grangeasse, C., Vaganay, E., Cozzone, A. J., & Duclos, B. (1999). Cells of Escherichia coli contain a protein-tyrosine kinase, Wzc, and a phosphotyrosine-protein phosphatase, Wzb. *Journal of Bacteriology, 181*(11), 3472–3477.

Wang, Y.-H., Ayrapetov, M. K., Lin, X., & Sun, G. (2006). A new strategy to produce active human Src from bacteria for biochemical study of its regulation. *Biochemical and Biophysical Research Communications, 346*(2), 606–611. https://doi.org/10.1016/j.bbrc.2006.05.180.

Wang, W., Marimuthu, A., Tsai, J., Kumar, A., Krupka, H. I., Zhang, C., et al. (2006). Structural characterization of autoinhibited c-met kinase produced by coexpression in bacteria with phosphatase. *Proceedings of the National Academy of Sciences of the United States of America, 103*(10), 3563–3568. https://doi.org/10.1073/pnas.0600048103.

Weijland, A., Neubauer, G., Courtneidge, S. A., Mann, M., Wierenga, R. K., & Superti-Furga, G. (1996). The purification and characterization of the catalytic domain of Src expressed in Schizosaccharomyces pombe. Comparison of unphosphorylated and tyrosine phosphorylated species. *European Journal of Biochemistry, 240*(3), 756–764.

Wilson, C., Agafonov, R. V., Hoemberger, M., Kutter, S., Zorba, A., Halpin, J., et al. (2015). Using ancient protein kinases to unravel a modern cancer drug's mechanism. *Science, 347*(6224), 882–886. https://doi.org/10.1126/science.aaa1823.

Wu, P., Nielsen, T. E., & Clausen, M. H. (2016). Small-molecule kinase inhibitors: An analysis of FDA-approved drugs. *Drug Discovery Today, 21*(1), 5–10. https://doi.org/10.1016/j.drudis.2015.07.008.

CHAPTER NINE

Design, cloning and characterization of transcription factor-based inducible gene expression systems

Erik K.R. Hanko, Nigel P. Minton, Naglis Malys*
BBSRC/EPSRC Synthetic Biology Research Centre (SBRC), Centre for Biomolecular Sciences, School of Life Sciences, The University of Nottingham, Nottingham, United Kingdom
*Corresponding author: e-mail address: n.malys@gmail.com

Contents

Abstract

Cellular functions are often controlled by small molecular weight molecules such as metabolites. Microorganisms, mainly prokaryotes, have evolved sensing and regulatory mechanisms based on transcriptional regulators (TRs) that are able to activate gene expression in response to changes in intra- and extracellular metabolite (ligand) concentrations. To understanding control mechanisms and cell factory development in synthetic biology applications, high throughput analytical procedures are required. In this chapter, we outline a methodological pipeline to design and build reporter constructs enabling the characterization of metabolite-responsive inducible gene

Methods in Enzymology, Volume 621
ISSN 0076-6879
https://doi.org/10.1016/bs.mie.2019.02.018

expression systems. As an example, we present the design, cloning and characterization of the itaconate-inducible system which is composed of the LysR-type transcriptional regulator ItcR and the promoter P_{ccl} from *Yersinia pseudotuberculosis*. Fluorescence-based plate reader and flow cytometry assays are described and the steps for performing data analysis are provided.

1. Introduction

The adjustment of RNA and protein levels through the control of gene expression is fundamental to the regulation of cellular functions. Living organisms have evolved a variety of sensing and regulatory mechanisms to monitor and respond to changes in their local environment by adjusting the expression of their genes. Transcriptional regulators (TRs) are DNA-binding proteins that control gene expression at the level of transcription in response to physical parameters (e.g., temperature, light or pH), ions, or small effector (ligand) molecules, also referred to as inducers. Coupling gene expression to physical conditions or the availability of specific metabolites in their environment gives cells the advantage of limiting the expenditure of energy and materials to the synthesis of enzymes that are required under these growth conditions.

Microorganisms, prokaryotes in particular, possess a large number of ligand-binding TRs, at present accounting for more than 400 unique ligand-TR pairs. Depending on their mode of action, TRs are classed as activators, repressors, or dual-function transcription factors. In the presence of ligand, activators bind to their target DNA sequence within the inducible promoter and recruit the RNA polymerase (RNAP) holoenzyme to initiate transcription. In contrast, repressors bind to their target sequence in the absence of ligand, preventing binding of RNAP to the promoter. Upon ligand binding, the repressor dissociates from the DNA allowing transcription to initiate. Dual-function transcription factors serve as both activators and repressors. For example, AraC, the TR of the arabinose utilization operon *araBAD* in *Escherichia coli*, represses transcription of *araBAD* in the absence of its ligand by forming a DNA loop, but acts as activator in the presence of arabinose through conformational change (Ogden, Haggerty, Stoner, Kolodrubetz, & Schleif, 1980). As a consequence of their ability to activate or repress the transcription of genes, transcription factor-based inducible gene expression systems are frequently employed in synthetic biology to build complex synthetic regulatory circuits, such as logic gates, toggle switches, or oscillators (Brophy & Voigt, 2014). Their application as genetically encoded biosensors has advanced the field of metabolic engineering by providing a solution to the bottleneck of construct evaluation (Rogers, Taylor, & Church, 2016). While the throughput of the design- and build-steps of the typical design-build-test cycle has increased by several orders of magnitude due to advances in *in silico* pathway design and automated genome engineering, evaluation of the generated genetic variants has been lagging behind. Linking a metabolite-responsive TR-based inducible gene expression system to a reporter (e.g., fluorescent protein or selection marker), however, enables each cell to measure the intracellular concentration of a target molecule and convert it into a detectable and quantifiable output. The same strategy can also be applied to evaluate basic cellular functions that are not easily accessible by other techniques. For example, inducible gene expression systems linked to a reporter may be used to determine the kinetics of ligand uptake and metabolism, or to discover new enzymatic functions by screening structurally similar secondary effectors for their ability to activate gene expression. Furthermore, reporter constructs enable the parameterization of fundamental system characteristics, including the promoter activity in the absence or presence of the ligand. Quantitative studies like these indicate whether an inducible system is suitable for the analysis of cellular functions, such as the effect of essential genes on downstream gene functions.

Novel transcription factor-based inducible gene expression systems are continuously added to the list of available systems. Online databases such as DBD (www.transcriptionfactor.org (Wilson, Charoensawan, Kummerfeld, & Teichmann, 2007)), RegPrecise (regprecise.lbl.gov/RegPrecise (Novichkov et al., 2013)), or PRODORIC (www.prodoric.de (Münch et al., 2003)) provide useful information on prokaryotic TRs and their corresponding target DNA sequences. Despite the availability of a considerable number of inducible systems, relatively few have been fully characterized. Rogers and co-workers established a workflow for the analysis of inducible systems, yielding key parameters that can be used to describe their overall performance (Rogers et al., 2015). Recently, this strategy has been further extended to the identification of novel systems and the characterization of their specificity (Hanko, Minton, & Malys, 2017, 2018).

In this chapter, we provide a guidance for the design, cloning, and characterization of inducible gene expression systems using the example of the recently identified itaconate-inducible system from *Yersinia pseudotuberculosis*.

2. Generation of reporter constructs

The choice of inducible system is often predetermined by the preference for the effector (ligand) or system parameters. After a potential inducible system has been identified, tests need to be performed to establish whether it is suitable for its intended application. One of the most effective ways of undertaking the necessary evaluation, and in particular to measure its response to appropriate ligand molecules, the TR-promoter pair is cloned into a reporter vector. This enables the quantitative evaluation of fundamental system parameters and characteristics, including the promoter activity in the absence or presence of ligand, the dose-response, as well as the specificity toward the ligand. The use of a reporter system also enables a more in-depth characterization of the individual components making up an inducible system. This part characterization may be crucial to aid in understanding functional mechanisms, which then can be implemented in synthetic circuit design.

2.1 Architecture of transcription factor-based inducible gene expression systems

A transcription factor-based inducible gene expression system is usually composed of two individual modules (Fig. 1A) (Mannan, Liu, Zhang, & Oyarzún, 2017). The regulation module is the DNA sequence that interacts with the TR to mediate the expression of a target gene or operon. It comprises the inducible promoter, including transcriptional regulator binding site(s) (TRBs), -35 and -10 RNAP holoenzyme recognition sequences, as well as a ribosome binding site (RBS). The sensing module is required for the interaction between the TR and the ligand. The associated DNA sequence contains the gene encoding the TR and the functional genetic elements, including the promoter and RBS needed for its expression. If the TR is endogenous to the organism it is tested in, the translational product of the chromosomal copy might be sufficient to control gene expression from its corresponding inducible promoter on a multicopy episomal vector. If the TR originates from a different organism, it needs to be expressed as part of the heterologous inducible system. This is often accomplished by employing a constitutive promoter. In some cases, however, the native promoter may be used to maintain TR-mediated autoregulation. Examples of inducible systems with TRs controlling their own expression include the widely used arabinose- and rhamnose-inducible systems (Ogden et al., 1980;

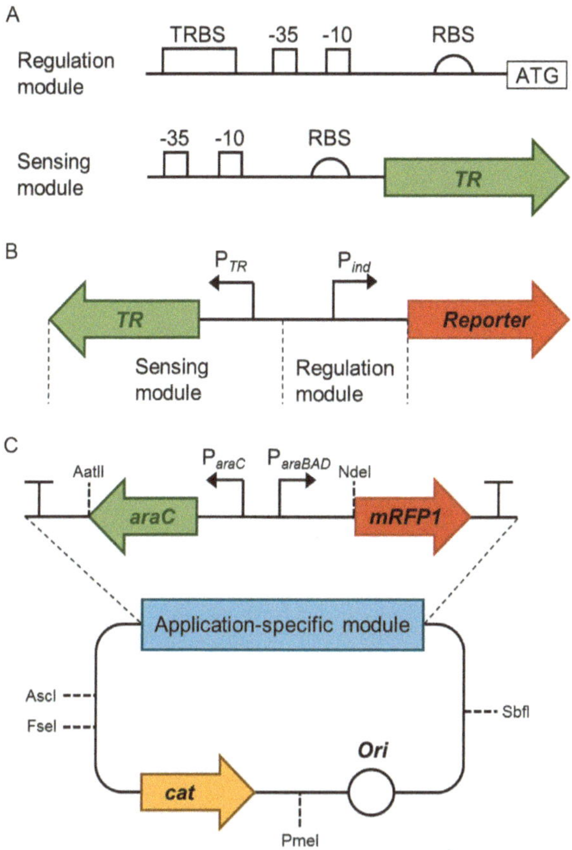

Fig. 1 Generation of inducible system-reporter constructs. (A) Schematic illustration of the regulation and sensing module of a transcription factor-based inducible gene expression system, including functional genetic elements. (B) Arrangement of the sensing and regulation module in a reporter construct to minimize background reporter gene expression. (C) Schematic illustration of the modular reporter vector pEH006. The application-specific module contains the arabinose-inducible system AraC/P_{araBAD}. Important restriction sites are highlighted.

Tobin & Schleif, 1987). Bioinformatics tools that aid in the identification of promoter sequences and transcriptional start sites include BPROM and NNPP, respectively (Reese, 2001; Solovyev & Salamov, 2011). Furthermore, codon-optimization of the TR coding sequence may be required to ensure synthesis of the regulator in a heterologous chassis.

After the two modules have been identified, they are cloned into a reporter vector for the inducible system to be characterized. Ideally, the promoter of the TR and the inducible promoter, which has been transcriptionally fused to a reporter gene, mediate transcription divergently (Fig. 1B). This arrangement minimizes the chance of transcriptional read-through, which may add to the basal activity of the inducible promoter in the absence of ligand.

2.2 Selection of reporter

The transcriptional output mediated by the inducible promoter in response to a ligand needs to be converted into a signal that can easily be detected and quantified. This is accomplished by coupling the inducible system to a reporter gene encoding a reporter protein. The activity or amount of reporter protein synthesized can be correlated to the intracellular ligand concentration. Commonly used reporter proteins include fluorescent proteins, luciferases, or proteins that cause a color change in the presence of an appropriate substrate, such as β-galactosidase in the presence of X-gal (5-bromo-4-chloro-3-indolyl-β-D-galactoside) (Van Der Meer & Belkin, 2010). For the characterization of inducible gene expression systems, fluorescent proteins are most often employed as reporters. They offer the advantage that they can be directly quantified *in vivo* with great specificity and sensitivity, without the necessity of external substrate addition. Furthermore, due to the variety of fluorescent proteins with nonoverlapping excitation and emission spectra, multiple reporters can be used in combination to monitor independent output signals simultaneously.

One important factor that needs to be taken into account when choosing a fluorescent reporter protein is the natural fluorescence of chemical species that are present in biological systems. These can result in autofluorescence which may mask small changes in fluorescent reporter protein synthesis and leading to low signal-to-noise ratios. In light of this, here we chose mRFP1 (monomeric red fluorescent protein) as the reporter protein (Campbell et al., 2002). It demonstrates considerably lower autofluorescence than other widely used reporters such as eGFP or eYFP (enhanced green or yellow fluorescent protein, respectively) (Telford, Hawley, Subach, Verkhusha, & Hawley, 2012). Another advantage of mRFP1 is its rapid maturation time of less than 1 h (Campbell et al., 2002). In our experiments, we found the time required to detect mRFP1 fluorescence above the level of basal promoter activity at 30 °C to be about 30 min after the ligand had been added (Alagesan et al., 2018). We therefore defined 30 min as the time needed from the activation of reporter gene expression to the maturation of mRFP1, which is used for the calculation of the protein synthesis rate. Based on mRFP1, several monomeric fluorescent proteins have been developed that emit in a similar wavelength range and therefore result in comparable low levels of autofluorescence. The new variants may perform better than mRFP1 due to improved extinction coefficients, quantum yields, and photostability (Shaner et al., 2004).

2.3 Vector design

Vectors for the characterization of inducible gene expression systems should be designed in such a way that the influence of sequence/genomic context is minimized and allow individual components to be easily exchanged. Ideally, the reporter vector design should incorporate the following features: (i) unique rare-cutter restriction sites flanking the selective marker gene, the origin of replication, and the application-specific module containing the inducible gene expression-reporter system, similar to the pMTL80000 modular vector series (Heap, Pennington, Cartman, & Minton, 2009); (ii) unique restriction sites within the application-specific module to allow the exchange of reporter gene and the insertion of an inducible system; (iii) if required, a unique restriction site separating the regulation- and sensing module of the inducible system to facilitate the replacement of functional genetic elements, such as promoters and RBSs; and (iv) transcriptional terminators flanking the application-specific module to prevent transcriptional read-through from the vector backbone. Following this design, the modular reporter vector pEH006 was developed (Fig. 1C) (Hanko et al., 2017). It harbors the L-arabinose-inducible system from *E. coli* and was used to evaluate the 3-hydroxypropionate-, L-rhamnose-, acrylate-, cumate-, and itaconate-inducible systems (Alagesan et al., 2018; Hanko et al., 2018).

Importantly, to determine the insulation of the application-specific module, vector pEH006E was constructed. This variant of pEH006 lacks the L-arabinose-regulation- and sensing module, leaving a promoterless *rfp* reporter gene flanked by transcriptional terminators. Cells of *E. coli* harboring pEH006E demonstrated fluorescence at the level of culture medium autofluorescence and therefore confirmed the suitability of pEH006 to serve as insulated modular reporter vector for the construction and analysis of inducible gene expression systems (Hanko et al., 2017).

2.4 Example: Cloning of an itaconic acid-inducible gene expression system

Recently, itaconic acid-responsive TRs have been identified in *Y. pseudotuberculosis* and *Pseudomonas aeruginosa* (Hanko et al., 2018). They control the expression of clusters of genes involved in itaconate catabolism. Interestingly, *in vitro* studies have shown that the first enzyme of the itaconate degradation pathway of *Yersinia pestis*, the itaconate-CoA transferase Ict, is able to use metabolites other than itaconate as CoA acceptors (Sasikaran, Ziemski, Zadora, Fleig, & Berg, 2014). By employing a reporter construct which harbors the itaconate-inducible system ItcR/P_{ccl} from *Y. pseudotuberculosis*, it could be demonstrated that the catalytic efficiency of Ict toward these secondary substrates directly correlates with their ability to activate gene expression. This suggested that there might be a structural evolutionary link between the itaconate-CoA transferase Ict and the regulator ItcR, where both proteins have co-evolved enabling a hierarchical ranking of metabolites as enzyme substrates and TR activators (Hanko et al., 2018). Here, we describe the cloning of the itaconate-inducible system ItcR/P_{ccl} from *Y. pseudotuberculosis* YPIII.

ItcR/P_{ccl} is composed of the ItcR regulator coding sequence and the *itcR-ccl* intergenic region (Fig. 2). The intergenic region harbors a bidirectional promoter mediating transcription of the TR gene and the cluster of three genes involved in itaconate catabolism (*ccl*, *ich* and *ict*). ItcR/P_{ccl} was amplified from *Y. pseudotuberculosis* YPIII genomic DNA by PCR using Phusion High-Fidelity DNA polymerase from New England Biolabs (NEB) in a 50 μL reaction according to the manufacturer's protocol. Oligonucleotide primers that were used to obtain the DNA fragment are EH191_forward (5′-gctactcgccatatggttcctcctccaacttcgct-3′) and EH190_reverse (5′-aggctcagtc gaaagactgggcctttcgttttatgacgtctcaaggaaacacggtcaggaca-3′). The primer overhangs were designed to contain AatII/NdeI restriction sites (underlined). The PCR product and the modular reporter vector pEH006 (Fig. 1C) were digested overnight using restriction enzymes AatII and NdeI under recommended conditions (NEB). Subsequently, the digested PCR product and reporter vector were purified by gel electrophoresis using a 1% (w/v) agarose gel. The fragments of correct size were extracted by employing the NEB Monarch Gel Extraction Kit, ligated overnight using T4 DNA ligase (NEB) and transformed into *E. coli* TOP10 (Invitrogen) by heat shock as described

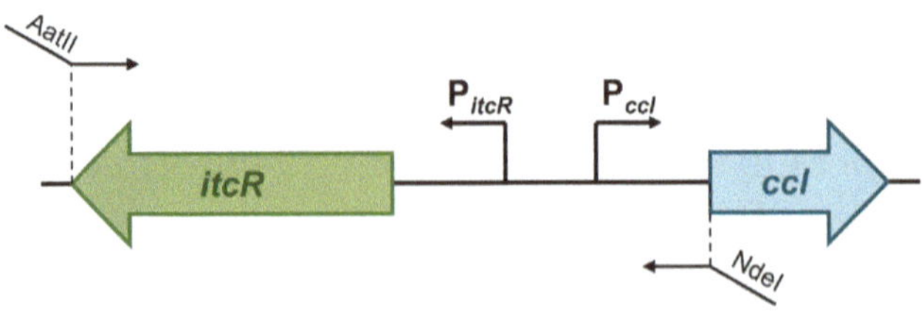

Fig. 2 The itaconate-inducible system ItcR/P_{ccl} from *Y. pseudotuberculosis*. The restriction sites within the oligonucleotide primers are highlighted.

previously (Sambrook & Russell, 2001). The transformed cells were spread on an LB agar plate containing 25 μg/mL chloramphenicol and incubated at 37 °C for 18 h. Clones harboring the correct insert were identified by colony PCR using Dream Taq Green PCR Master Mix (Thermo Fisher Scientific) and oligonucleotide primers EH191_forward and EH190_reverse in 25 μL reactions according to the manufacturer's protocol. The resulting plasmid pEH086 was obtained by miniprep using the NEB Monarch Plasmid Miniprep Kit from an *E. coli* overnight culture and the construct was verified by DNA sequencing (Source BioScience, Nottingham, UK).

Tip: The screening process for successfully ligated constructs can be facilitated by supplementing the selection plate with L-arabinose at a final concentration of 0.1% (w/v). Colonies harboring the undigested original vector pEH006 will turn visibly pink due to expression of the mRFP1 reporter gene from the arabinose-inducible promoter. White colonies contain either the successfully ligated insert or, less likely, the religated vector backbone without an insert. Depending on the inducible system, colonies with successfully ligated inserts may turn pink because of the leakiness of the cloned inducible promoter or cross-induction by medium components.

3. Fluorescence reporter assays

Following the assembly of the reporter construct, the inducible system can be evaluated for its ability to control gene expression. The fluorescence output can be visualized and quantified using specialized equipment, such as fluorescence spectrophotometers, plate readers, flow cytometers, or fluorescence microscopes. Plate readers offer the advantage of allowing the determination of system characteristics, such as the specificity toward the ligand and the dose-response, which might require a large number of samples, in a high throughput format. Moreover, modern plate readers enable on-line measurement of fluorescence and absorbance during the time course of cultivation, allowing the calculation of rates of change rather than steady-state values. Flow cytometers are useful to determine whether the bulk fluorescence measured using a plate reader reflects the behavior of individual cells.

3.1 Fluorescence measurement using plate reader

3.1.1 Equipment

1. 0.22 μm pore size filter
2. 96-well plate (flat and clear bottom, black; Greiner One International, Germany; catalog no. 655090)

3. UV/visible spectrophotometer
4. Infinite M1000 Pro plate reader (Tecan, Switzerland) or any other plate reader suitable for fluorescence and absorbance measurements as described below
5. Multichannel pipette for dispensing culture and inducers (optional)

3.1.2 Single time point fluorescence measurement

1. Inoculate a fresh single colony of *E. coli* MG1655 harboring pEH086 into 2mL of M9 minimal medium (containing the following per liter: 200mL of 5×M9 salts [64g/L $Na_2HPO_4{\cdot}7H_2O$, 15g/L KH_2PO_4, 2.5g/L NaCl, 5g/L NH_4Cl], 2mL of 1*M* $MgSO_4$, 0.1mL of 1*M* $CaCl_2$) (Sambrook & Russell, 2001) supplemented with 1μg/mL thiamine, 20μg/mL uracil (Jensen, 1993), 25μg/mL chloramphenicol, and 0.4% (w/v) glucose.
2. Grow cells for 18h in a 50-mL conical centrifuge tube with orbital shaking at 37°C and 200rpm.
3. Dilute the preculture 1:20 into 1.96mL of fresh medium and incubate at 37°C and 200rpm.
4. At an OD_{600} of 0.5, add 40μL of stock inducer to the desired concentration. To prepare the stock inducer, dissolve itaconic acid (purity ≥99%, Sigma-Aldrich) in distilled water to a final concentration of 1*M*. Filter-sterilize it using a 0.22-μm pore size filter. If required, further dilute the stock inducer in sterile water to the desired concentration. The volume of inducer added was chosen to be 1/50th of the cell culture. For example, add 40μL of 0.25*M* itaconate to achieve a final concentration of 5m*M* in the cell culture.
5. Six hours after incubation at 37°C and 200rpm, transfer 100μL of cell culture into a 96-well plate.
6. Using a plate reader, measure the RFP fluorescence using 585nm as excitation wavelength and 620nm as emission wavelength both with a 10nm bandwidth. Set the gain to 100% unless the fluorescence reaches the detection limit. Set the flash frequency to 100Hz and the integration and lag time to 20μs and 0μs, respectively. At the same time, measure the absorbance using a wavelength of 600nm with a 5nm bandwidth. For both fluorescence and absorbance measurements, set additional acquisition parameters to: number of flashes of 10 and settle time of 10ms.

3.1.3 Time course fluorescence measurement

1. Inoculate a fresh single colony of *E. coli* MG1655 harboring pEH086 into 2mL of M9 minimal medium (Sambrook & Russell, 2001)

supplemented with 1 μg/mL thiamine, 20 μg/mL uracil (Jensen, 1993), 25 μg/mL chloramphenicol, and 0.4% (w/v) glucose.
2. Grow cells for 18 h in a 50-mL conical centrifuge tube with orbital shaking at 37 °C and 200 rpm.
3. Dilute the preculture 1:40 into 2 mL of fresh medium and incubate at 37 °C and 200 rpm.
4. At an OD_{600} of 0.2, transfer 142.5 μL of cell culture into a 96-well plate.
5. Add 7.5 μL of stock inducer (prepared as in 3.1.2) at the desired concentration. For example, add 7.5 μL of 100 m*M* itaconate to achieve a final concentration of 5 m*M* in the cell culture.
6. Using a plate reader, measure RFP fluorescence and absorbance with the same wavelength settings as for the single time point fluorescence measurement (Section 3.1.2 step 6). The temperature of the plate reader was set to 37 °C with orbital shaking at a frequency of 582 rpm and an amplitude of 1 mm. Measurements were taken in intervals of 5 min for a time course of 16 h.

3.2 Fluorescence measurement using flow cytometer

3.2.1 Equipment

1. UV/visible spectrophotometer
2. Astrios EQ flow cytometer (Beckman Coulter) or any other flow cytometer suitable for fluorescence measurements as described below

3.2.2 Procedure

1. Grow *E. coli* MG1655 harboring pEH086 and perform induction as described for the single time point fluorescence measurement.
2. Collect cells by centrifugation at 5000 × *g* for 4 min at 4 °C.
3. Take off the supernatant and resuspend the cells in ice-cold sterile filtered phosphate buffered saline (PBS) to an OD_{600} of approximately 0.01. Keep the sample on ice.
4. Measure fluorescence using a flow cytometer equipped with a 561 nm laser and a 614/20 nm emission band-pass filter. Set the voltage of the photomultiplier tube at 400 V. Adjust the area and height gain to 1.0. Collect at least 100,000 events.

4. Data analysis

4.1 Induction level and dose-response

The data from the fluorescence reporter assays can be used to characterize different aspects of an inducible gene expression system. As cellular functions have evolved to be regulated to different extents, the ligand concentration-dependent output of an inducible gene expression system can be expressed as induction level and dose-response.

Although the induction level only provides limited information, it is generally used as a first indicator of the inducible system's performance. It is defined as the level of reporter gene expression relative to the basal promoter activity at a certain ligand concentration and at a specific time point. The induction level can be obtained using the data from the single time point, the time course, or the flow cytometer fluorescence measurement. Due to differences in the cultivation and data acquisition method used, however, induction levels may be comparable on a qualitative, but not on a quantitative level.

Before the reporter output from the induced sample can be compared to the output of the uninduced cell culture, fluorescence values need to be normalized to account for variability in different samples which can be attributed to either or both biological or/and technical variation. Fluorescence is normalized by cell density of the whole population or given as intensity per cell in case of flow cytometry. When using a plate reader to collect fluorescence data, the measured absolute fluorescence and absorbance values (RFP_{raw} and OD_{raw}, respectively) are corrected by subtracting the absolute fluorescence and absorbance value of the culture medium (RFP_{medium} and OD_{medium}). The absolute normalized fluorescence value (RFP/OD) is subsequently obtained by dividing the corrected absolute fluorescence value (RFP) by its corresponding corrected absorbance value (OD).

$$\frac{RFP}{OD} = \frac{RFP_{raw} - RFP_{medium}}{OD_{raw} - OD_{medium}} \tag{1}$$

The single time point fluorescence and absorbance values of the culture medium, an uninduced cell culture of *E. coli* MG1655 pEH086, and a culture of *E. coli* MG1655 pEH086 in the presence of 5 m*M* itaconate are given as examples in Table 1. The induction level is subsequently calculated by dividing the absolute normalized fluorescence value of the induced sample by the absolute normalized fluorescence value of the uninduced sample. For the given example this would result in an induction level of 193.2.

The dose-response curve provides more information than the induction factor, as it encompasses the full range of ligand concentration that results in a measurable change in reporter output—from the minimum compound concentration required to elicit an induction of gene expression, to the

Table 1 Absolute fluorescence and absorbance values used to calculate the induction level of a cell culture of *E. coli* MG1655 pEH086 in the presence of 5 m*M* itaconate relative to the uninduced cells.

	RFP (A.U., arbitrary unit)		OD at 600 nm	
	RFP_{medium}	RFP_{raw}	OD_{medium}	OD_{raw}
Medium	13.2	N/A	0.103	N/A
Culture uninduced	N/A	26.4	N/A	0.210
Culture induced	N/A	3087.2	N/A	0.232

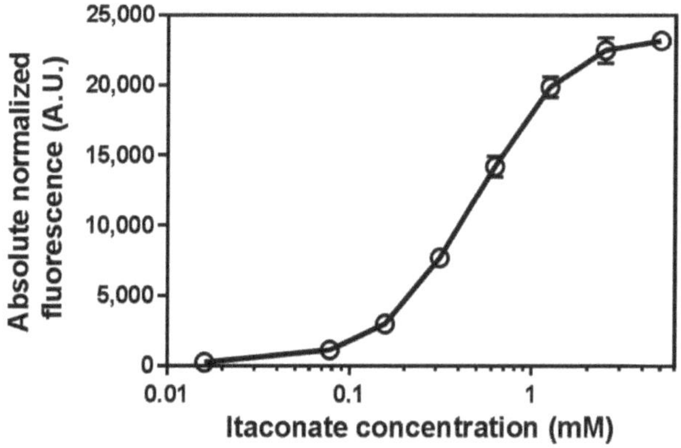

Fig. 3 Example: dose-response curve of the ItcR/P_{ccl} inducible system in *E. coli* MG1655 pEH086 8 h after supplementation with different concentrations (0.0195, 0.0781, 0.1563, 0.3125, 0.625, 1.25, 2.5, and 5 m*M*) of itaconate. Absolute normalized fluorescence values were calculated using the data from the time course fluorescence measurement using formula (1). Error bars represent standard deviations of three biological replicates.

maximum ligand concentration that saturates the system. It correlates absolute normalized fluorescence values and ligand concentrations at a specific time point. The dose-response curve of the itaconate-inducible system ItcR/P_{ccl} from a culture of *E. coli* MG1655 pEH086, 8 h after addition of ligand, is given as example (Fig. 3). When used as a whole-cell biosensor, the dose-response curve of the pure compound may be used to approximate the unknown ligand concentration in a sample by interpolating its absolute normalized fluorescence value.

4.2 Estimation of protein synthesis rate and hill coefficient

In recent years, numerous transcription factor-based inducible gene expression systems have been identified and added to the list of available systems to be applied in microbial biotechnology. To understand kinetic characteristics of inducible gene expression systems, and to aid in part selection and the design of complex synthetic circuits, key parameters of individual systems need to be determined. Since inducible systems are used to fine-tune the expression of genes, it is important to estimate how the addition of ligand translates into the synthesis of protein and impacts on cellular functions. The rate at which the protein is produced can be calculated for each ligand concentration at any time *t* using the normalized absolute fluorescence values from the time course fluorescence measurements with formula (2) (Rogers et al., 2015):

$$\Delta \mathrm{RFP}_{n,t} = \frac{\mathrm{RFP}_t}{\mathrm{OD}_t} - \frac{\mathrm{RFP}_{t-1}}{\mathrm{OD}_{t-1}} \tag{2}$$

The time that is required from the activation of reporter gene expression to the maturation of the fluorescent protein was defined as Δt ($\Delta t \approx 30$ min for mRFP1). The rate of mRFP1 degradation was neglected in Eq. (2) as it was reported to demonstrate slow protein unfolding kinetics (Stepanenko et al., 2004). As protein synthesis rates depend not only on the genomic context of the inducible system, but also on growth rate-dependent parameters, such as the plasmid copy number and the abundance of RNAP and ribosomes (Klumpp, Zhang, & Hwa, 2009), it is not surprising that it changes over the time course of experiment (Fig. 4A). The maximum rate of RFP synthesis $v_{\max}$ can subsequently be fit to the corresponding ligand

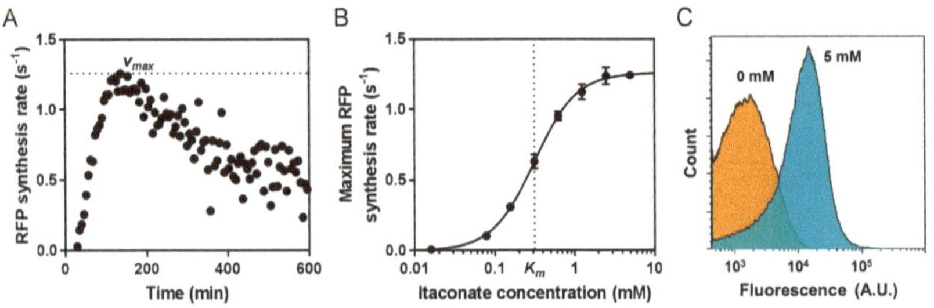

Fig. 4 (A) RFP synthesis rate over the time course of experiment for the itaconate-inducible system ItcR/P_{ccl} in a culture of *E. coli* MG1655 pEH086 after supplementation with 5 m*M* itaconate calculated using formula (2). (B) The maximum rate of RFP synthesis as a function of inducer concentration. The line reflects the Hill function fit to the data using Eq. (3). Error bars represent standard deviations of three biological replicates. (C) Flow cytometry analysis of *E. coli* MG1655 pEH086 in the absence of inducer and 6h after addition of 5 m*M* itaconate. For the induced sample, fluorescence from >98% of cells are displayed in the histogram, whereas for the uninduced sample, <25% of cells are below the 429 A.U. fluorescence threshold.

concentration using a Hill function, taking into account the basal rate of protein synthesis of the uninduced cells $v_{\min}$ (Fig. 4B) (Rogers et al., 2015):

$$\left(\frac{\Delta RFP_n}{\Delta t}\right)_{\max} = v_{\max} \cdot \frac{I^h}{I^h + K_m{}^h} + v_{\min} \tag{3}$$

The parameters correspond to the ligand concentration I, the Hill coefficient h, and the ligand concentration that results in half-maximal RFP synthesis K_m. In this case, h is an indicator of the range of ligand concentration over which the system results in a change in protein synthesis (Madar, Dekel, Bren, & Alon, 2011). For example, a high Hill coefficient results in a steeper and more sigmoidal curve indicating a small range of ligand over which gene expression can be fine-tuned. In other words, a small change in ligand concentration is required for system activation.

4.3 Evaluation of induction homogeneity

The collected data from the flow cytometer fluorescence measurement can be used to analyze the induction behavior on a single cell level. This aids in concluding whether a promoter can be gradually induced on a single-cell level or if it acts as an on/off switch. Systems that produce a unimodal induction across the whole range of the ligand dose-response are often preferred in synthetic biology as they facilitate a more predictable circuit design. A bimodal distribution of reporter output on the other hand may indicate limitations of ligand uptake, resulting in an "all-or-nothing" induction response (Khlebnikov, Risa, Skaug, Carrier, & Keasling, 2000).

The data was analyzed using the software Kaluza 1.5 (Beckman Coulter). To generate a histogram of mRFP1 fluorescence intensity, gating was performed to include the main cell population of the uninduced cells on a dot plot of forward (FSC1 Height) and side scatter (SSC2 Height). The same gate was applied to the induced cells. Subsequently, an overlay histogram was created illustrating the mRFP1 fluorescence intensity (561-614/20 Area) of the gated uninduced and gated induced cells (Fig. 4C).

5. Concluding remarks

There is a need to develop tools and high throughput methods to study metabolic processes *in vivo* to provide a better understanding of cellular functions. Genetic elements derived from inducible gene expression systems can provide new parts to build synthetic circuits for controlling and improving cellular functions to enable metabolic product focused applications. Whereas inducible system-based technologies, such as genetically encoded biosensors, aid in the efficient and high throughput strain construction and evaluation in synthetic biology approaches.

Here, we have described the architecture of transcription factor-based inducible gene expression systems, as well as provided information on necessary steps and considerations for selection of the reporter, vector design and inducible system's cloning. We have placed special emphasis on discussing standard assays for evaluation of inducible systems using fluorescence-based equipment, such as plate reader and flow cytometer. Special attention has been dedicated to describing the data analysis. Application of these methodologies generates quantitative and dynamic data, which help to evaluate characteristics and parameters of inducible gene expression systems and monitor cellular metabolite changes. This can be further advanced by combining the acquired data with a computational model that can simulate and verify the experimental results suggesting new hypotheses and directions for the study of metabolic cellular functions or applications in synthetic biology.

Acknowledgments

This work was supported by the Biotechnology and Biological Sciences Research Council [grant number BB/L013940/1] (BBSRC); and the Engineering and Physical Sciences Research Council (EPSRC) under the same grant number. We thank University of Nottingham for providing SBRC-DTProg PhD studentship to E.K.R H., Amy Slater for gifting *Y. pseudotuberculosis* genomic DNA and all members of SBRC who helped in carrying out this research.

References

Alagesan, S., Hanko, E. K. R., Malys, N., Ehsaan, M., Winzer, K., & Minton, N. P. (2018). Functional genetic elements for controlling gene expression in *Cupriavidus necator* H16. *Applied and Environmental Microbiology*, *84*(19), e00878-18.

Brophy, J. A., & Voigt, C. A. (2014). Principles of genetic circuit design. *Nature Methods*, *11*(5), 508.

Campbell, R. E., Tour, O., Palmer, A. E., Steinbach, P. A., Baird, G. S., Zacharias, D. A., et al. (2002). A monomeric red fluorescent protein. *Proceedings of the National Academy of Sciences of the United States of America*, *99*(12), 7877–7882.

Hanko, E. K. R., Minton, N. P., & Malys, N. (2017). Characterisation of a 3-hydroxypropionic acid-inducible system from Pseudomonas putida for orthogonal gene expression control in *Escherichia coli* and *Cupriavidus necator*. *Scientific Reports*, 7(1), 1724.

Hanko, E. K. R., Minton, N. P., & Malys, N. (2018). A transcription factor-based biosensor for detection of itaconic acid. *ACS Synthetic Biology*, 7(5), 1436–1446.

Heap, J. T., Pennington, O. J., Cartman, S. T., & Minton, N. P. (2009). A modular system for *Clostridium* shuttle plasmids. *Journal of Microbiological Methods*, *78*(1), 79–85.

Jensen, K. F. (1993). The *Escherichia coli* K-12" wild types" W3110 and MG1655 have an *rph* frameshift mutation that leads to pyrimidine starvation due to low *pyrE* expression levels. *Journal of Bacteriology*, *175*(11), 3401–3407.

Khlebnikov, A., Risa, Ø., Skaug, T., Carrier, T. A., & Keasling, J. (2000). Regulatable arabinose-inducible gene expression system with consistent control in all cells of a culture. *Journal of Bacteriology*, *182*(24), 7029–7034.

Klumpp, S., Zhang, Z., & Hwa, T. (2009). Growth rate-dependent global effects on gene expression in bacteria. *Cell*, *139*(7), 1366–1375.

Madar, D., Dekel, E., Bren, A., & Alon, U. (2011). Negative auto-regulation increases the input dynamic-range of the arabinose system of *Escherichia coli*. *BMC Systems Biology*, *5*(1), 111.

Mannan, A. A., Liu, D., Zhang, F., & Oyarzún, D. A. (2017). Fundamental design principles for transcription-factor-based metabolite biosensors. *ACS Synthetic Biology*, *6*(10), 1851–1859.

Münch, R., Hiller, K., Barg, H., Heldt, D., Linz, S., Wingender, E., et al. (2003). PRODORIC: Prokaryotic database of gene regulation. *Nucleic Acids Research*, *31*(1), 266–269.

Novichkov, P. S., Kazakov, A. E., Ravcheev, D. A., Leyn, S. A., Kovaleva, G. Y., Sutormin, R. A., et al. (2013). RegPrecise 3.0—A resource for genome-scale exploration of transcriptional regulation in bacteria. *BMC Genomics*, *14*(1), 745.

Ogden, S., Haggerty, D., Stoner, C. M., Kolodrubetz, D., & Schleif, R. (1980). The *Escherichia coli* L-arabinose operon: Binding sites of the regulatory proteins and a mechanism of positive and negative regulation. *Proceedings of the National Academy of Sciences of the United States of America*, 77(6), 3346–3350.

Reese, M. G. (2001). Application of a time-delay neural network to promoter annotation in the *Drosophila melanogaster* genome. *Computers & Chemistry*, *26*(1), 51–56.

Rogers, J. K., Guzman, C. D., Taylor, N. D., Raman, S., Anderson, K., & Church, G. M. (2015). Synthetic biosensors for precise gene control and real-time monitoring of metabolites. *Nucleic Acids Research*, *43*(15), 7648–7660.

Rogers, J. K., Taylor, N. D., & Church, G. M. (2016). Biosensor-based engineering of biosynthetic pathways. *Current Opinion in Biotechnology*, *42*, 84–91.

Sambrook, J., & Russell, D. W. (2001). *Molecular cloning: A laboratory manual* (3rd ed.). New York: Cold Spring Harbor Laboratory Press.

Sasikaran, J., Ziemski, M., Zadora, P. K., Fleig, A., & Berg, I. A. (2014). Bacterial itaconate degradation promotes pathogenicity. *Nature Chemical Biology*, *10*(5), 371–377.

Shaner, N. C., Campbell, R. E., Steinbach, P. A., Giepmans, B. N., Palmer, A. E., & Tsien, R. Y. (2004). Improved monomeric red, orange and yellow fluorescent proteins derived from *Discosoma* sp. red fluorescent protein. *Nature Biotechnology*, *22*(12), 1567.

Solovyev, V., & Salamov, A. (2011). Automatic annotation of microbial genomes and metagenomic sequences. In R. W. Li (Ed.), *Metagenomics and its applications in agriculture, biomedicine and environmental studies* (pp. 61–78). New York: Nova Science Publishers.

Stepanenko, O. V., Verkhusha, V. V., Kazakov, V. I., Shavlovsky, M. M., Kuznetsova, I. M., Uversky, V. N., et al. (2004). Comparative studies on the structure and stability of fluorescent proteins EGFP, zFP506, mRFP1, "dimer2", and DsRed1. *Biochemistry*, *43*(47), 14913–14923.

Telford, W. G., Hawley, T., Subach, F., Verkhusha, V., & Hawley, R. G. (2012). Flow cytometry of fluorescent proteins. *Methods*, *57*(3), 318–330.

Tobin, J., & Schleif, R. (1987). Positive regulation of the *Escherichia coli* L-rhamnose operon is mediated by the products of tandemly repeated regulatory genes. *Journal of Molecular Biology*, *196*(4), 789–799.

Van Der Meer, J. R., & Belkin, S. (2010). Where microbiology meets microengineering: Design and applications of reporter bacteria. *Nature Reviews. Microbiology*, *8*(7), 511.

Wilson, D., Charoensawan, V., Kummerfeld, S. K., & Teichmann, S. A. (2007). DBD—Taxonomically broad transcription factor predictions: New content and functionality. *Nucleic Acids Research*, *36*(Suppl. 1), D88–D92.

CHAPTER TEN

Design, construction, and validation of optogenetic proteins

Colin P. O'Banion[a], Anwesha Goswami[a], David S. Lawrence[b,c,*]

[a]Division of Chemical Biology and Medicinal Chemistry, UNC Eshelman School of Pharmacy, University of North Carolina, Chapel Hill, NC, United States
[b]Department of Chemistry, University of North Carolina, Chapel Hill, NC, United States
[c]Department of Pharmacology, University of North Carolina, Chapel Hill, NC, United States
*Corresponding author: e-mail address: lawrencd@email.unc.edu

Contents

Abstract

Cellular optogenetics employs light-regulated, genetically encoded protein actuators to perturb cellular signaling with unprecedented spatial and temporal control. Here, we present a potentially generalized approach for transforming a given protein of interest (POI) into an optogenetic species. We describe the rational and methods by which we developed three different optogenetic POIs utilizing the Cry2-Cib photodimerizing pair. The process pipeline is highlighted by (1) developing a low level, constitutively active

Methods in Enzymology, Volume 621
ISSN 0076-6879
https://doi.org/10.1016/bs.mie.2019.02.019

POI that is independent of endogenous regulation, (2) fusion of the mutant protein of interest to an optogenetic photodimerizing system, and (3) light-mediated recruitment of the light-responsive POI to specific subcellular regions.

Abbreviations

C PKA catalytic subunit
CFP cyan fluorescent protein
Cib cryptochrome-interacting basic helix-loop-helix protein
Cry2 cryptochrome 2 photolyase homology region
FAD flavin adenine dinucleotide
FITC fluorescein isothiocyanate
FMN flavin mononucleotide
GFP green fluorescent protein
LOV light, oxygen, voltage domain
mCh mCherry
PKA cAMP dependent protein kinase
POI protein of interest
R PKA regulatory subunit
ROI region of interest
TxRed Texas red

1. Introduction

1.1 Optogenetics

Cellular optogenetics employs genetically encoded, light-regulated proteins to drive changes in cellular signaling. Pairing molecular biology and light mediated control of protein activity yields systems that can act globally on whole cells or locally at subcellular sites on timescales from seconds to hours. We have developed optogenetic analogs of endogenous proteins that control cell migration (Hughes & Lawrence, 2014), apoptosis (Hughes et al., 2015), and kinase signaling (O'banion et al., 2018).

Desired characteristics of an optogenetic protein of interest (POI) is that it is (1) silent in the dark and (2) activated in response to a minimally invasive light. Light-activated proteins have been developed by introducing light sensitive motifs into the POI. However, generating these constructs generally requires a significant engineering effort and may only provide partial control over protein activity. A key challenge is to identify the precise location on the POI to position the photoreceptor so that activity is compromised in the dark but liberated upon illumination.

We have developed an alternative and potentially generalizable approach for generating optogenetic POIs. The two-step strategy requires (1) the acquisition of a low level, constitutively active, analog of the POI. This analog is (2) fused to a photoreceptor which, for the constructs described herein, is the cryptochrome photolyase homology region (Cry2). The latter, upon excitation at the appropriate wavelength (vide infra), associates with its binding partner (Cib). The POI-Cry2 construct is designed to be cytoplasmic and functionally silent in the dark. Upon illumination, the POI-Cry2 conjugate binds to Cib, which is sequestered at a specific subcellular region. This generates a dramatic increase in the local concentration of the POI furnishing spatially focused activity.

The inspiration for the design of our optogenetic constructs is derived from the Michaelis-Menten equation ($v_0 = k_{cat}[E][S]/(K_M + [S])$). The reaction rate for an enzyme-catalyzed reaction is dependent on the turnover number of an enzyme (k_{cat}), the substrate concentration at which the reaction rate is half maximal (K_M) and the concentrations of both enzyme [E] and substrate [S]. A dramatic light-driven increase in [E] at a specific subcellular location should result in an increase in the overall reaction rate. We've applied this concept to cofilin, which severs F-actin, and the cAMP-dependent protein kinase (PKA), which catalyzes the phosphorylation of an array of proteins. In addition, an analogous strategy has been used to generate an optogenetic analog of the pro-apoptotic protein Bax, which forms pore-generating oligomers at the outer mitochondrial membrane.

1.2 An optogenetic engineering strategy

Each of the three optogenetic POIs described herein has a specific subcellular location of activity. Cofilin mediates cellular migration via modulation of the actin cytoskeleton, Bax triggers apoptosis at the outer mitochondrial membrane, and PKA is a highly compartmentalized enzyme with discrete activity profiles at the plasma membrane, mitochondria, and other cellular locations. We note that many effector proteins require an appropriate interaction with a target, be it a subcellular location or a protein/protein interaction, to induce an action and/or transmit a signal. The primary engineering strategy that we've developed is based on the ability to bring the POI close to its intended site of activity in a light-dependent fashion. Consequently, the optogenetic engineering approach should prove applicable to POIs that act upon substrates positioned at distinct subcellular sites.

As described in the introduction, we've taken an approach based on the Michealis-Menten equation ($v_0 = k_{cat}[E][S]/(K_M + [S])$) in order to generate optogenetic constructs that are silent and diffusely distributed in the dark. Illumination furnishes a concentration ([E]) jump at designated intracellular sites sufficient to drive the desired protein activity. A key strategic element associated with our approach is the design of an optogenetic POI that is functionally silent when disseminated throughout the cytoplasm. In order to achieve this, we take into account (1) the mechanism of POI regulation, (2) the location of native POI activity, and (3) protein-protein interactions between the POI and its substrates. Our strategy requires the generation of POI mutants that are (1) diffuse in the cytosol by disrupting subcellular localization of the native protein, (2) impervious to up- or downregulation by endogenous signaling, and (3) exhibit reduced activity at concentrations comparable to that of the wild type POI. Fortunately, site-directed mutagenesis studies constitute one of the very first experiments performed on newly described proteins, which offer insight into structure function activity relationships. Consequently, in many instances, the strategy for generating low level, constitutively active proteins is straightforward. In the examples described below, all optogenetic constructs contain an appended fluorescent protein to assist in visualizing the location of the POI.

2. Preparing low level, constitutively active POIs

Cofilin mutants: Cofilin activity is dramatically reduced by phosphorylation at S3. Consequently, a constitutively active (i.e., unregulated) cofilin is readily generated via a S3A mutation. In order to achieve low level cofilin activity so that F-actin is not severed in the dark, we identified three mutations (S120A, S94D, and K96Q) from the literature (Moriyama & Yahara, 2002), which have been shown to compromise activity by interfering with the ability of cofilin to bind to F-actin. Three low level constitutively active cofilin double mutants were prepared: cofilin$^{S3A/S120A}$, cofilin$^{S3A/S94D}$, and cofilin$^{S3A/K96Q}$. We sought to identify a double mutant that displays a significant concentration-dependent increase in activity at ~0.5 μM, which is approximately the intracellular concentration of cofilin. An in vitro F-actin severing assay was used to assess the activity of these constructs at various concentrations (Hughes & Lawrence, 2014). The cofilin$^{S3A/S120A}$ analog displays an order of magnitude change in F-actin severing activity from 0.2 to 1.0 μM, suggesting that a sudden concentration jump of fivefold could transform low activity cofilin$^{S3A/S120A}$ into an efficient F-actin severing protein. The double mutant cofilin$^{S3A/S120A}$ was appended to the Cry2 photoreceptor: cofilin$^{S3A/S120A}$-Cry2-mCh (optoCofilin; where mCh = mCherry for visualization purposes). A second construct, LifeAct-Cib-GFP, positions Cib (the binding partner of Cry2) at the cytoskeleton via an F-actin binding element (LifeAct) (Riedl et al., 2008). Simultaneous transfection of cells with optoCofilin and LifeAct-Cib-GFP furnishes an optogenetic system that recruits cofilin to the cytoskeleton.

Bax mutants: Like cofilin, Bax is regulated via phosphorylation. Phosphorylation of Bax at S184 sequesters the protein in the cytoplasm, which prevents Bax from associating with the outer mitochondrial membrane where it triggers apoptosis (Nechushtan, Smith, Hsu, & Youle, 1999). In response to an apoptotic stimulus, Bax is dephosphorylated, migrates to the mitochondria, oligomerizes, and generates pores in the outer mitochondrial membrane. An optogenetic version of Bax was designed by first removing the ability of native Bax to migrate to the mitochondria. The phosphomimetic S184E mutant has previously been shown to be cytoplasmic (Nechushtan et al., 1999). Based on this observation, we constructed Cry2-mCh-Bax^{S184E} (optoBax) and demonstrated that it is cytoplasmic in the dark in cells expressing Cib at the outer mitochondrial membrane (Hughes et al., 2015). Illumination provokes migration of the optogenetic Bax to the mitochondria (vide infra).

PKA mutants: The cAMP dependent protein kinase (PKA) holoenzyme exists as a tetramer composed of two catalytic subunits (C) and two regulatory (R) subunits. cAMP binds to the R subunits, which releases the C subunits, allowing them to phosphorylate local protein substrates. First, based on literature precedent, a W196R mutation/substitution was introduced in the C subunit, where it abolishes binding of the R subunit (Gibson & Taylor, 1997). This generates a cAMP-independent PKA. Second, the catalytic activity of the C subunit was dramatically reduced in order to prevent widespread catalytic activity of the cAMP-independent (i.e., constitutively active) enzyme. Three mutations were screened that are known to either impair substrate binding or catalytic activity. The C^{E203A}, C^{Y204A}, and C^{F327A} mutants display reduced activity that is 43%, 5.6%, and 1.2% as active, respectively, of wild type C subunit. As in the examples of optogenetic cofilin and Bax, the C subunit mutants were appended to the Cry2 photoreceptor, which can then be directed to specific subcellular sites in a light-dependent fashion using the Cry2 binding partner Cib.

2.1 Protocol for generating site-directed mutants

A general protocol for generating site-directed mutants is adapted from the QuickChange mutagenesis kit:

1. Design primers for site-directed mutagenesis so that they have a minimum T_m of 72 °C.
   a. *Note: Primers for site-directed mutagenesis should be designed using the Agilent QuickChange primer design web tool: https://www.chem.agilent.com/store/primerDesignProgram.jsp*
2. Combine the following: 125 ng of each primer, 50 ng of template DNA, 4 μL of 2.5 mM dNTPs, 5 μL of 10 × PfuUltra buffer, 1 μL of PfuUltra Hotstart DNA polymerase, and ddH_2O to a final volume 50 μL.
3. Treat the PCR products obtained from this reaction with *Dpn*I for 1 h at 37 °C and use 1 μL of this solution for the transformation of 50 μL of Invitrogen Max Efficiency DH5α *E. coli* cells.
4. The next day, choose individual bacterial colonies to be amplified in selective LB media and scale up to miniprep DNA extraction for sequence verification.

3. Fusion of the POI to Cry2

In order to drive local light-dependent concentration jumps for a given POI, fusion to a photodimerizing system is necessary (Fig. 1). Two photoreceptors comprise the entirety of blue light regulated optogenetic dimerization: cryptochrome 2 (Cry2) and LOV. Both are flavin binding photoreceptors

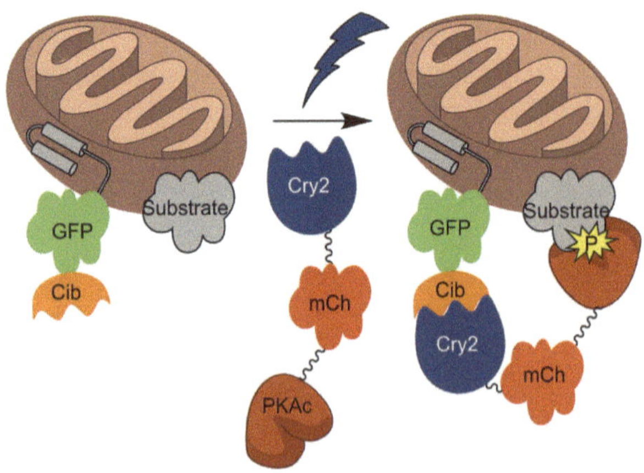

Fig. 1 Schematic for light-induced recruitment of a POI fused to Cry2-mCh to the OMM. *Reproduced with permission from O'Banion, C. P., Priestman, M. A., Hughes, R. M., Herring, L. E., Capuzzi, S. J., & Lawrence, D. S. (2018). Design and profiling of a subcellular targeted optogenetic cAMP-dependent protein kinase.* Cell Chemical Biology, 25*(1), 100–109 e108. doi:10.1016/j.chembiol.2017.09.011.*

with Cry2 binding to FAD and LOV to FMN and absorb light between 400 and 500 nm (O'banion & Lawrence, 2018). The best characterized are Cry2 from *Arabidopsis thaliana* and LOV2 from *Avena sativa*. As noted above, Cry2 associates with its binding partner, Cib, upon excitation with blue light.

There are several choices available with respect to the design of fused POI/photoreceptor construct, especially if a fluorescent protein is part of the optogenetic species. First, in the case of the Cry2/Cib binding partner, Cry2 is generally recruited best if Cib is immobilized at a subcellular location of interest and the POI is fused to the Cry2 construct. Second, a determination of whether a POI is best placed at the N- or C-terminus of the construct (or inserted between: Cry2-POI-fluorescent protein) is dependent, in part, on the POI itself. For example, the POI may require a free N-terminus for activity. In the absence of strict POI structural requirements, we have found that some experimentation is necessary to identify the best-behaved constructs. For example, the C subunit of PKA, when fused to the N-terminus of Cry2, does not recruit to Cib upon illumination. By contrast, fusion of the C subunit to the C-terminus of Cry2 provides constructs that are efficiently recruited to Cib in a light-dependent fashion (O'banion et al., 2018). Interestingly, in the case of cofilin, the exact opposite behavior is observed (Hughes & Lawrence, 2014).

3.1 Protocol for the construction of POI/Cry2/fluorescent protein constructs

The following general protocol describes our strategy for fusing the POI to the N- or C-termini of Cry2-mCh (in a pmCherry-N1 plasmid backbone) to generate the following constructs: Bax-Cry2-FLAG-mCh, Cry2-mCh-Bax, Cofilin-Cry2-FLAG-mCh, Cry2-mCh-Cofilin and Cry2-mCh-C (PKA). The protocol is as follows:

Cloning of POI-Cry2-FLAG-mCherry into pmCherry-N1 vector by splice overlap extension PCR:

1. Design and generate primers that contain: (1) an appropriate gene specific region for amplification of POI, (2) any linkers to be appended, and (3) an additional region containing 30–40 homologous bases to the other DNA fragment to be spliced together (see examples for cloning below).
   a. Forward primer for POI: 5′-*Xho*I-POI gene specific region-3′
   b. Reverse primer for POI: 5′-gene specific region-3′
   c. Forward primer for Cry2: 5′-POI homology region-linker region-Cry2 gene specific region-3′.
   d. Reverse primer for Cry2: 5′-*Xma*I-FLAG-Cry2 gene specific region-3′

2. Separately PCR amplify the POI and Cry2 gene constructs with Phusion polymerase and primers encoding linkers and homology regions.
3. After subsequent gel purification of the individual isolates, use aliquots (6 μL of each PCR isolate) of the obtained DNA fragments to combine with a cocktail of 1.5 μL each of the forward primer for the POI (a. from primer list above) and reverse primer for Cry2 (d. from primer list above) (10 μM stocks), 4 μL dNTPS (2.5 mM stock), 3 μL of 10× Phusion HF buffer, 8 μL ddH_2O, and 0.5 μL of the Phusion polymerase enzyme and subject this mix to splice overlap extension PCR. Standard thermocycling conditions can be applied for this reaction.
4. Purify the resultant PCR fragments via agarose gel and subject them to restriction digestion. The resultant fragments should be ligated into a plasmid backbone containing pmCherry-N1 with *Xho*I and *Xma*I sites to yield 5′-*Xho*I-POI-GGSGGS-Cry2-GGSDYKDDDKARDPPVAT-*Xma*I-mCh-3′.

Cloning of Cry2-mCh-POI in pCry2-mCherry-N1 vector:

1. PCR amplify POI with primers appending *Bsr*GI and *Not*I restriction digestion sites as well as a six amino-acid N-terminal linker, followed by gel purification and isolation of the appropriately sized PCR amplicon.
2. The pCry2-mCherry-N1 should be analogously linearized via *Bsr*GI/*Not*I restriction digestion, and both digested PCR and vector fragments should be ligated via T4 DNA ligase using standard procedures. The resultant construct will yield Cry2-mCh-GGSGGS-POI.
3. In the case of optoPKA, inclusion of an additional 16 amino acid tether as an alternative to the six amino acid tether, followed by *Bsr*GI/*Not*I restriction and ligation should generate Cry2-mCh-SAGGSAGGSAGGSAGG-C (PKA).

For generating the corresponding Cib constructs fused to the respective anchoring peptide sequences, the following protocol can be employed to generate Tom20MLS-Cib-GFP and LifeAct-Cib-GFP in a phCMV-GFP vector. Tom20MLS is an N-terminal fragment of the Tom20 protein that is anchored to the outer mitochondrial membrane. The protocol entails the following steps:

1. PCR amplify Tom20MLS, LifeAct (anchoring peptides) and Cib (Kennedy et al., 2010) separately by PCR Phusion polymerase. After gel isolation and purification of the fragments, combine aliquots of the purified DNA and subject them to overlap extension PCR, as before.

2. Digest the resultant constructs with *Xho*I and *Hind*III enzymes, followed by ligation into an analogously digested phCMV-GFP vector backbone to generate the final construct(s) expressed as (anchoring peptide)-GGSGGS-Cib-GGSGGSRSFEF-GFP.
3. A somewhat modified protocol can be used to generate Cib-CAAX (lacking GFP) from Cib-GFP-CAAX (Addgene 26867) via PCR-mediated site-directed mutagenesis to insert a poly-lysine sequence CAAX box [GGSGKKKKKKSKTK-CVIM] and a stop codon between Cib and GFP.

4. Characterization and validation of light mediated translocation

4.1 Experimental setup and considerations

A characteristic displayed by many genetically encoded light-activated proteins, such as the Cry2/Cib system, is functional reversibility. The latter transpires by exposing the optogenetic species to a series of light and dark cycles. Under these circumstances, optogenetic proteins display a characteristic recruitment or activity half-life following illumination. For example, an optogenetic Bax mutant fused to the C-terminus of Cry2-mCh (i.e., Cry2-mCh-Bax) displays a longer residence time at mitochondria than its N-terminal counterpart (i.e., Bax-Cry2-mCh) (Hughes et al., 2015). Mitochondrial recruitment serves as a direct means for quantifying light mediated translocation. Mitochondria have clear morphology that allow for obvious qualitative assessment of translocation as well as simple quantitative measurements of translocation from the cytoplasm to the mitochondria. In addition, imaging mitochondrial recruitment can be performed on either a widefield or confocal microscope whereas plasma membrane recruitment (another popular region of interest) requires the use of confocal microscopy for precise measurement of light-mediated recruitment.

4.2 Experimental design

Short single pulse recruitment vs. *multiple pulse recruitment.* Recruitment of an optogenetic protein to a subcellular location can be imaged on a microscope and analyzed by initiation with either a single pulse of light or sustained light pulses. Short, single pulses (100 ms) at 488 nm provide information about the sensitivity of an optogenetic system as well as qualitative and quantitative information about reversibility (Fig. 2). On the other hand, multiple short

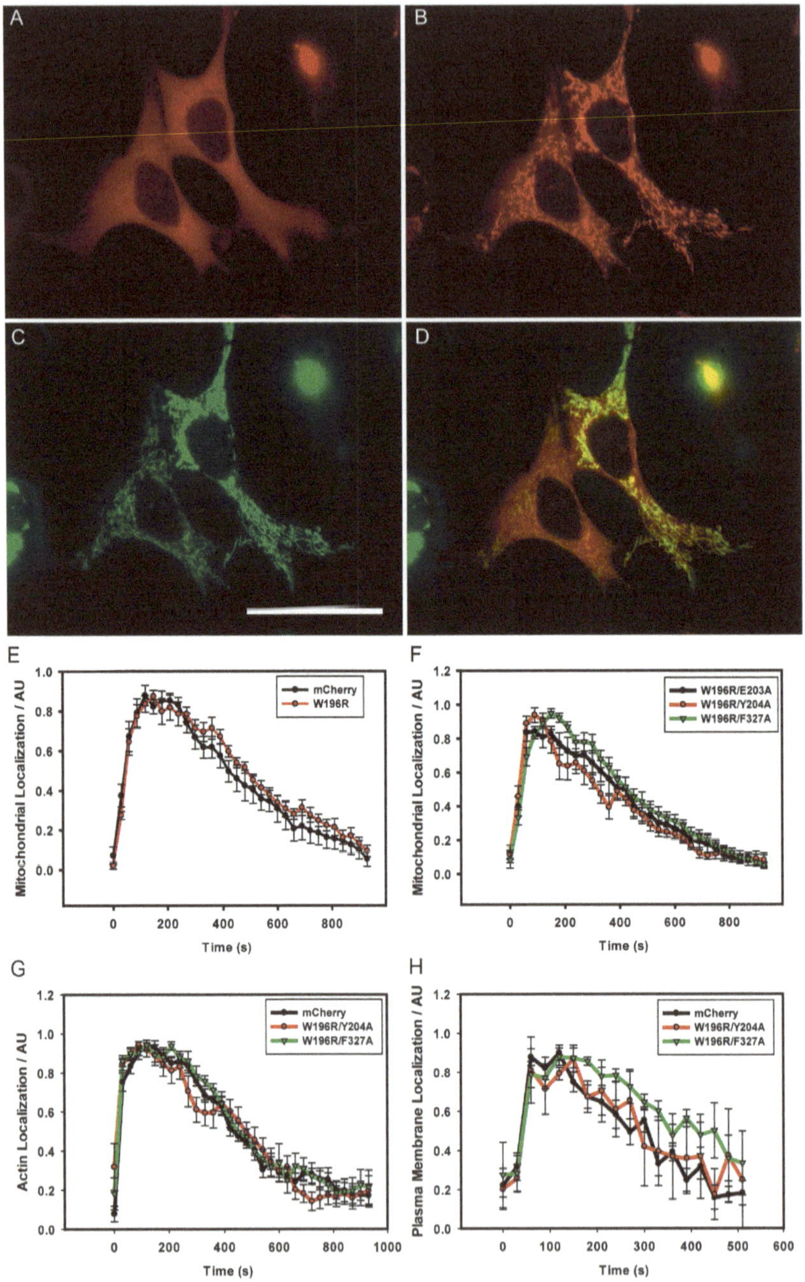

Fig. 2 See legend on opposite page.

pulses of light establish the maximal possible recruitment of the POI. The latter defines the dynamic range of recruitment and assists in identifying an appropriate light dosing regimen for the biological system to be studied.

4.3 Light-mediated translocation of opto-POIs

A general protocol for the characterization and validation of light mediated translocation of the opto-POI constructs is outlined here:

4.3.1 Equipment and materials

1. Fluorescent microscopy (widefield) imaging is performed with an inverted Olympus IX81 microscope equipped with a Hamamatsu C848 camera, 60× oil immersion Plan S-Apo objective and FITC and TxRed filter cubes (Semrock).
2. Metamorph Imaging Suite
3. Opaque heat, humidity, and atmosphere (5% CO_2) controlled microscope enclosure
4. 37°C tissue culture incubator with 5% CO_2 and 95% humidity
5. 35mm glass bottom dishes (Mattek)
6. Fluorophore-labeled purified proteins
7. DMEM (Gibco) supplemented with 10% FBS, 100U/mL penicillin, 100μg/mL streptomycin.
8. Lebowitz L-15 media without phenol red (Gibco).

Fig. 2 Light-Triggered optoPKA association with, and subsequent dissociation from, the OMM, cytoskeleton, and PM. All optoPKA constructs contain the mCh fluorescent protein: Cry2-mCh-$C^{W196R/E203A}$ (A–D and F), Cry2-mCh (E), control, Cry2-mCh-C^{W196R} (E), Cry2-mCh-$C^{W196R/Y204A}$ (F–H), and Cry2-mCh-$C^{W196R/F327A}$ (F–H). The following Cib constructs were employed to recruit optoPKA to specific intracellular sites: Tom20MLS-Cib-GFP (OMM-Cib in A–F) at the OMM; LifeAct-GFP-Cib (LifeAct-Cib) in (G) at the actin cytoskeleton; Cib-GFP-CAAX (PM-Cib) in (H) at the PM. Visualization of the mCh label in $PKA^{196R/E203A}$ (A) before and (B) 1min after stimulation with a 100ms, 488nm light pulse. (C) Visualization of the GFP label in OMM-Cib (Tom20MLS-Cib-GFP), where (D) is an overlay of (B and C). (E–H) Association and subsequent dissociation of optoPKA with and from the PM, OMM, and the cytoskeleton were monitored via mCh fluorescence. A single 100ms pulse (FITC cube) was employed to initiate recruitment of the optoPKA constructs to designated sites. Experiments were performed on either a wide-field (OMM-Cib, LifeAct-Cib) or confocal (PM-Cib) microscope. $N=3$ cells per group. Scale bar, 50μm. Data expressed as mean±SEM. *Reproduced with permission from O'Banion, C. P., Priestman, M. A., Hughes, R. M., Herring, L. E., Capuzzi, S. J., & Lawrence, D. S. (2018). Design and profiling of a subcellular targeted optogenetic cAMP-dependent protein kinase.* Cell Chemical Biology, 25*(1), 100–109 e108. doi:10.1016/j.chembiol.2017.09.011.*

9. Mammalian expression vector containing the sequence for (photoswitch)-fused to fluorescent protein spectrally compatible with our system
10. Cell lines of interest
11. Transfection reagent (Polyplus transfection 114-07)

4.3.2 Protocol

Day 1

1. Plate cells in 35 mm glass bottomed MatTek dishes at 75,000 cells per dish and return to incubator for 16–24 h.

Day 2:

2. Using DMEM and Polyplus-transfection reagent, transfect cells with the plasmids containing photoswitch according to transfection protocol (each plasmid in a 1:1 ratio with a total of 2 μg DNA per well; 1:2 ratio of DNA: jetPrime transfection cocktails should be incubated for 10 min at room temperature (RT) before being added to each well and allowed to incubate).
3. 4–6 h later, aspirate media and exchange for fresh DMEM, followed by overnight incubation.

Day 3:

4. On the day of imaging, aspirate and exchange media for complete Lebowitz L-15 media without phenol red for live cell imaging.

Day 4:

5. Allow for dishes to equilibrate in heated incubators (32 °C, 45% humidity) for at least 30 min prior to imaging.
6. In live cell experiments, initiate whole cell recruitment of either N- or C-termini fused Cry2-mCh constructs to the respective peptide-anchored-Cib-GFP with a 100 ms pulse of blue light (488 nm) using FITC filter cube.
7. Monitor time-lapse images via the TxRed filter cube (500 ms pulse) and acquire images every 30 s over 15 min during the course of the experiment; collected images should be processed by ImageJ.

4.4 Data analysis

Data analysis is performed in ImageJ (or other analysis software). Generally, data are presented as the fold change in localization relative to the initial timepoint. For mitochondrial recruitment, we first calculate the mitochondrial:cytoplasmic ratio of the POI and then normalize to the first timepoint (F/F_0). Generally, we define three regions of interest (ROIs) within the mitochondrial area and three ROIs in adjacent cytoplasmic regions as well as a number of background measurements from regions containing either nontransfected cells or no cells. Background is subtracted from cellular ROIs and the ratio of a mitochondrial ROI to its adjacent cytoplasmic ROI is calculated.

The FIJI (ImageJ) software package is used for data analysis. A general protocol for the analysis of image data using ImageJ consists of the following steps:

1. For each cell, collect images from two channels: Cib-GFP (acts as label for either mitochondria, cytoskeleton or the plasma membrane) and Cry2-mCh.
2. Define three mitochondrial ROIs and three adjacent cytoplasmic ROIs for each cell, as well as three background ROIs.
3. Record the mean fluorescent intensity for each ROI and subtract background fluorescence from each ROI.
4. The ratio of mitochondrial:cytoplasmic fluorescence is obtained by dividing background corrected mitochondrial ROI fluorescence by adjacent background corrected cytoplasmic ROI fluorescence.
5. Steps 1–5 should then be repeated for every image in the series.
6. Normalize values between 0 and 1 for comparison or as fold change relative to the initial time point.

5. Characterization and validation of light-mediated biological activity of optogenetic constructs

A critical element of optogenetic protein design is characterization of the optoPOI using a well-defined cellular assay.

5.1 Filopodia induction and cell spreading by light activation of optoCofilin

optoCofilin: The cofilin$^{S3A/S120A}$-Cry2-mCh (optoCofilin) and LifeAct-Cib-GFP constructs were expressed in MTLn3 cells. The light-induced (488 nm, 100 ms pulse) association of optoCofilin with the cytoskeleton was imaged and quantified. optoCofilin does not associate with the cytoskeleton in the dark, which is in marked contrast to the wild type (cofilin-Cry2-mCh) and high constitutively active (cofilin^{S3A}-Cry2-mCh) analogs.

Initial characterization of optogenetic cofilin construct activity was assessed by whole-cell spreading after whole-cell illumination by measuring cell area during and after multiple 488 nm pulses over 10 min. Cofilin activation by

EGF is known to induce cell spreading (Song et al., 2006). optoCofilin induces both cell spreading and filopodial formation, another phenotype of increased cofilin activity (Lichtner, Wiedemuth, Noeske-Jungblut, & Schirrmacher, 1993), in response to illumination. In contrast, a Cry2-mCh control recruited in the same manner does not generate filopodia (Hughes & Lawrence, 2014).

A protocol for assessing filopodia formation by light activation of optoCofilin is described here:

5.1.1 Equipment and materials

1. Fluorescent microscopy (widefield) imaging is performed with an inverted Olympus IX81 microscope equipped with a Hamamatsu C848 camera, 60 × oil immersion Plan S-Apo objective and FITC and TxRed filter cubes (Semrock).
2. Metamorph Imaging Suite.
3. Opaque heat, humidity, and atmosphere (5% CO_2) controlled microscope enclosure.
4. 37 °C tissue culture incubator with 5% CO_2 and 95% humidity.
5. 35 mm glass bottom dishes (Mattek).
6. MEM alpha (Gibco) supplemented with 10% FBS and 1% Pen-Strep.
7. MEM alpha without phenol red (Gibco) supplemented with 10% FBS and 1% Pen-Strep.
8. Mammalian expression plasmids: $Cofilin^{S3A}$-Cry2-mCh, $Cofilin^{S3A/S120A}$-Cry2-mCh, Cry2-mCh, LifeAct-Cib-GFP.
9. MTLn3 cells.
10. Transfection reagent (Polyplus transfection 114-07).

5.1.2 Protocol

1. Grow MTLn3 cells to confluency in MEM alpha medium supplemented with 5% FBS and 1% Pen-Strep.
2. After plating on 35 mm MatTek dishes, transfect cells with $Cofilin^{S3A/S120A}$-Cry2-mCh, $Cofilin^{S3A}$-Cry2-mCh (or Cry2-mCh) along with LifeAct-Cib-GFP for initiating the association of optoCofilin to the F-actin cytoskeleton.
3. Select individual cells for filopodia-forming experiments by assessing their adherence and brightness; Image cells using both the TxRed and FITC channels for each time point (60 s over 10 min). The exposure time for the TxRed channel should be 300 ms, and 100 ms for the FITC channel.

4. Image analysis: Manually count the filopodia at both the beginning and end of each time-lapse image series for each cell using the FITC channel (via the LifeAct-Cib-GFP construct). The fold change in filopodia (10 min vs. 0 min) should then be quantified for each cell and the net fold change calculated for each individual experimental setup.

5.2 Cellular collapse by the recruitment of optoBax to the outer mitochondrial membrane

optoBax: The Cry2-mCh-Bax^{S184E} (optoBax) construct was expressed in HeLa cells and is retained in the cytoplasm in the dark. Cells transfected with optoBax and Tom20MLS-Cib-GFP (positioned on the outer mitochondrial membrane) are likewise healthy in the dark. However, upon pulsed illumination at 488 nm for 1 h, optoBax is recruited to the outer mitochondrial membrane, resulting in cellular collapse and subsequent death. By contrast, when Cib is positioned on the plasma membrane (Cib-GFP-CAAX), optoBax recruitment to the membrane has no effect on cell viability (Hughes et al., 2015).

A protocol for the characterization and validation of cellular collapse in HeLa cells transfected with optoBax is described here:

5.2.1 Equipment and materials

1. Fluorescent microscopy (widefield) imaging is performed with an inverted Olympus IX81 microscope equipped with a Hamamatsu C848 camera, 60 × oil immersion Plan S-Apo objective and FITC and TxRed filter cubes (Semrock)
2. Metamorph Imaging Suite
3. Opaque heat, humidity, and atmosphere (5% CO_2) controlled microscope enclosure
4. 37 °C tissue culture incubator with 5% CO_2 and 95% humidity
5. 35 mm glass bottom dishes (Mattek)
6. DMEM (Gibco) supplemented with 10% FBS and 1% Pen-Strep
7. DMEM without phenol red (Gibco) supplemented with 10% FBS and 1% Pen-Strep
8. Mammalian expression plasmids: Cry2-mCh-Bax^{S184E}, Cry2-mCh and Tom20-Cib-GFP
9. HeLa cells
10. Transfection reagent (Polyplus transfection 114-07)

5.2.2 Protocol

1. Grow HeLa cells to confluency in DMEM media supplemented with 5% FBS and 1% Pen-Strep
2. Plate cells in 35 mm MatTek dishes as described followed by transfection with Cry2-mCh-Bax^{S184E}, Cry2-mCh and Tom20-Cib-GFP for initiating the association of optoBax to outer mitochondrial membrane.
3. Select individual cells for cellular collapse experiments by assessing their adherence and brightness; Image cells using both the TxRed and FITC channels for each time point (every 120 s for the duration of the experiment, typically an average of 60 min). The exposure time for the TxRed channel should be between 100 to 300 ms, and FITC channel should be kept at a minimum (10 ms) to reduce light induced toxicity.
4. For image analysis, the loss of adherent cell morphology and the appearance of membrane blebbing can be used as direct indicators of cell collapse. Time-lapse images should be collected for each individual cell in each field of view followed by processing using ImageJ. A timestamp is appended to each image series and thereby manually viewed over an appropriate image interval to inspect the loss of adherent morphology and/or membrane blebbing.

5.3 Phosphorylation of mitochondrial PKA reporter by optoPKA

optoPKA: We evaluated the behavior of three different PKA C subunit mutants: $C^{W196R/E203A}$, $C^{W196R/Y204A}$, and $C^{W196R/F327A}$, which were inserted into the construct of the general form Cry2-mCh-C^{mutant} (optoPKA). Desired optoPKA properties include low kinase activity in the dark, observable kinase activity upon exposure to 470 nm light, and inhibition by known PKA inhibitors. MV^{D7} cells were simultaneously transfected with one of the three optoPKA constructs and a Cib construct targeted to the outer mitochondrial membrane (Tom20MLS-Cib). Cells were also transfected with a PKA reporter described in O'banion et al. (2018). The levels of mitochondrial PKA reporter phosphorylation were compared for cells in the dark, illuminated (470 nm) for 1 min, or illuminated in the presence of the PKA inhibitor H89. We found Cry2-mCh-$C^{W196R/Y204A}$ displays robust kinase activity (reporter phosphorylation) upon exposure to 1 min of 470 nm light and insignificant kinase activity in the dark. Furthermore, the presence of H89 interferes with light-triggered kinase activity (O'banion et al., 2018).

5.3.1 Equipment and materials

1. Fluorescent microscopy (widefield) imaging was performed with an inverted Olympus IX81 microscope equipped with a Hamamatsu C848 camera, 60× oil immersion Plan S-Apo objective and CFP, TxRed, and Cy5.5 filter cubes (Semrock)
2. Blue LED light source (470 nm, Home Depot PAR38 Blue LED Flood Lamp Model #469072)
3. 37 °C tissue culture incubator with 5% CO_2 and 95% humidity
4. DMEM (Gibco) supplemented with 10% FBS, 1% Pen-Strep, 50 UI recombinant mouse interferon-γ, and 2 mM Glutamax
5. Optimem (Gibco)
6. DPBS with calcium and magnesium (Gibco)
7. Trypsin-EDTA (0.05%) (Gibco)
8. MatTek 12-well glass bottom plates or 35 mm # 1.5 glass bottom dishes
9. Mammalian expression plasmids: Cry2-mCh, Cry2-mCh-C^{W196R}, Cry2-mCh-$C^{W196R/Y204A}$, Cry2-mCh-$C^{W196R/F327A}$, Cry2-mCh-$C^{W196R/E203A}$, OMM-Cib, and mitochondrial PKA-reporter
10. JetPrime transfection reagent (PolyPlus transfection 114-07)
11. Mouse anti-phospho$VASP^{S157}$ antibody (abcam ab58555)
12. Anti-mouse Alexa 647 antibody (Life Technologies)
13. MV^{D7} cells (gift from Dr. James E. Bear, double knockout VASP and Mena embryonic mouse fibroblast clonal cells with no detectable expression of the EVL protein)
14. Forskolin (Sigma; 10 mM stock solution in DMSO)
15. H-89 (Sigma; 10 mM stock solution in DMSO)

5.3.2 Protocol

1. Grow MV^{D7} cells to confluency in DMEM media supplemented with 5% FBS and 1% Pen-Strep
2. Plate cells in 35 mm MatTek dishes as described, followed by transfection with the plasmids containing optoPKA, Tom20MLS-Cib, and mitochondrial PKA reporter at a ratio of 2:1:1 with a total of 2 μg DNA per well using jetPrime transfection reagent (Polyplus Transfection).
3. The following day, exchange media for 1 mL L-15/10% FBS/1× GlutaMAX/50 UI interferon γ without phenol red. L-15 is phosphate and amino acid buffered and is to be used without CO_2. If a CO_2 incubator is to be used, then bicarbonate buffered media without phenol red is required.
4. Pre-treat positive control cells with 100 μM forskolin, an cAMP generator, for 30 min. Equilibrate cells expressing optoPKA mutants,

Tom20MLS-Cib (lacking GFP) and mitochondrial PKA reporter for 10 min at RT. They are either to be left in the dark, exposed to 1 min blue light (470 nm) or exposed to blue light in the presence of 10 μM PKA inhibitor H89.

5. Cells treated with H89 should be preequilibrated with inhibitor for 10 min at RT before light exposure.
6. Post-light treatment, fix cells by adding 1 mL of 2 × paraformaldehyde (8% v/v) in PBS yielding a final concentration of 4% paraformaldehyde; incubate for 10 min at RT before washing twice with PBS.
7. Block fixed cells with PBS containing 5% BSA + 0.1% Triton X-100 for 1 h at RT.
8. Incubate blocked cells with primary antibody (1:100 dilution of mouse anti-phosphoVASP^{S157} in PBS with 1% BSA + 0.1% Triton X-100) overnight at 4 °C.
9. The following day, wash cells three times with PBS for 5 min each and then incubate with secondary antibody (1:1000 anti-mouse Alexa fluor 647) in blocking buffer at RT for 2 h.
10. Repeat wash steps with PBS three times for 5 min each before imaging.
11. Three channels are used for imaging on the widefield microscope: CFP (cyan fluorescent protein; total mitochondrial PKA reporter), TxRed (mCh), and Cy5.5 (phosphorylated mitochondrial PKA reporter via immunostaining with anti phospho-VASP^{S157} and detecting with anti-mouse Alexa fluor 647)
12. At least 10 frames containing one or more cells per frame should be captured. Analyze cells that express both mCh and CFP fluorescence.
13. Draw ROIs around CFP fluorescence and calculate mean and minimum fluorescent intensity for both CFP and Alexa fluor 647.
14. Use the minimum fluorescent intensity as background signal and subtract from respective channels.
15. The ratio of phosphorylated reporter/total reporter can be determined by dividing background corrected Alexa fluor 647 mean fluorescent intensity by CFP mean fluorescent intensity. Data can be presented as a percentage of forskolin treated positive control.

6. Summary

We've described the process of developing an optogenetic POI using a concentration jump approach in conjunction with the Cry2-Cib system. Three optogenetic POIs across three protein families have been developed:

optoCofilin, an actin remodeler, optoBax, an apoptosis inducer, and optoPKA, a protein kinase. These optogenetic species are functionally silent in the dark but can be recruited upon illumination to specific intracellular sites where they are functionally active. Mutants of the POI are generated cytoplasmically disseminated low activity analogs of the wild type protein. These mutants are fused to a photoreceptor (Cry2), and along with the photoreceptor's binding partner (Cib), the constructs are expressed in cells. Although our work has, to this point, primarily utilized the Cry2-Cib photodimerizing pair, we anticipate that the concentration jump approach may be applied to other photodimerizing systems such as iLid.

Acknowledgments

Eshelman Institute for Innovation at the University of North Carolina at Chapel Hill (R1040 RX03712116) and the National Institutes of Health (1R21NS093617, 1R01NS103486, 1U01CA207160).

References

Gibson, R. M., & Taylor, S. S. (1997). Dissecting the cooperative reassociation of the regulatory and catalytic subunits of camp-dependent protein kinase. Role of Trp-196 in the catalytic subunit. *The Journal of Biological Chemistry*, *272*(51), 31998–32005.

Hughes, R. M., Freeman, D. J., Lamb, K. N., Pollet, R. M., Smith, W. J., & Lawrence, D. S. (2015). Optogenetic apoptosis: Light-triggered cell death. *Angewandte Chemie (International Ed. in English)*, *54*(41), 12064–12068. https://doi.org/10.1002/Anie.201506346.

Hughes, R. M., & Lawrence, D. S. (2014). Optogenetic engineering: Light-directed cell motility. *Angewandte Chemie (International Ed. in English)*, *53*(41), 10904–10907. https://doi.org/10.1002/Anie.201404198.

Kennedy, M. J., Hughes, R. M., Peteya, L. A., Schwartz, J. W., Ehlers, M. D., & Tucker, C. L. (2010). Rapid blue-light-mediated induction of protein interactions in living cells. *Nature Methods*, 7(12), 973–975. https://doi.org/10.1038/Nmeth.1524.

Lichtner, R. B., Wiedemuth, M., Noeske-Jungblut, C., & Schirrmacher, V. (1993). Rapid effects of EGF on cytoskeletal structures and adhesive properties of highly metastatic rat mammary adenocarcinoma cells. *Clinical & Experimental Metastasis*, *11*(1), 113–125.

Moriyama, K., & Yahara, I. (2002). The actin-severing activity of cofilin is exerted by the interplay of three distinct sites on cofilin and essential for cell viability. *Biochemical Journal*, *365*(1), 147–155. https://doi.org/10.1042/Bj20020231.

Nechushtan, A., Smith, C. L., Hsu, Y. T., & Youle, R. J. (1999). Conformation of the Bax C-terminus regulates subcellular location and cell death. *The EMBO Journal*, *18*(9), 2330–2341. https://doi.org/10.1093/Emboj/18.9.2330.

O'banion, C. P., & Lawrence, D. S. (2018). Optogenetics: A primer for chemists. *Chembiochem*, *19*(12), 1201–1216. https://doi.org/10.1002/Cbic.201800013.

O'banion, C. P., Priestman, M. A., Hughes, R. M., Herring, L. E., Capuzzi, S. J., & Lawrence, D. S. (2018). Design and profiling of a subcellular targeted optogenetic CAMP-dependent protein kinase. *Cell Chemical Biology*, *25*(1). https://doi.org/10.1016/J.Chembiol.2017.09.011. 100–109.e18.

Riedl, J., Crevenna, A. H., Kessenbrock, K., Yu, J. H., Neukirchen, D., Bista, M., et al. (2008). Lifeact: A versatile marker to visualize F-actin. *Nature Methods*, *5*(7), 605–607. https://doi.org/10.1038/Nmeth.1220.

Song, X., Chen, X., Yamaguchi, H., Mouneimne, G., Condeelis, J. S., & Eddy, R. J. (2006). Initiation of cofilin activity in response to EGF is uncoupled from cofilin phosphorylation and dephosphorylation in carcinoma cells. *Journal of Cell Science*, *119*(Pt. 14), 2871–2881. https://doi.org/10.1242/Jcs.03017.

CHAPTER ELEVEN

Overcoming component limitations in synthetic biology through transposon-mediated protein engineering

Joshua T. Atkinson[a], Bingyan Wu[b,c], Laura Segatori[c,d,e], Jonathan J. Silberg[c,d,e,*]

[a]Systems, Synthetic, and Physical Biology Graduate Program, Rice University, Houston, TX, United States
[b]Biochemistry and Cell Biology Graduate Program, Rice University, Houston, TX, United States
[c]Department of Biosciences, Rice University, Houston, TX, United States
[d]Department of Chemical and Biomolecular Engineering, Rice University, Houston, TX, United States
[e]Department of Bioengineering, Rice University, Houston, TX, United States
*Corresponding author: e-mail address: joff@rice.edu

Contents

Abstract

Protein fission and fusion can be used to create biomolecules with new structures and functions, including circularly permuted proteins that require post-translational modifications for activity, split protein AND gates that require multiple inputs for activity, and fused domains that function as chemical-dependent protein switches. Herein

Methods in Enzymology, Volume 621
ISSN 0076-6879
https://doi.org/10.1016/bs.mie.2019.02.014

we describe how transposon mutagenesis can be used for protein design to create libraries of permuted, split, or domain-inserted proteins. When coupled with a functional screen or selection, these approaches can rapidly diversify the topologies and functions of natural proteins and create useful protein components for synthetic biology.

1. Component limitations in synthetic biology

A limited repertoire of well-defined genetically encoded components is available for constructing genetic circuits. These simple parts have enabled the construction of programs with complex dynamic phenotypes, such as toggle switches (Gardner, Cantor, & Collins, 2000), oscillators (Elowitz & Leibler, 2000), logic gates (Gao, Chong, Kim, & Elowitz, 2018; Tamsir, Tabor, & Voigt, 2010), and edge-detectors (Tabor et al., 2009). As synthetic biologists program more complex phenotypes, an increasingly larger set of orthogonal components are required. One-way researchers have diversified the parts list for programming cells is by looking to nature. Libraries of candidate genes identified through metagenomic studies have been synthesized and characterized to identify natural parts with desired functions (van der Helm, Genee, & Sommer, 2018). However, synthetic biology frequently requires genetically encoded components that have not yet been observed in nature, limiting the utility of metagenomic mining as a strategy to identify useful components.

Protein engineering strategies have also been used to overcome component limitations in synthetic biology (Roberta, 2010). Biomolecules with altered functions have been created by screening or selecting libraries of mutated proteins for variants with desired functions. A wide range of methods have been used to create sequence diversity in libraries, such as error-prone PCR (Bloom et al., 2005; Brandsen, Mattheisen, Noel, & Fields, 2018), site-saturation mutagenesis (Öling et al., 2018), homologous recombination (Drummond, Silberg, Meyer, Wilke, & Arnold, 2005), structure-guided recombination (Ho, Adler, Torre, Silberg, & Suh, 2013; Meyer et al., 2003; Otey et al., 2004), and computational design (Dou et al., 2018; Thompson, Bashor, Lim, & Keating, 2012). By altering protein primary structure while conserving protein topology, these approaches have successfully tuned protein substrate specificities, catalytic activities, and stabilities (Arnold, 2009; Dougherty & Arnold, 2009; Renata, Wang, & Arnold, 2015). While these mutagenesis methods have been successfully used to improve preexisting properties in proteins, they are limited in their potential to create new allosteric properties because they are typically designed to conserve protein topologies.

2. Overcoming component limitations with transposon mutagenesis

In cases where there is a need to control cellular processes on fast timescales, natural proteins can be engineered to display switch-like behaviors through topological modifications. For example, circular permutation has been used to create proteins with altered contact order that are inactive upon expression until they are cut into a pair of fragments by a protease, at which point the resulting fragments associate and cooperatively function (Mitrea, Parsons, & Loh, 2010; Plainkum, Fuchs, Wiyakrutta, & Raines, 2003). Protein fission and fusion can also be used to endow natural proteins with chemical-dependent activities. This latter approach has been achieved by identifying pairs of protein fragments that do not associate and function unless they are fused to a pair of proteins whose interaction is stabilized by chemical binding (Hoff et al., 2009; Pelletier, Campbell-Valois, & Michnick, 1998; Pu, Zinkus-Boltz, & Dickinson, 2017). This approach was recently leveraged to create a split protein sensor that links aggregation of the fused protein to the activity of a transcriptional repressor (Zeng et al., 2018). Protein fragments have also been fused to the termini of ligand-binding domains, such that the activity of the split protein is dependent upon ligand binding. This latter domain insertion approach has been used to achieve post-translational control over fluorescent reporters (Baird, Zacharias, & Tsien, 1999), transcriptional regulation (Younger et al., 2018), antibiotic degradation (Guntas, Mansell, Kim, & Ostermeier, 2005), gene editing (Oakes et al., 2016), metabolic labeling of proteins with non-natural amino acids (Thomas, Pandey, Knudsen, Ball, & Silberg, 2017), and electron transfer (Atkinson et al., 2019).

In vitro transposon mutagenesis (Haapa, Taira, Heikkinen, & Savilahti, 1999) represents a simple strategy to create libraries of vectors that express proteins with altered topologies (Fig. 1). With this approach, a transposase randomly inserts a synthetic transposon into a vector containing a gene of interest, and the product of this reaction is subjected to molecular biology manipulations to create a library of expression vectors with well-defined sequence diversity. This design strategy was initially leveraged to create

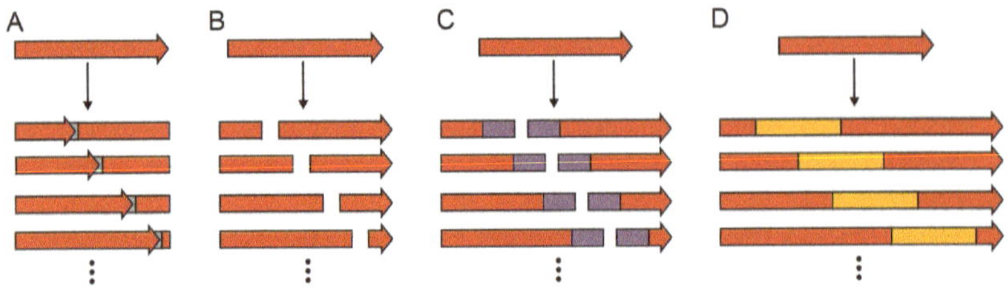

Fig. 1 Topological mutations sampled using transposon mutagenesis. Libraries of vectors can be created that encode different topological variants of a native gene (red), including: (A) circularly permuted genes where the first and last codons of a native gene have been fused through a linker (gray), (B) a pair of gene fragments where two distinct open reading frames are created through fission of a native gene, (C) a pair of gene fragments in which each fragment is fused to genes encoding different proteins (purple), and (D) a pair of gene fragments that are fused to ends of a gene encoding a different protein (yellow), such as a ligand-binding protein.

libraries of vectors that express proteins with pentapeptide insertions (Hoeller, Reiter, Abad, Graze, & Glieder, 2008; Poussu, Vihinen, Paulin, & Savilahti, 2004), hexahistidine insertions (Hoeller et al., 2008; Koerber, Jang, Yu, Kane, & Schaffer, 2007), deletions (Jones, 2005), truncations (Poussu, Jäntti, & Savilahti, 2005), amino acid substitutions (Baldwin, Busse, mm, & Jones, 2008), and non-natural amino acid substitutions (Daggett, Layer, & Cropp, 2009). More recently, transposon mutagenesis has been used to create vector libraries that express proteins with altered topologies, such as randomly split proteins (Segall-Shapiro, Meyer, Ellington, Sontag, & Voigt, 2014; Segall-Shapiro et al., 2011), split proteins fused to pairs of proteins that associate (Mahdavi et al., 2013; Pandey, Nobles, Zechiedrich, Maresso, & Silberg, 2015; Zeng et al., 2018), split proteins fused to the termini of ligand-binding domains (Edwards, Busse, Allemann, & Jones, 2008; Nadler, Morgan, Flamholz, Kortright, & Savage, 2016; Oakes et al., 2016; Thomas et al., 2017), and circularly permuted proteins (Jones et al., 2016; Mehta, Liu, & Silberg, 2012; Pandey et al., 2016). In contrast to early DNAse methods developed for creating libraries of proteins with topological changes (Graf & Schachman, 1996; Hennecke, Sebbel, & Glockshuber, 1999), which are enriched in variants with deletions and insertions of varying lengths, transposon-based methods create well-defined sequence diversity that avoid deletions. This library feature allows for facile analysis using deep mutational scanning (Atkinson, Jones, Zhou, & Silberg, 2018; Higgins & Savage, 2017; Nadler et al., 2016; Oakes et al., 2016).

3. Overview of the library construction workflow

Herein, we describe how transposon mutagenesis can create topological changes in proteins. We highlight how the use of different synthetic transposons in these reactions enables the creation of libraries of vectors that contain genes encoding proteins with different topological changes (Fig. 1), including circularly permuted proteins, split proteins, split proteins fused to other proteins, and split proteins fused to the ends of ligand-binding domains. Finally, we describe how these libraries can be subjected to deep mutational scanning before and after functional analysis to comprehensively characterize protein tolerance to topological changes.

Regardless of the design goal, the first step of all transposon mutagenesis methods involves the insertion of a synthetic transposon into a gene of interest using MuA transposase (Fig. 2). When creating libraries of vectors that encode different topological mutations, slightly different protocols are required. Some methods require that the target gene be encoded within an expression vector, while others require that this gene be circularized. Additionally, some methods yield the desired library in a single step, while others require further manipulation through standard molecular biology

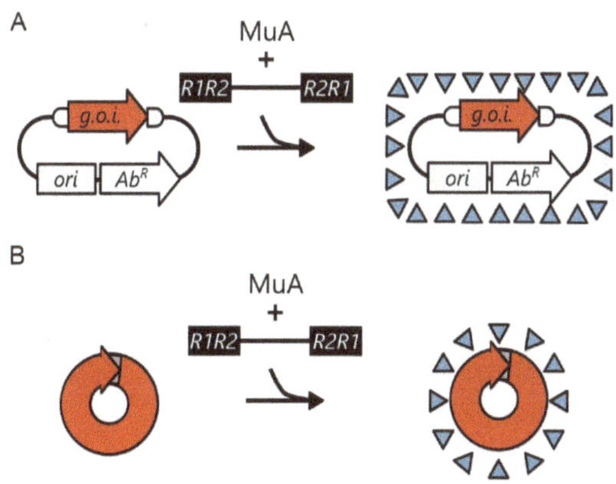

Fig. 2 Transposon insertions occur randomly in the target DNA. Transposition reactions are performed in vitro by mixing an artificial transposon and MuA transposase with (A) a plasmid containing the gene of interest (g.o.i.) or (B) a circularized gene. MuA binds the transposase recognition sequences (R1R2 and R2R1) within the synthetic transposon and randomly inserts it at different locations (triangles) in the target DNA. When performing this reaction with vectors, only transposon insertions within the g.o.i. are desired. All other products of the MuA reaction yield undesired products because they leave the target gene unmodified.

methods, such as replacement of the synthetic transposon with DNA that encodes desired regulatory elements and/or genes that become fused to the target gene. In all cases, careful attention must be paid to maintain stop codons in the correct frame. MuA-mediated DNA insertion results in a 5 base pair duplication at the insertion site, which can affect the relative frame of the open reading frames and the stop codons (Haapa-Paananen, Rita, & Savilahti, 2001).

4. Choosing an artificial transposon

Three classes of synthetic transposons have been reported for building libraries that encode proteins with topological changes. We only discuss previously described transposons below. However, these can be easily modified as long as the transposase recognition sites (R1R2 and R2R1) remain functional within the synthetic transposons. Table 1 provides transposase recognition site sequences that have been shown to function with the transposase MuA.

4.1 Miniature transposons that insert a unique restriction site

Libraries of vectors expressing split proteins and domain-inserted proteins were initially generated using miniature transposons (Edwards et al., 2008; Segall-Shapiro et al., 2011). These synthetic linear DNA sequences contain a selectable cassette, transposase recognition sequences (R1R2 and R2R1), and unique restriction enzyme recognition sequences on the periphery of the R1 sites, such as those cleaved by the Type IIP restriction enzyme NotI (Fig. 3A) or the Golden Gate compatible Type IIS restriction enzyme BsaI (Fig. 3B). When this type of transposon is inserted into a plasmid, it creates an ensemble of vectors that contain the transposon inserted at different sites. Since the transposon is flanked on both sides by the same restriction site, digestion of that site yields sticky ends that can be ligated to compatible synthetic DNA, provided that the plasmid used in the original reaction lacks the restriction sites. A variety of synthetic DNAs have been developed for subcloning in place of the miniature transposon. These include synthetic DNAs that encode regulatory elements required to express split proteins (Fig. 3C), split proteins fused to a pair of proteins (Fig. 3D), split proteins fused to a single protein (Fig. 3E), and split proteins fused to the N- and C-termini of a domain (Fig. 3F).

Table 1 Transposase recognition sequences that can serve as MuA substrates.

Transposon	5′	R1 (5′)	R1 (3′)	3′	Availability
WT Mu	TGAAG	CGGCGCACGAAAAACGCGAAAG	CTTTCGCGTTTTTCGTGCGCCG	CTTCA	–
M1-CamR (NotI)	TGCGG	CCGCGCACGAAAAACGCGAAAG	CTTTCGCGTTTTTCGTGCGCGG	CCGCA	ThermoFisher F-760
Permuteposon P1	TTGAG	CGGCGCACGAAAAACGCGAAAG	CTTTCGCGTTTTTCGTGCGCCG	CTCAA	Addgene #120863
Permuteposon P2	TGAAG	CGGCGCACGAAAAACGCGAAAG	CTTTCGCATTTTGAGTGAGGTA	TATGA	Addgene #120864
Permuteposon P3	TGATT	GATTGAACGAAAAACGCGAAAG	CTTTCGCATTTTGAGTGAATTA	TATGA	Addgene #120865
Splitposon	TGATT	GATTGAACGAAAAACGCGAAAG	CTTTCGCATTTTGAGTGAGGTA	TATGA	Addgene #59947
Mu-BsaI	TGCAT	CGGAGACCGAAAAACGCGAAAG	CTTTCGCGTTTTTCGGTCTCCG	CGTCA	Addgene #79769

Synthetic transposons harbor different mutations within the R1 recognition sites. These mutations generate restriction enzyme sites (M1-CamR and Mu-BsaI), stop codons (permuteposon P3 and the splitposon), or ribosomal binding sites (permuteposons P2, P3, and the splitposon). Highlighted nucleotides are mutations relative to wildtype (WT). The Mu transposase recognition sequence is shown as a frame of reference. Bolded nucleotides represent restriction enzyme sites introduced into the recognition sequences through mutation of WT. Underlined nucleotides represent stop (5′) or start codons (3′). Italicized nucleotides represent 4 bp overhangs used for downstream Golden Gate ligations (Engler, Kandzia, & Marillonnet, 2008). All of the transposons have wildtype R2 sequences. In the case of the R2 that follows the R1(5′), the sequence is CGTTTCACGATAAATGCGAAAA. In the case of the R2 that precedes the R1(3′), the sequence is TTTTCGCATTTATCGTGAAACG.

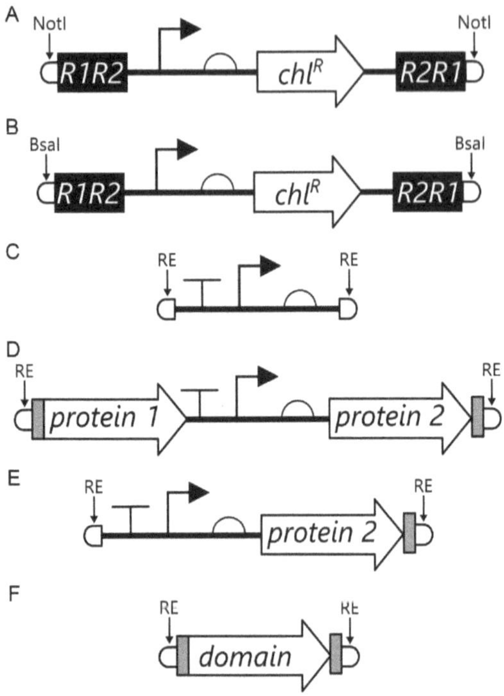

Fig. 3 Transposons that insert unique restriction sites. (A, B) Miniature transposons contain a chloramphenicol resistance cassette (Chl^R) and transposase recognition sequences (R1R2 and R2R1) flanked by a unique restriction site. If this restriction site is absent from the target vector and g.o.i., then the small artificial transposons can be rapidly removed from the product of the insertion reaction through restriction digestion and replaced with synthetic DNA encoding regulatory elements, peptides, or domains. Previously described synthetic DNA include inserts encoding: (C) regulatory elements required to express a split protein without protein fusions (Mahdavi et al., 2013; Segall-Shapiro et al., 2011), (D) a pair of genes with linkers (gray) that become fused to the fragments of the split gene (Pandey et al., 2015; Thomas et al., 2017), (E) a single gene that becomes fused to only one of the fragmented genes (Pandey et al., 2015; Thomas et al., 2017), and (F) a protein domain (Nadler et al., 2016; Oakes et al., 2016; Thomas et al., 2017).

4.2 Splitposons

This synthetic transposon also creates vectors that express different split variants of proteins (Segall-Shapiro et al., 2014). Similar to a miniature transposon, the splitposon contains a selection cassette and transposase recognition sequences (Fig. 4). This synthetic transposon also contains a conditional promoter, a ligand-dependent transcriptional regulator that controls expression from that promoter, and a modified R2R1 that has been mutated to contain a ribosome-binding site (RBS). This hybrid

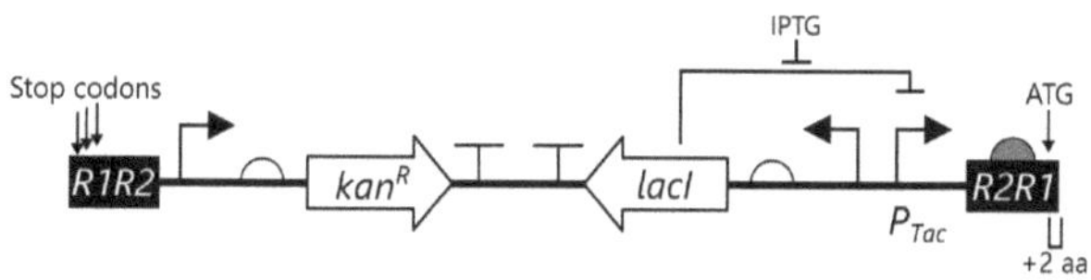

Fig. 4 Splitposon for constructing libraries of fragmented proteins. This synthetic transposon contains a kanamycin resistance cassette (*kan*^{R}), a gene encoding a ligand-dependent transcriptional regulator (*lacI*) that controls transcription from an internal promoter (P_{Tac}), and transposon binding sites (R1R2 and R2R1). The R2R1 site is mutated to contain an RBS (green). This hybrid transposase/ribosome binding site minimizes the number residues added to the N-terminus of the second protein fragment that is under control of P_{Tac} (Segall-Shapiro et al., 2014).

transposase/ribosome binding site, which ultimately controls expression of the C-terminal protein fragment, minimizes the number of residues that are amended to the protein fragment.

4.3 Permuteposons

These synthetic transposons were developed to enable a one-pot synthesis of vectors that express different circularly permuted variants of proteins. When inserted into a circularized gene of interest, permuteposons generate an ensemble of expression vectors since they contain all of the attributes of an expression vector, including a selection cassette, origin of replication, and regulatory elements required to transcribe and translate the circularly permuted genes. Several permuteposons have been developed (Jones et al., 2016; Mehta et al., 2012; Pandey et al., 2016). Some generate a polycistronic mRNA that encodes the selectable marker and the permuted genes (e.g., permuteposons P1, P2, and P3), while others generate a unique mRNA encoding the permuted genes (e.g., P4). The first permuteposon reported (P1) uses an RBS that is located between the selectable cassette and R2R1 to initiate translation of the permuted proteins (Fig. 5A), while the others permuteposons (P2, P3, and P4) use a RBS that is embedded within the transposase binding site (Fig. 5B–D), like the splitposon. P1 adds an 18 amino acid peptide to the N-terminus of all permuted proteins that are expressed (Mehta et al., 2012); the others add 2 amino acids due to the 5 base pair duplication of the MuA insertion site (Jones et al., 2016; Pandey et al., 2016). The advantage of the P1 permuteposon is that it maintains the RBS used to initiate translation in a similar genetic context and yields consistent translation initiation rates across all variants in a library. In contrast, the other permuteposons yield libraries where the circularly permuted protein variants do not experience the same translation initiation rates because of this variability in the genetic context of

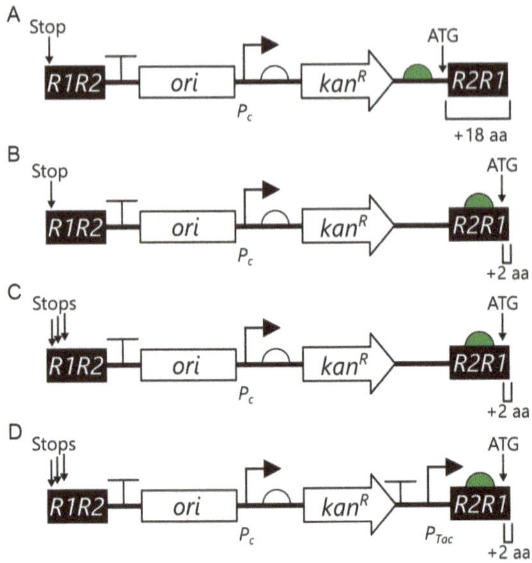

Fig. 5 Permuteposons for building libraries of circularly permuted proteins. (A) Permuteposon P1 contains an origin of replication (ori), a kanamycin resistance cassette (*kan*^{R}) that is constitutively expressed using a promoter (P_c), a transcriptional terminator, transposase recognition sequences (R1R2 and R2R1) at the 5′ and 3′ ends (Mehta et al., 2012), and a stop codon to terminate translation of in frame permuted proteins. With P1, the RBS (green) and start codon used to initiate translation of the permuted proteins adds an 18 amino acid peptide to the N-terminus of each permuted variant that is encoded by the R2R1 site. (B) Permuteposon P2 contains a hybrid transposase/ribosome binding site to minimize the number of residues fused to the N-terminus of each permuted variant (Jones et al., 2016). (C) Permuteposon P3 contains a hybrid transposase/ribosome binding site with a different RBS strength from P2. Additionally, P3 includes cascading stop codons that terminate translation occurring in the +1 and −1 reading frames in addition to the in frame stop codon (Jones et al., 2016). (D) Permuteposon P4 uses the same hybrid R2R1 sequence as P2 but includes a terminator and a LacI regulated promoter, P_{Tac}, to control expression of permuted variants (Pandey et al., 2016). P4 also contains cascading stop codons.

the RBS. However, these permuteposons minimize the peptide scar amended to the N-terminus of circularly permuted proteins like splitposons (Jones et al., 2016), which may be desired for certain applications.

5. Library construction

To generate libraries of vectors that express proteins with topological changes, the following materials are required: (1) target DNA encoding the gene of interest, (2) linear synthetic transposon, (3) MuA transposase,

(4) MuA buffer, (5) library-quality electrocompetent *Escherichia coli*, (6) DNA Clean and Concentrator Kit, (7) LB medium and LB-agar plates containing antibiotics, and (8) DNA Miniprep Kit. The first step for generating libraries requires incubating target DNA (500 ng), a synthetic transposon (100 ng), and MuA (0.22 μg; Thermo Fischer F-750) in a 20 μL reaction containing 1× MuA buffer (25 m*M* Tris-HCl pH 8.0 at 25 °C, 10 m*M* $MgCl_2$, 110 m*M* NaCl, 0.05% Triton X-100, and 10% glycerol) as previously outlined for circular permutation (Jones, Atkinson, & Silberg, 2017). This mixture is incubated at 30 °C for 16 h before heat inactivating the MuA at 75°C for 10 min. Following heat inactivation, the reaction is desalted using a DNA Clean and Concentrator Kit. The DNA mixture is then electroporated into *E. coli* cells, such as MegaX DH10B T1R Electrocompetent Cells (Thermo Fisher Scientific), which can yield $>3 \times 10^{10}$ cfu/μg of plasmid DNA. Transformed cells are plated on multiple LB-agar plates that select for the antibiotic resistance encoded by the synthetic transposon, and the plates are incubated overnight at 37 °C. The colony forming units (cfu) on plates are counted, and all colonies are harvested from plates by adding 1 mL LB to the surface of each plate, scraping colonies into a slurry with a sterile spreader, and transferring the cell slurry from all plates into a single 50 mL sterile tube. This cell slurry is mixed, and the "initial vector library" is purified from the cell slurry using a DNA Miniprep Kit (400 μL of cell slurry/DNA Miniprep column). For each design goal listed below, we describe the target DNA, the synthetic transposon, and any additional steps that are required to synthesize a library of expression vectors. Note that if the desired colony counts are not obtained from the initial reaction, then the MuA reaction can be scaled up prior to the electroporation step.

5.1 Split protein libraries created using miniature transposons

When using miniature transposons for library construction (Fig. 6A), the initial vector library created by the MuA reaction requires further manipulation to yield expression vectors. The first manipulation requires digesting the initial library with a restriction enzyme that cuts adjacent to the gene of interest, separating the genes containing inserted transposons from native genes using agarose gel electrophoresis, subcloning the purified gene-transposon hybrids back into fresh vector backbone, transforming the product of the ligation reaction into electrocompetent cells, and plating the transformed cells onto LB-agar containing antibiotic that selects for the vector. After an overnight incubation, the colonies on plates are

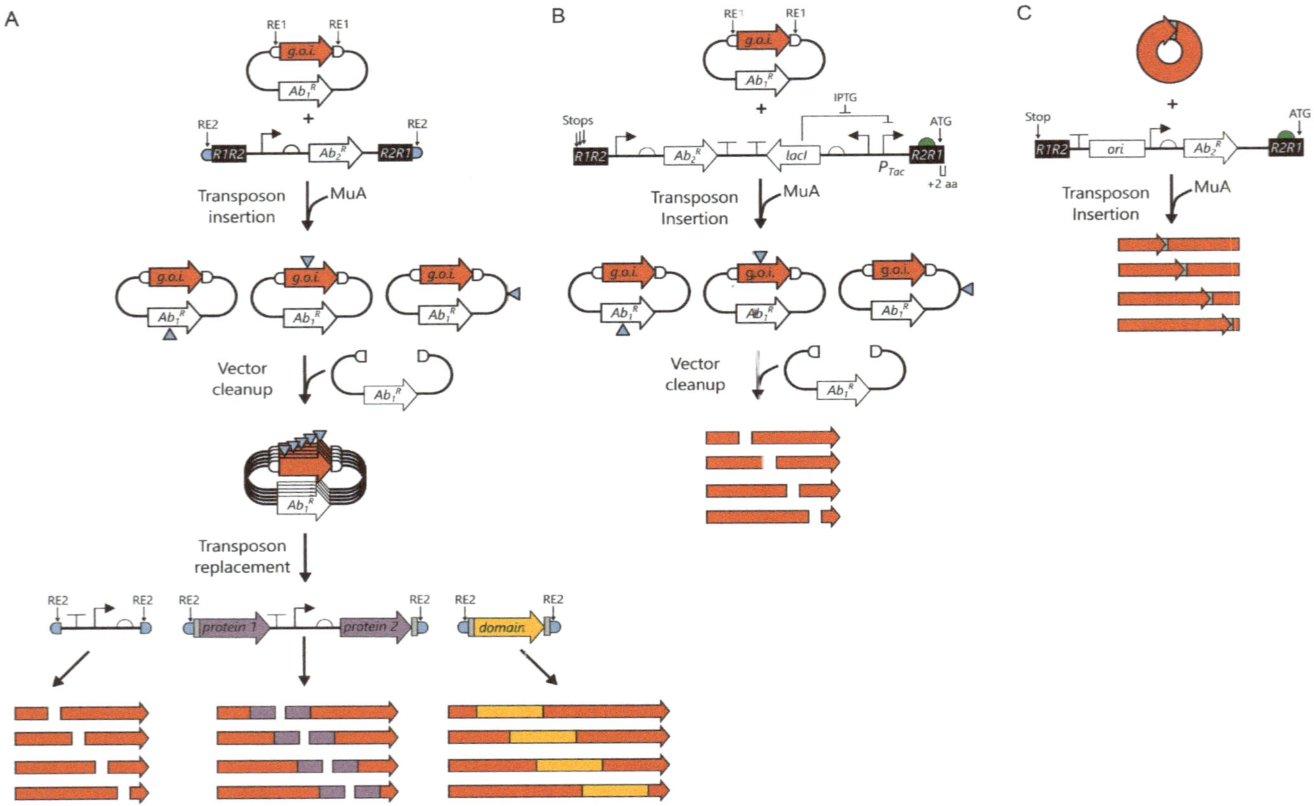

Fig. 6 See legend on opposite page.

harvested as described for the initial vector library, and the resulting vectors are purified from the cell slurry using a DNA Miniprep Kit. This results in a "staged vector library" that has unique restriction sites inserted at different locations within the gene of interest. In contrast to the initial vector library, the staged vector library lacks plasmids containing native genes that are unmodified and vectors having the transposon inserted in the vector backbone.

Once the staged vector library is purified, a variety of split protein libraries can be generated in parallel using this vector ensemble. For all design goals, the synthetic transposons are removed from the staged vector library using restriction enzyme digestion, the linearized vectors lacking the transposons are separated from the transposons using agarose gel electrophoresis, the linear vector ensemble is purified using a DNA Gel Purification Kit, and synthetic DNA is ligated to the vector ensemble to create a final library. The synthetic DNAs for creating split protein libraries are shown in Fig. 3C-E. These yield different kinds of libraries, including libraries of vectors that express split proteins lacking protein fusions (Fig. 3C), split proteins having

Fig. 6 Workflow for creating libraries that express proteins with different classes of topological mutations. (A) When creating vectors that express split proteins or domain-inserted proteins, three steps are required (Pandey et al., 2015; Segall-Shapiro et al., 2011; Thomas et al., 2017). First, transposition reactions are performed in vitro by mixing purified plasmids containing the g.o.i., a miniature transposon, MuA transposase, and MuA buffer. MuA inserts the transposon (triangles) within the gene of interest at different locations and throughout the plasmid. Typically, MuA insertion creates plasmids containing a single inserted transposon. Second, the product of the reaction is digested with a restriction enzyme (RE1) that cuts adjacent to the gene of interest. This yields genes containing or lacking a synthetic transposon, and vector backbone containing or lacking the synthetic transposon. The gene-transposon hybrids are purified and ligated into fresh expression vector to yield a staging library through this vector cleanup process. Third, the synthetic transposons are removed by restriction digestion with a second restriction enzyme (RE2), and a synthetic DNA is subcloned in place of the transposon. By varying the DNA inserted at this step, different topological mutations can be generated. (B) When using splitposons to generate vectors expressing split proteins, only two steps are required for library construction (Segall-Shapiro et al., 2014). First, transposition reactions are performed in vitro by mixing purified plasmids containing a g.o.i., a splitposon, MuA transposase, and MuA buffer. Second, the product of the reaction is digested with a restriction enzyme (RE1) that cuts adjacent to the gene of interest. The purified gene-transposon hybrids are then ligated into fresh expression vector. (C) When using permuteposons to create vectors that express circularly permuted proteins, only one step is required for library construction. Transposition reactions are performed in vitro by mixing a circularized g.o.i., a permuteposon, MuA transposase, and MuA buffer (Atkinson et al., 2018).

the fragments fused to a pair of proteins (Fig. 3D), and split proteins having only the second fragment fused to a protein (Fig. 3E). During the design of these types of synthetic DNA, it is critical to remove stop codons from the genes being fused to the split genes so that open reading frames are generated that express the protein fragments fused to the desired proteins.

5.2 Domain insertion libraries created using miniature transposons

The protocol for creating libraries that express domain inserted proteins is identical to the protocol described for creating split protein libraries with one exception (Fig. 6A). During the last step of library construction, a synthetic DNA encoding a gene that lacks a stop codon (Fig. 3F) is subcloned in place of the synthetic transposon in the staged vector library. It is critical that the synthetic DNA be designed so that the open reading frames encoding the fragments of the split gene are in frame with the inserted gene encoding the domain.

5.3 Split protein libraries created using splitposons

The major benefit of generating libraries with splitposons is that the initial library created by the MuA reaction requires only one processing step to generate the final library (Fig. 6B). With this approach, the initial library contains vectors: (1) having the splitposon inserted within the gene of interest, (2) having the splitposon inserted at other locations, and (3) lacking inserted splitposon. To generate the final library, the gene of interest containing a splitposon inserted in different locations must be subcloned into a fresh vector backbone. This is achieved by digesting the initial vector library with a restriction enzyme that cuts adjacent to the gene of interest, separating this gene-splitposon hybrid from the other digestion products using agarose gel electrophoresis, gel purifying the gene-splitposon hybrids, and subcloning the purified DNA into fresh vector backbone.

5.4 Circularly permuted protein libraries created using permuteposons

Libraries that express circularly permuted variants of a protein can be created in a single step as shown in Fig. 6C. When creating this type of library, the target gene must initially be circularized through ligation prior to mixing with MuA and a permuteposon. The circular gene must lack a stop codon, and the first and last codons in the gene must be linked with DNA consisting of in frame codons that encode for a linker peptide.

6. Sampling considerations when screening versus selecting

Libraries generated by transposon mutagenesis create well-defined sequence diversity that is two times larger than the length of the target gene. This occurs because synthetic transposons are inserted in two orientations. Among the variants created by transposon mutagenesis, only one-sixth of the variants express the desired protein variants with topological changes. This occurs because transposons are randomly inserted in all three frames within the gene of interest and in the two orientations (Fig. 7). The transposon insertion frequency can vary by many orders of magnitude, since MuA does not insert transposons with uniform efficiency across different DNA sequences (Haapa-Paananen et al., 2001). A recent study examined

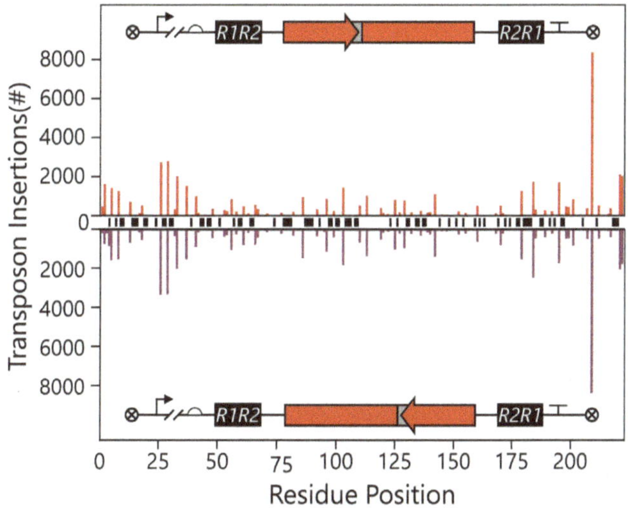

Fig. 7 Positional biases are observed with transposon mutagenesis. Sequencing data from an unselected library of permuted adenylate kinase genes generated by mixing a circular g.o.i. with permuteposon P1 and MuA. When counting insertions in both the forward (red; top) and reverse (purple; bottom) orientations, we observed that a single variant comprised 7% of the total library. Black bars in the center represent the subset of the 223 positions ($n = 64$) that were unsampled in one or both orientations. A total of 114,754 in frame insertions were observed in this library that would theoretically have sampled both orientations in every position of the protein over 100 times. However, due to the bias of MuA, 29% of the positions went unsampled. *Replotted from Atkinson, J. T., Jones, A. M., Zhou, Q., & Silberg, J. J. (2018). Circular permutation profiling by deep sequencing libraries created using transposon mutagenesis.* Nucleic Acids Research, 46*(13), e76.*

the sequence bias when constructing a library of circularly permuted adenylate kinases using a permuteposon (Atkinson et al., 2018). As shown in Fig. 7, this study observed that abundance of individual variants could by vary from 0 to 8377 copies (Atkinson et al., 2018). Even though this library had this level of bias, >70% of the possible vectors that express the different circularly permuted protein variants could be sampled using a functional selection (Atkinson et al., 2018). The bias that occurs with transposase reactions also requires that the initial library construction steps achieve large colony counts from transformations to avoid loss of sequence diversity during the library construction process. The sequence bias observed with libraries is primarily positional. Transposons inserted in the forward and reverse orientations occur with similar frequencies at the same insertion location (Atkinson et al., 2018).

A wide range of methods can be used to mine libraries for functional protein variants (Fig. 8). While low throughput methods such as colony screening on agar plates (or 96-well plates) are limited in their throughput, these have been successfully used to discover proteins with topological mutations that display the desired functional properties (Pandey et al., 2016, 2015). Higher throughput functional assays, such as flow cytometry and bacterial selections, are required when the design goal is to generate a comprehensive profile of a protein's functional tolerance to a topological mutation (Atkinson et al., 2018; Nadler et al., 2016; Oakes et al., 2016).

7. Profiling protein tolerance to topological changes

One of the great advantages of creating libraries through transposon mutagenesis is that this approach generates well-defined sequence diversity that can be analyzed using deep mutational scanning (Atkinson et al., 2018; Nadler et al., 2016; Oakes et al., 2016). Recent studies have illustrated how deep mutational scanning of libraries before and after functional analysis can provide insight into a protein's tolerance to topological changes (Atkinson et al., 2018; Nadler et al., 2016; Oakes et al., 2016). These studies, which are named domain-insertion profiling with DNA sequencing (DIP-seq) and circular permutation profiling with DNA sequencing (CPP-seq), generate profiles that map topological changes that yield functional proteins onto the primary structure of the protein encoded by the target gene. These profiles allow comparison of the sequence abundance of each variant to the

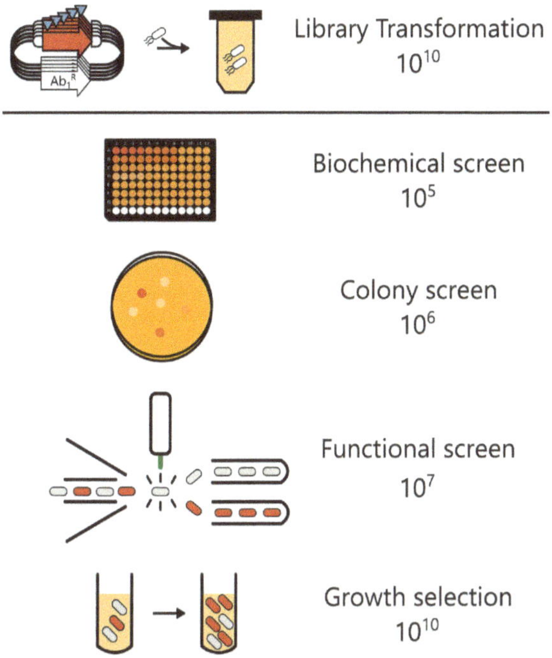

Fig. 8 Strategies for mining libraries for functional protein components. Plasmid libraries created using transposon mutagenesis should be transformed into cells using library-grade competent cells with transformation efficiencies $>10^{10}$ cfu/μg. A variety of strategies can be used to link protein fitness to sequence. Each strategy has a different practical throughput that sets the limit for its ability to comprehensively sample transposon-generated libraries, especially when compounded with the transposon insertion bias (Fig. 7). Biochemical screens of purified proteins are the lowest throughput technique requiring expression, purification, and in vitro activity assay. Screening individual colonies from agar plates (or 96-well plates) using a visual output (e.g., colorimetric substrate or protein) allows for about an order of magnitude increase in throughput. Coupling protein activity to a functional screen in liquid media where cell sorting based on fluorescence can further enhance screening rates. Finally, linking protein activity to cell fitness through a growth selection can greatly increase the number of variants analyzed. These final two strategies are best for generating mutational profiles through deep sequencing as shown in Fig. 9.

abundance of variants that cannot express functional proteins before and after functional enrichment (Fig. 9A). For each variant encoded in the library, the ratio of sequence reads after and before the functional enrichment is calculated and compared to the ratios obtained for vectors that cannot express functional proteins (Fig. 9B). This comparison is used to determine which topological mutants are biologically active. Functional

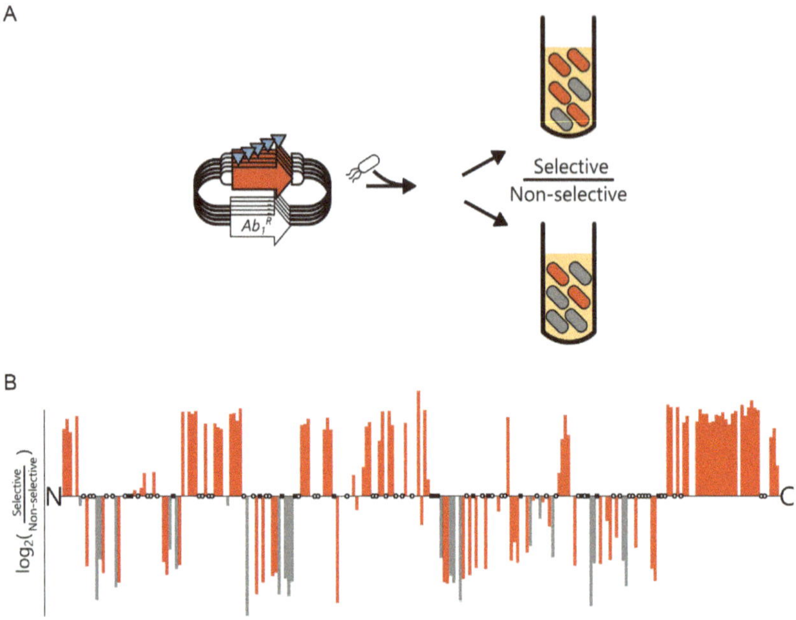

Fig. 9 A profile illustrating protein tolerance to a topological mutation. (A) A plasmid library is subjected to deep mutational scanning where cultures are sequenced before and after subjecting the library to a functional enrichment using a growth selection or functional screen with cell sorting. (B) The log_2-fold ratio of the abundances of each variant observed in the selected versus the unselected conditions is calculated for every native position in the primary structure. These data enable an experimenter to identify positions in the protein primary sequence that retain function following the topological change; only functional variants show significant increases in the selected condition when compared to the unselected condition. Variants having the transposon inserted in the reverse orientation (Fig. 7, bottom panel) are used as a frame of reference for plasmids that cannot express functional variants. CPP-seq data of an adenylate kinase subjected to growth selection after permutation by permuteposon P1. Positions that are significantly enriched relative to vectors that do not express a functional protein are red, while non-significant are in gray. Positions that were not represented in the library are noted with a small circle on the *x* axis. *Replotted from a previously reported data set Atkinson, J. T., Jones, A. M., Zhou, Q., & Silberg, J. J. (2018). Circular permutation profiling by deep sequencing libraries created using transposon mutagenesis.* Nucleic Acids Research, 46*(13), e76.*

mutants are defined as those having sequence ratios that are significantly enriched over the ratios of those variants in the library that cannot express functional proteins. Previous studies have used two different statistical methods for this analysis, including DESeq and Fisher's Exact Test (Atkinson et al., 2018; Nadler et al., 2016; Oakes et al., 2016).

8. Conclusions

Transposon-mediated protein engineering can be used to create combinatorial libraries that are rich in folded and functional proteins with topological mutations, including circularly permuted proteins, split proteins, and domain-inserted proteins. Libraries encoding each of these topological mutations can be rapidly generated with virtually any protein using this methodology. Because transposon mutagenesis libraries are rapid to generate, and they are frequently enriched in proteins with novel switching behaviors, they represent an excellent way to design protein components for synthetic biology applications that require fast, post-translational regulation. Additionally, the libraries generated by transposon mutagenesis contain well-constrained sequence diversity. This feature allows for their analysis through deep mutational scanning before and after functional enrichment and the generation of profiles that comprehensively map a protein's functional tolerance to topological mutations to its primary structure. To date, this approach has only been applied to a handful of proteins. In future studies that apply these methods to a wider range of proteins, we will certainly extend our understanding of the relationship between topological mutations and the evolution of protein allostery.

Acknowledgments

This work was supported by the Office of Naval Research (N00014-17-1-2639), Department of Energy (DE-SC0014462), the National Science Foundation (1805317 and 1615562), and the Welch Foundation (C-1824). J.T.A. was supported by the National Science Foundation Graduate Research Fellowship Program under grant number (R3E821) and Department of Energy Office of Science Graduate Student Research Program.

References

Arnold, F. H. (2009). How proteins adapt: Lessons from directed evolution. *Cold Spring Harbor Symposia on Quantitative Biology*, *74*, 41–46.

Atkinson, J. T., Campbell, I. J., Thomas, E. E., Bonitatibus, S. C., Elliot, S. J., Bennett, G. N., et al. (2019). Metalloprotein switches that display chemical-dependent electron transfer in cells. *Nature Chemical Biology*, *15*(2), 189–195.

Atkinson, J. T., Jones, A. M., Zhou, Q., & Silberg, J. J. (2018). Circular permutation profiling by deep sequencing libraries created using transposon mutagenesis. *Nucleic Acids Research*, *46*(13), e76.

Baird, G., Zacharias, D., & Tsien, R. (1999). Circular permutation and receptor insertion within green fluorescent proteins. *Proceedings of the National Academy of Sciences of the United States of America*, *96*(20), 11241–11246.

Baldwin, A. J., Busse, K., mm, A., & Jones, D. D. (2008). Expanded molecular diversity generation during directed evolution by trinucleotide exchange (TriNEx). *Nucleic Acids Research*, *36*(13), e77.

Bloom, J. D., Silberg, J. J., Wilke, C. O., Drummond, A. D., Adami, C., & Arnold, F. H. (2005). Thermodynamic prediction of protein neutrality. *Proceedings of the National Academy of Sciences of the United States of America*, *102*(3), 606–611.

Brandsen, B. M., Mattheisen, J. M., Noel, T., & Fields, S. (2018). A biosensor strategy for *E. coli* based on ligand-dependent stabilization. *ACS Synthetic Biology*, 7(9), 1990–1999.

Daggett, K. A., Layer, M., & Cropp, A. T. (2009). A general method for scanning unnatural amino acid mutagenesis. *ACS Chemical Biology*, *4*(2), 109–113.

Dou, J., Vorobieva, A. A., Sheffler, W., Doyle, L. A., Park, H., Bick, M. J., et al. (2018). De novo design of a fluorescence-activating β-barrel. *Nature*, *561*(7724), 485–491.

Dougherty, M. J., & Arnold, F. H. (2009). Directed evolution: New parts and optimized function. *Current Opinion in Biotechnology*, *20*(4), 486–491.

Drummond, A. D., Silberg, J. J., Meyer, M. M., Wilke, C. O., & Arnold, F. H. (2005). On the conservative nature of intragenic recombination. *Proceedings of the National Academy of Sciences of the United States of America*, *102*(15), 5380–5385.

Edwards, W. R., Busse, K., Allemann, R. K., & Jones, D. D. (2008). Linking the functions of unrelated proteins using a novel directed evolution domain insertion method. *Nucleic Acids Research*, *36*(13), e78.

Elowitz, M. B., & Leibler, S. (2000). A synthetic oscillatory network of transcriptional regulators. *Nature*, *403*(6767), 335.

Engler, C., Kandzia, R., & Marillonnet, S. (2008). A one pot, one step, precision cloning method with high throughput capability. *PLoS One*, *3*(11), e3647.

Gao, X. J., Chong, L. S., Kim, M. S., & Elowitz, M. B. (2018). Programmable protein circuits in living cells. *Science*, *361*(6408), 1252–1258.

Gardner, T., Cantor, C., & Collins, J. (2000). Construction of a genetic toggle switch in *Escherichia coli*. *Nature*, *403*(6767), 339–342.

Graf, R., & Schachman, H. (1996). Random circular permutation of genes and expressed polypeptide chains: Application of the method to the catalytic chains of aspartate transcarbamoylase. *Proceedings of the National Academy of Sciences of the United States of America*, *93*(21), 11591–11596.

Guntas, G., Mansell, T. J., Kim, J., & Ostermeier, M. (2005). Directed evolution of protein switches and their application to the creation of ligand-binding proteins. *Proceedings of the National Academy of Sciences of the United States of America*, *102*(32), 11224–11229.

Haapa, S., Taira, S., Heikkinen, E., & Savilahti, H. (1999). An efficient and accurate integration of mini-Mu transposons in vitro: A general methodology for functional genetic analysis and molecular biology applications. *Nucleic Acids Research*, *27*(13), 2777–2784.

Haapa-Paananen, S., Rita, H., & Savilahti, H. (2001). DNA transposition of bacteriophage Mu. A quantitative analysis of target site selection in vitro. *The Journal of Biological Chemistry*, *277*(4), 2843–2851.

Hennecke, J., Sebbel, P., & Glockshuber, R. (1999). Random circular permutation of DsbA reveals segments that are essential for protein folding and stability. *Journal of Molecular Biology*, *286*(4), 1197–1215.

Higgins, S., & Savage, D. F. (2017). Protein science by DNA sequencing: How advances in molecular biology are accelerating biochemistry. *Biochemistry*, *57*(1), 38–46.

Ho, M. L., Adler, B. A., Torre, M. L., Silberg, J. J., & Suh, J. (2013). SCHEMA computational design of virus capsid chimeras: Calibrating how genome packaging, protection, and transduction correlate with calculated structural disruption. *ACS Synthetic Biology*, *2*(12), 724–733.

Hoeller, B. M., Reiter, B., Abad, S., Graze, I., & Glieder, A. (2008). Random tag insertions by transposon integration mediated mutagenesis (TIM). *Journal of Microbiological Methods*, *75*(2), 251–257.

Hoff, K. G., Culler, S. J., Nguyen, P. Q., McGuire, R. M., Silberg, J. J., & Smolke, C. D. (2009). In vivo fluorescent detection of Fe-S clusters coordinated by human GRX2. *Chemistry & Biology*, *16*(12), 1299–1308.

Jones, D. D. (2005). Triplet nucleotide removal at random positions in a target gene: The tolerance of TEM-1 beta-lactamase to an amino acid deletion. *Nucleic Acids Research*, *33*(9), e80.

Jones, A. M., Atkinson, J. T., & Silberg, J. J. (2017). PERMutation using transposase engineering (PERMUTE): A simple approach for constructing circularly permuted protein libraries. *Methods in Molecular Biology (Clifton, N.J.)*, *1498*, 295–308.

Jones, A. M., Mehta, M. M., Thomas, E. E., Atkinson, J. T., Segall-Shapiro, T. H., Liu, S., et al. (2016). The structure of a thermophilic kinase shapes fitness upon random circular permutation. *ACS Synthetic Biology*, *5*(5), 415–425.

Koerber, J. T., Jang, J.-H., Yu, J. H., Kane, R. S., & Schaffer, D. V. (2007). Engineering adeno-associated virus for one-step purification via immobilized metal affinity chromatography. *Human Gene Therapy*, *18*(4), 367–378.

Mahdavi, A., Segall-Shapiro, T. H., Kou, S., Jindal, G. A., Hoff, K. G., Liu, S., et al. (2013). A genetically encoded AND gate for cell-targeted metabolic labeling of proteins. *Journal of the American Chemical Society*, *135*(8), 2979–2982.

Mehta, M. M., Liu, S., & Silberg, J. J. (2012). A transposase strategy for creating libraries of circularly permuted proteins. *Nucleic Acids Research*, *40*(9), e71.

Meyer, M. M., Silberg, J. J., Voigt, C. A., Endelman, J. B., Mayo, S. L., Wang, Z.-G., et al. (2003). Library analysis of SCHEMA-guided protein recombination. *Protein Science*, *12*(8), 1686–1693.

Mitrea, D. M., Parsons, L. S., & Loh, S. N. (2010). Engineering an artificial zymogen by alternate frame protein folding. *Proceedings of the National Academy of Sciences of the United States of America*, *107*(7), 2824–2829.

Nadler, D. C., Morgan, S.-A., Flamholz, A., Kortright, K. E., & Savage, D. F. (2016). Rapid construction of metabolite biosensors using domain-insertion profiling. *Nature Communications*, 7, 12266.

Oakes, B. L., Nadler, D. C., Flamholz, A., Fellmann, C., Staahl, B. T., Doudna, J. A., et al. (2016). Profiling of engineering hotspots identifies an allosteric CRISPR-Cas9 switch. *Nature Biotechnology*, *34*(6), 646–651.

Öling, D., Lawenius, L., Shaw, W., Clark, S., Kettleborough, R., Ellis, T., et al. (2018). Large scale synthetic site saturation GPCR libraries reveal novel mutations that alter glucose signaling. *ACS Synthetic Biology*, 7(9), 2317–2321.

Otey, C. R., Silberg, J. J., Voigt, C. A., Endelman, J. B., Bandara, G., & Arnold, F. H. (2004). Functional evolution and structural conservation in chimeric cytochromes p450: Calibrating a structure-guided approach. *Chemistry & Biology*, *11*(3), 309–318.

Pandey, N., Kuypers, B. E., Nassif, B., Thomas, E. E., Alnahhas, R. N., Segatori, L., et al. (2016). Tolerance of a knotted near-infrared fluorescent protein to random circular permutation. *Biochemistry*, *55*(27), 3763–3773.

Pandey, N., Nobles, C. L., Zechiedrich, L., Maresso, A. W., & Silberg, J. J. (2015). Combining random gene fission and rational gene fusion to discover near-infrared fluorescent protein fragments that report on protein–protein interactions. *ACS Synthetic Biology*, *4*(5), 615–624.

Pelletier, J., Campbell-Valois, F., & Michnick, S. (1998). Oligomerization domain-directed reassembly of active dihydrofolate reductase from rationally designed fragments. *Proceedings of the National Academy of Sciences of the United States of America*, *95*(21), 12141–12146.

Plainkum, P., Fuchs, S. M., Wiyakrutta, S., & Raines, R. T. (2003). Creation of a zymogen. *Nature Structural Biology*, *10*(2), 115–119.

Poussu, E., Jäntti, J., & Savilahti, H. (2005). A gene truncation strategy generating N- and C-terminal deletion variants of proteins for functional studies: Mapping of the Sec1p binding domain in yeast Mso1p by a Mu in vitro transposition-based approach. *Nucleic Acids Research*, *33*(12), e104.

Poussu, E., Vihinen, M., Paulin, L., & Savilahti, H. (2004). Probing the alpha-complementing domain of *E. coli* beta-galactosidase with use of an insertional pentapeptide mutagenesis strategy based on Mu in vitro DNA transposition. *Proteins*, *54*(4), 681–692.

Pu, J., Zinkus-Boltz, J., & Dickinson, B. C. (2017). Evolution of a split RNA polymerase as a versatile biosensor platform. *Nature Chemical Biology*, *13*(4), 432–438.

Renata, H., Wang, J. Z., & Arnold, F. H. (2015). Expanding the enzyme universe: Accessing non-natural reactions by mechanism-guided directed evolution. *Angewandte Chemie International Edition*, *54*(11), 3351–3367.

Roberta, K. (2010). Five hard truths for synthetic biology. *Nature*, *463*(7279), 288–290.

Segall-Shapiro, T. H., Meyer, A. J., Ellington, A. D., Sontag, E. D., & Voigt, C. A. (2014). A 'resource allocator' for transcription based on a highly fragmented T7 RNA polymerase. *Molecular Systems Biology*, *10*(7), 742.

Segall-Shapiro, T. H., Nguyen, P. Q., Santos, E. D., Subedi, S., Judd, J., Suh, J., et al. (2011). Mesophilic and hyperthermophilic adenylate kinases differ in their tolerance to random fragmentation. *Journal of Molecular Biology*, *406*(1), 135–148.

Tabor, J. J., lis, H., Simpson, Z., Chevalier, A. A., Levskaya, A., Marcotte, E. M., et al. (2009). A synthetic genetic edge detection program. *Cell*, *137*(7), 1272–1281.

Tamsir, A., Tabor, J. J., & Voigt, C. A. (2010). Robust multicellular computing using genetically encoded NOR gates and chemical 'wires'. *Nature*, *469*(7329), 212.

Thomas, E. E., Pandey, N., Knudsen, S., Ball, Z. T., & Silberg, J. J. (2017). Programming post-translational control over the metabolic labeling of cellular proteins with a non-canonical amino acid. *ACS Synthetic Biology*, *6*(8), 1572–1583.

Thompson, K., Bashor, C. J., Lim, W. A., & Keating, A. E. (2012). SYNZIP protein interaction toolbox: In vitro and in vivo specifications of heterospecific coiled-coil interaction domains. *ACS Synthetic Biology*, *1*(4), 118–129.

Van der Helm, E., Genee, H. J., & Sommer, M. O. (2018). The evolving interface between synthetic biology and functional metagenomics. *Nature Chemical Biology*, *14*(8), 752–759.

Younger, A. K., Su, P. Y., Shepard, A. J., Udani, S. V., Cybulski, T. R., Tyo, K. E., et al. (2018). *Development of novel metabolite-responsive transcription factors via transposon-mediated protein fusion, Protein Engineering, Design, and Selection*, *31*(2), 55–63. https://doi.org/10.1093/protein/gzy001.

Zeng, Y., Jones, A. M., Thomas, E. E., Nassif, B., Silberg, J. J., & Segatori, L. (2018). A split transcriptional repressor that links protein solubility to an orthogonal genetic circuit. *ACS Synthetic Biology*, 7(9), 2126–2138.

CHAPTER TWELVE

Chemical biology of glycoproteins: From chemical synthesis to biological impact

Yaohao Li[a,b], Amy H. Tran[b], Samuel J. Danishefsky[c,*], Zhongping Tan[a,b,*]

[a]Institute of Materia Medica, Chinese Academy of Medical Sciences and Peking Union Medical College, Beijing, China
[b]Department of Biochemistry and BioFrontiers Institute, University of Colorado, Boulder, CO, United States
[c]Chemical Biology Program, Sloan Kettering Institute, Memorial Sloan Kettering Cancer Center, New York, NY, United States
*Corresponding authors: e-mail address: danishes@mskcc.org; zhongping.tan@imm.pumc.edu.cn

Contents

Abstract

Recent advances have demonstrated the feasibility and robustness of chemical synthesis for the production of homogeneously glycosylated protein forms (glycoforms). By taking advantage of the unmatchable flexibility and precision provided by chemical synthesis, the quantitative effects of glycosylation were obtained using chemical glycobiology approaches. These findings greatly advanced our fundamental knowledge of glycosylation. More importantly, analysis of these findings has led to the development of glycoengineering guidelines for rationally improving the properties of peptides and proteins. In this chapter, we present the key experimental steps for chemical biology studies of protein glycosylation, with the aim of facilitating and promoting research in this important but significantly underexplored area of biology.

Methods in Enzymology, Volume 621
ISSN 0076-6879
https://doi.org/10.1016/bs.mie.2019.02.030

1. Introduction

Protein glycosylation is the most common form of posttranslational modification (PTM) on excreted and extracellular membrane-associated proteins (Spiro, 2002). It involves the covalent attachment of many different types of glycans (also called carbohydrates, saccharides, or sugars) to a protein. Based on the amino acid side-chain atoms to which glycans are linked, protein glycosylation can be divided into two major categories: N-glycosylation and O-glycosylation. In N-glycosylation, glycans are attached to the side-chain nitrogen atoms of Asn residues in a conserved consensus sequence Asn-Xaa-Ser/Thr (Xaa≠Pro), whereas in O-glycosylation, glycans are attached to the side-chain oxygen atoms of hydroxyl amino acids, primarily serine and threonine residues (Fig. 1).

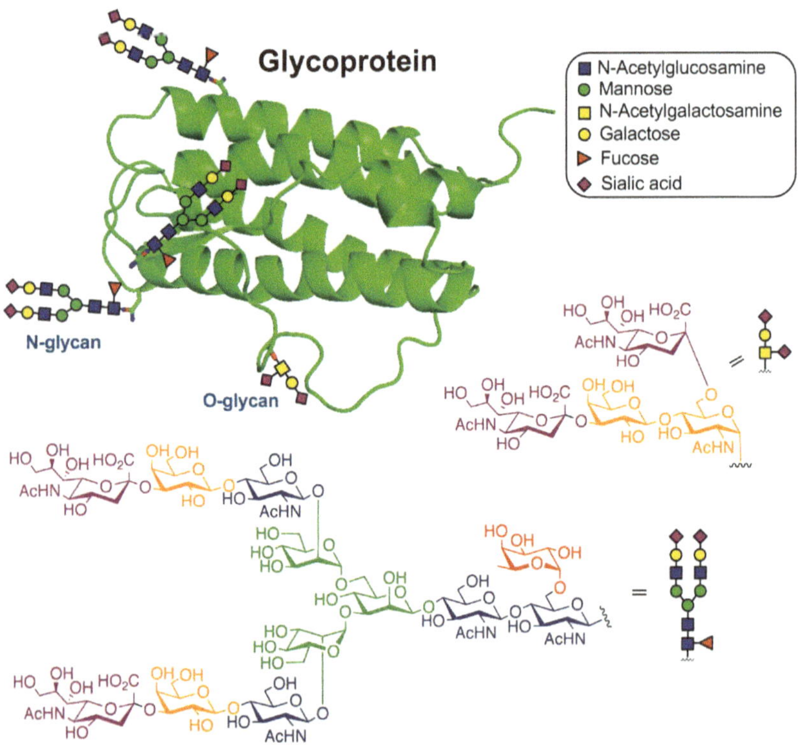

Fig. 1 The structure of a typical glycoprotein, human EPO. The ribbon diagram of the tertiary structure of human EPO protein is shown in green. It contains three N-linked glycosylation sites, N24, N38, and N83, and one O-linked glycosylation site, S126.

Over the past several decades, research from many disciplines has established the importance of protein glycosylation (Varki, 2017). More and more evidence has pointed to the fact that glycosylation can greatly diversify the functions of therapeutically important proteins (Anthony, Wermeling, & Ravetch, 2012) and industrial enzymes (Greene, Himmel, Beckham, & Tan, 2015), possibly achieved by adjusting their physicochemical and/or biological properties (Moremen, Tiemeyer, & Nairn, 2012). Aberrant glycosylation has been implicated as a significant contributor to many human diseases and to reduced (or even completely abolished) enzyme activity (Dwek, 1995; Meany & Chan, 2011). However, despite its significance, protein glycosylation has long remained understudied, and thus underutilized. This lagging behind is mainly due to the lack of efficient strategies to obtain large collections of glycosylated protein forms (glycoforms) with defined, systematically varied structures in a sufficient amount for characterization (Rich & Withers, 2009). The absence of such strategies is a result of the complexity of glycosylation biosynthetic pathways and the inseparable nature of glycoforms (Rudd & Dwek, 1997).

Depending on the exact glycan structure, glycosylation is initiated and proceeds variously at different cellular locations and/or different stages of protein synthesis (Ohtsubo & Marth, 2006). Like most enzymatic processes, it is regulated by many different host and environmental factors, such as conformation at potential glycosylation sites, local amino acid sequence, and availability of activated sugar donors and enzymes. Under the combined influence of these highly variable factors, glycoproteins synthesized by cells usually exist as complex mixtures of up to hundreds of glycoforms (or glycosylated variants, glyco-variants). These variants differ in glycosylation site occupancy and glycan structures (giving rise to macroheterogeneity and microheterogeneity, respectively) and are often impossible to separate from one another. As a result, most previous studies had to rely on mixture samples with unknown compositions (Rudd & Dwek, 1997). Not surprisingly, the uncertainty in sample composition has led to inconsistent and even conflicting conclusions about protein glycosylation, which hampered further progress in this area (Varki, 1993, 2017).

The study of homogenous glycoforms with defined structures is key to drawing strong and valid conclusions about protein glycosylation (Wang et al., 2013). Similar to the studies of other biopolymers like nucleic acids and proteins, the effects of glycosylation are expected to be precisely established by comparing the properties of a series of representative glycoforms with systematically varied glycans (Chaffey, Guan, Li, & Tan, 2018). Since it is currently not possible to separate and isolate individual glycoforms from biological sources, alternative methods must therefore be developed. Many biochemical methods, including glycosylation pathway engineering, glycan remodeling and enzymatic glycosylation, have been explored for the purpose of generating homogeneous glycoforms (Rich & Withers, 2009). However, although these methods are more practical means to prepare large glycoproteins, and to produce glycoproteins on preparative scales, they are severely limited with respect to the generation of large collections of glycoforms with structurally closely related glycan structures. Their intrinsic lack of flexibility thus makes them not suitable to address the issues in protein glycosylation research.

In recent years, chemical synthesis has evolved to become a core method for preparing homogeneous glycoforms (Wilson, Dong, Wang, & Danishefsky, 2013). Unlike biosynthesis, chemical glycosylation is not subject to the restrictions imposed by substrate specificity. Therefore, chemical synthesis is a highly flexible and precise means by which to access glycoforms with diverse structural variations. Such advantages make chemical synthesis an attractive method to help reach a comprehensive and definitive understanding of protein glycosylation. While useful and promising, chemical synthesis also has its drawbacks. Chemical synthesis generally involves labor-intensive, sometimes tedious preparation of starting materials and intermediates. In order to obtain a large collection of homogenous glyco-variants in a reasonable amount of time through chemical synthesis, innovation in the form of synthetic strategy, tactics and method invention is required.

Chemical synthesis of glycoproteins generally involves the ligation of a set of chemically prepared peptide and glycopeptide fragments into a long polyglycopeptide chain, which then folds up to form targeted glycoproteins (Fig. 2). Peptide fragments can be readily prepared using well-developed Solid-Phase Peptide Synthesis (SPPS) methods. The most challenging tasks in the synthesis of glycoproteins are the preparation of glycopeptide fragments and their assembly. In order to achieve highly effective synthesis of glycoproteins, the Danishefsky group has systematically optimized the synthetic strategies and methodologies in these two key areas (Tan, Shang, & Danishefsky, 2011; Wang et al., 2012). The success of this effort paves the way for the production of glycoforms with diverse structural variations and forms the essential basis for developing a molecular-level understanding of protein glycosylation.

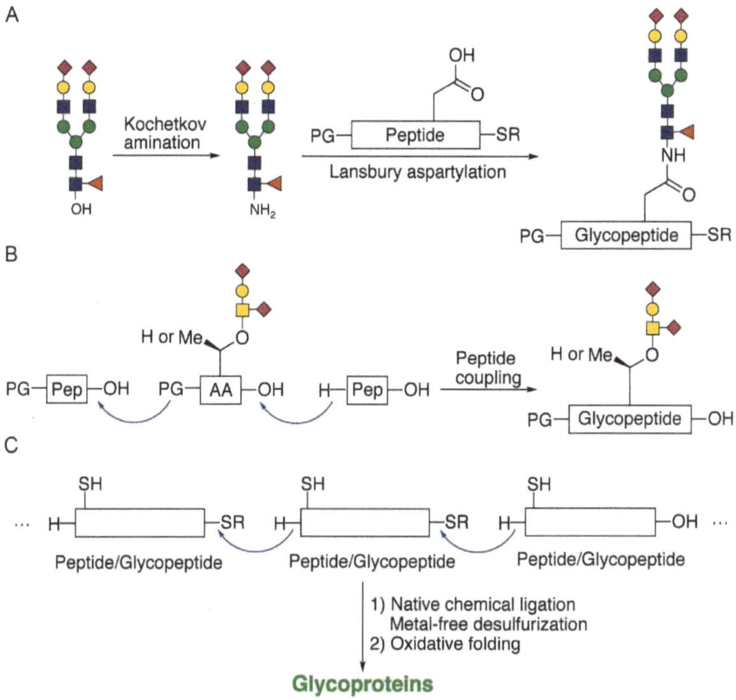

Fig. 2 Chemical synthesis of glycoproteins. (A) Convergent synthesis of N-linked glycopeptides; (B) Stepwise synthesis of O-linked glycopeptides; (C) Native chemical ligation/metal-free desulfurization-based strategy for the synthesis of glycopeptides. PG, protecting groups; SR, thioesters; Pep, peptides; AA, amino acids.

With the newly increased sophistication of glycoprotein synthesis, the Tan group was able to apply a chemical glycobiology approach to investigate the role and mechanism of glycosylation in regulating the properties and function of proteins (Chaffey et al., 2018). Their efforts established the general workflow of this approach as well as demonstrated its feasibility and effectiveness to provide new definitive insights into the biochemical and biological impacts of protein glycosylation. Most importantly, the results from these efforts have greatly advanced the application of rational glycoengineering to the optimization of therapeutic peptides and proteins.

This chapter will review the above-mentioned innovative developments in protein glycosylation research. We will first review the chemical synthesis of glycoproteins, and then describe the implementation of chemical glycobiology approaches to explore and exploit the beneficial effects of protein glycosylation.

2. Chemical synthesis of homogeneous glycoforms

There are two general strategies for the synthesis of glycopeptides: one is a convergent strategy, where sugars are directly coupled to appropriately-protected peptides to generate glycopeptides; the other is a cassette-based strategy, where pre-formed glycoamino acids (so-called cassette) are used as building blocks in synthesis (Fig. 2). Most N-linked glycopeptides can be prepared convergently by joining glycosylamines (glycans bearing anomeric amines) to aspartate residues (Fig. 2A), while most *O*-linked glycopeptides need to be prepared using the stepwise cassette approaches to ensure that the stereo-purity of the products is rigorously controlled (Fig. 2B) (Tan, Shang, Halkina, Yuan, & Danishefsky, 2009). However, regardless of the strategy selected, the chemical synthesis of glycans and glycoamino acids is the most challenging, but unavoidable step. The Danishefsky group did a significant amount of optimization and provided better synthetic routes for many N-linked glycans as well as O-GalNAc glycoamino acids (Sames, Chen, & Danishefsky, 1997; Wu et al., 2006). With these notable achievements, it became possible to conduct chemical synthesis of highly complex glycopeptide fragments.

The Danishefsky group exploited a number of synthetic advances toward the creation of glycopeptides (Wilson et al., 2013). Their endeavors along this direction led to the development of a range of broadly useful methods. They established that the unprotected oligosaccharides could undergo anomeric β-amination under Kochetkov conditions to produce glycosylamines, which in turn can be directly added to peptides using the Lansbury aspartylation protocol to yield N-glycopeptide fragments (Fig. 2A) (Wang, Zhang, Live, & Danishefsky, 2000). They also creatively addressed the challenges met in the synthesis of *O*-linked glycopeptides using an optimized "mask-off" approach by exploiting differences in reaction rates caused by steric effects of different carboxylic groups in unprotected *O*-glycoamino acids (Fig. 3) (Tan et al., 2009). Through these optimized preparation methods, they were able to complete the synthesis of many different glycopeptides, including both N- and *O*-linked glycopeptides (Wilson et al., 2013).

After the synthesis of glycopeptide fragments, the production of multiply glycosylated protein targets would require the development of capabilities in one more key research area. Through more than 10 years of efforts, the Danishefsky group introduced to this area a generally applicable and powerful new method: native chemical ligation coupled to metal-free

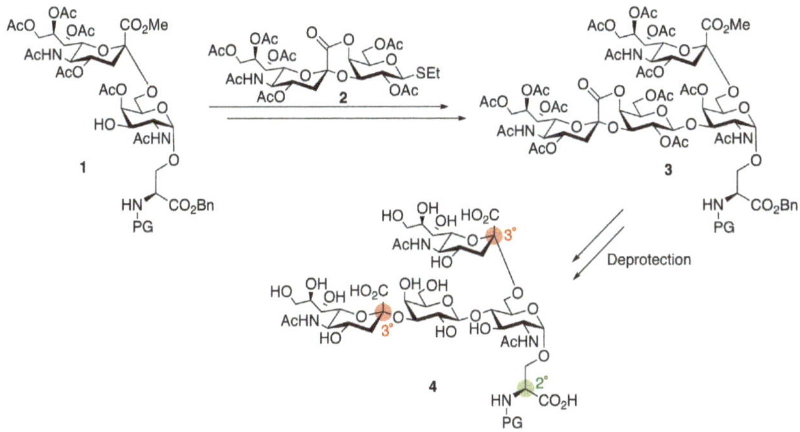

Fig. 3 Chemical synthesis of O-linked glycoamino acid with unmasked carbohydrate moiety. The reactivity of different carboxyl groups in **4** can be effectively controlled by the substitution level (steric hindrance) of their neighboring carbons. The one that is directly attached to a secondary carbon (2°, green colored) has better reactivity to the N-terminal amine groups of peptides. PG, protecting groups.

desulfurization (NCL/MFD) (Shang, Tan, Dong, & Danishefsky, 2011; Tan, Shang, & Danishefsky, 2010; Wan & Danishefsky, 2007). Traditionally, NCL is a process in which two peptide or glycopeptides fragments are chemoselectively joined together by reactions of C-terminal thioesters with N-terminal cysteine residues. One of its major limitations is the requirement of cysteine residues at ligation junctions, which are rare or absent in most proteins and glycoproteins. To address this challenge, the Danishefsky group greatly extended the application scope of NCL to many abundant amino acid residues (Chaffey et al., 2018; Shang, Tan, & Danishefsky, 2011; Wilson et al., 2013). To overcome the synthesis issues associated with the removal of thiol groups that were introduced into the side chains of N-terminal amino acids to catalyze the ligation reactions, they contributed one of the most impactful developments in protein/glycoprotein synthesis, the metal-free desulfurization strategy (Fig. 4) (Wan & Danishefsky, 2007).

The optimized methods described above have been applied to the synthesis of many different glycoproteins (Fernandez-Tejada et al., 2015). Depending on the structural features of the synthetic targets, there are small differences in experimental details. They can be easily found in previously published research. Here, we only briefly describe the most commonly used experimental techniques and tips that are critical for experimental success.

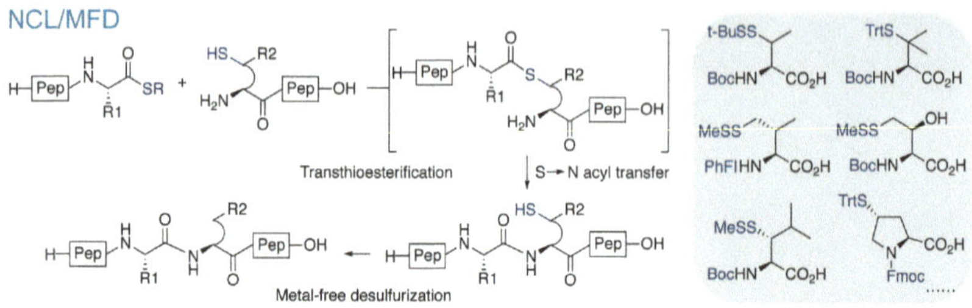

Fig. 4 Ligation at noncysteine amino acid residues by the native chemical ligation/metal-free desulfurization strategy. SR, thioester; Pep, peptides; R1 and R2, functional groups in amino acid side chains.

Equipment

1. NMR Instruments: Bruker Advance DRX 500 MHz and Bruker Advance II 600 MHz
2. Analytical TLC: Merck silica gel 60 F254
3. Flash Chromatography: Merck silica gel 60 (40–63 mm)
4. Solvent Drying System
5. Applied Biosystems Pioneer continuous flow peptide synthesizer
6. LC-MS System: Waters 2695 Separations Module, Waters Micromass ZQ mass spectrometer, Waters 996 Photodiode Array Detector, and Varian Microsorb columns 100–5, C18, 150 × 2.0 mm and 300–5, C4, 250 × 2.0 mm
7. UPLC-MS System: Waters Acquity™ Ultra Performance LC system and Waters Acquity UPLC® BEH columns C18, 1.7 μL, 2.1 × 100 mm, C8, 1.7 μL, 2.1 × 100 mm, and C4, 1.7 μL, 2.1 × 100 mm
8. Preparative HPLC system: Rainin HPLC solvent delivery system, Rainin UV-1 detector, and Varian Dynamax Microsorb columns 100–5, C18, 250 × 21.4 mm and 300–5, C4, 250 × 21.4 mm
9. VirTis Bench Top 4.0 K ZL Freeze Dryer (SP Industries, NY, USA)
10. Eppendorf 5430R Microcentrifuge
11. Analytical Balance: Mettler Toledo XS205DU
12. Bath Sonicator

Chemicals

1. Carbohydrate Building Blocks
2. Amino Acid Building Blocks
3. Pseudoproline Dipeptides
4. Preloaded NovaSyn®TGT resin
5. Peptide Coupling Reagents
6. Desulfurization Reagents

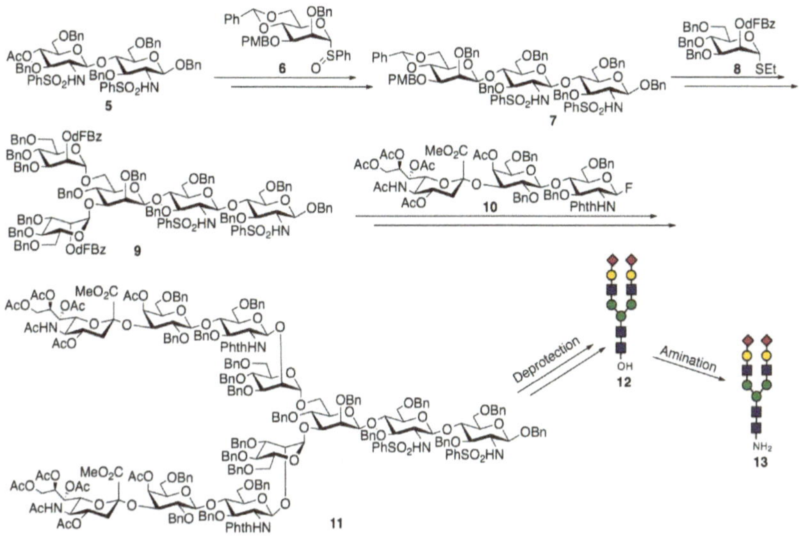

Fig. 5 Chemical synthesis of glycosyl amines.

2.1 Synthesis of N-linked glycopeptides

The preparation of N-linked glycopeptides began with the synthesis of N-glycans, such as a complex-type glycosylamine **13** (Walczak & Danishefsky, 2012; Wu et al., 2006). As shown in Fig. 5, this undecasaccharide can be synthesized convergently from a pentasaccharide **9** and a trisaccharide **10**. Briefly, the synthesis commenced with the preparation of a trisaccharide **7** through the application of the Crich–Kahne direct mannosylation protocol. Next, attachment of the mannose thioethyl donor **8** to **7** provided pentasaccharide **9**. This intermediate was then subjected to a double glycosylation with anomeric fluoride to afford the fully protected undecasaccharide **11**, which in turn was deprotected through a six-step sequence to generate the unprotected glycan **12**. The Kochetkov amination of **12** followed by repetitive lyophilization gave the glycosylamine **13**. Finally, the synthetic glycosyl amines were coupled to peptides in the presence of HATU to afford the desired glycopeptides (Fig. 2A).

Tip: The coupling efficiency of the sialylated trisaccharide **10** to the pentasaccharide or other oligosaccharides is dependent on the reactivity of the anomeric donor group. If the yield of this step is low, the first thing for troubleshooting should be to test more donor groups (Walczak & Danishefsky, 2012).

Tip: The deprotection of fucosylated N-linked oligosaccharides can result in partial cleavage of glycan chains. Special attention needs to be paid to the NaOMe/NaOH-mediated hydrolysis step (Nagorny et al., 2009).

Tip: The purity of glycosyl amine products is important for their reactivity in Lansbury aspartylation reaction. In order to completely remove impurities, the lyophilization needs to be performed many times until a constant weight is obtained.

Tip: Glycosyl amine products are sensitive to H_2O. When dissolved in H_2O and frozen for the next lyophilization cycle, they may decompose. Therefore, this step must be done very quickly, in less than 10 s.

Tip: The Lansbury aspartylation reaction is a bimolecular reaction and its rate is dependent on the concentrations of both reactants. Therefore, high reactant concentrations are crucial for the success of this reaction.

2.2 Synthesis of *O*-linked glycopeptides

The cassette approach was applied to prepare *O*-linked glycopeptides. In a typical synthesis, fully-protected glycoamino acids were convergently prepared under optimized reaction conditions (Fig. 3). After deprotection, the HATU-mediated peptide coupling reaction between unprotected glycoamino acids and short peptides proceeded very selectively to afford the desired *O*-linked glycopeptide products (Fig. 2B) (Tan et al., 2009).

Tip: The HATU-mediated peptide coupling reaction occurs quite rapidly. It is normally completed within 1 min.

Tip: Under HATU-mediated conditions, the carboxylic groups in glycoamino acids can be clearly differentiated on the basis of the substitution levels of their β-carbons. The formation of lactone between the sialic acid carboxyl group and the hydroxyl group in position 4 of the adjacent galactose is another reason for the selectivity.

Tip: The lactone formation can also beneficially prevent the cleavage of the labile α-2,3-glycosidic bond between sialic acid and galactose. The lactone ring can be opened by dilute sodium bicarbonate solution after the completion of glycoprotein synthesis.

Tip: The by-products formed by reacting with sialic acid carboxyl groups can be identified through the analysis of their fragmentation patterns in electrospray ionization-mass spectrometry.

2.3 Ligation and metal-free desulfurization

The chemical assembly of the synthetic fragments was carried out in the direction of C-terminus to N-terminus as shown in Fig. 2C. Briefly, the to-be-ligated peptides and glycopeptides were mixed together and lyophilized. The resulting powder was dissolved in buffer solution to initiate the ligation reaction. After ligation was completed, the thiol functional groups were removed under metal-free desulfurization conditions to yield the full-length glycoproteins, which were then oxidatively folded using a redox buffer (Shang, Tan, & Danishefsky, 2011; Wang et al., 2013).

Tip: Native chemical ligation reactions are bimolecular reactions. The reaction rates are influenced by concentration and other factors like the availability of free thiol groups at the N-terminal sides of peptides and glycopeptides, pH, and temperature. To ensure high concentrations of reactants, the lyophilized reactant powder mixtures should be centrifuged to the bottom of reaction vessels before a minimum required amount of buffer solution is added. Since the rate-limiting step in most NCL is the transthioesterification with the side-chain thiol groups of the N-terminal amino acids, it is important to include soluble reducing agents, like TCEP in degassed ligation buffer, to cleave disulfide bonds at the beginning of ligation and to prevent the oxidation of thiols to disulfides during the reaction. Furthermore, for intermolecular transthioesterification to occur, the proton on the thiol group must be transferred to a putative "base" or to the bulk solvent. Therefore, the desired pH of the buffer solution should be slightly alkaline (~7.5). Ligation rates also increase with increasing temperature. If the ligation is slow, the reaction temperature can be increased to 37 °C.

Tip: After ligation, the free thiol groups in the middle of the resulting peptides/glycopeptide products can further react with excess thioesters to form higher molecular weight species. The presence of these species can significantly complicate the analysis of reaction progress. This issue can be overcome by adding the alkyl thiolsodium 2-mercaptoethanesulfonate (MESNa) to reaction mixtures.

Tip: The presence of excess thiol additives and large amount of guanidine prevents direct desulfurization of ligation products. An efficient MFD can only be realized after removing these substances with a rapid and simple HPLC separation.

Tip: A large portion of ligation product can be lost during HPLC purification. Therefore, less purification steps are required for a more efficient and higher yielding synthesis.

Tip: Correctly folded proteins/glycoproteins normally eluted earlier on HPLC than other folding isomers. Correct folding can be verified by mass spectrometry analysis of enzymatically digested glyoprotein fragments.

3. Biological impact of protein glycosylation

The development of chemical synthesis of multiply glycosylated proteins by the Danishefsky group provided a sufficient scientific and technical base for overcoming challenges associated with research on protein glycosylation. It enabled the Tan group to apply chemical glycobiology approaches to further the understanding of the roles of glycans in regulating protein biophysical and biological properties (Chaffey, Chi, & Tan, 2017; Chaffey et al., 2018).

The predictability and flexibility of chemical synthesis allowed the generation of large collections of structurally well-defined glycoproteins with subtle differences in their glycan structures and amino acid sequences. The availability of such collections makes it possible to get a clearer and more in-depth view of the biological impact of glycosylation (Chen et al., 2014). By systematically analyzing the impact, the Tan group was able to dissect detailed structure-property relationships and to gain insights into the structural basis for the inferred relationships (Guan et al., 2015). More importantly, the knowledge developed from these studies helps to formulate a directional strategy that can guide future studies toward other posttranslational modifications, and to improve the properties and performance of peptides/proteins through rational glycoengineering (Guan, Chaffey, Chen, et al., 2018; Guan, Chaffey, Wei, et al., 2018; Li, Clark, & Tan, 2018).

In the following section, we briefly summarize these studies and the findings relating to the biological impact of protein glycosylation.

Equipment

1. Chirascan™-plus CD Spectrometer
2. Voyager-DE™ STR MALDI-TOF Mass Spectrometer
3. Waters Synapt G2 HDMS (q-TOF) Mass Spectrometer
4. Proteomelab XL-I Protein Characterization System
5. 48-Well Micro Chemotaxis Chamber
6. QuantaMaster-4/2005 SE Spectrofluorometer

Chemicals

1. Proteases
2. Chemicals for biochemical and biological buffers

3.1 Impact on folding

The process of protein folding is a notoriously complicated process. For proteins containing disulfide bonds, it is a combination of two distinct, but interrelated processes: conformational folding of the protein sequence into secondary and tertiary structures and formation of disulfide bonds. By taking advantage of the relatively slow kinetics in disulfide bond formation, the Tan group was able to confirm the importance of glycosylation in regulating the folding of a family 1 carbohydrate-binding module (CBM). Additionally, they were able to elucidate the glycosylation site-, glycan size- and structure-specific effects of O-glycosylation on the folding using a mass spectrometry-based approach (Chaffey, Guan, Chen, et al., 2017; Chaffey, Guan, Wang, et al., 2017).

Via comparison of the folding rates of a large collection of synthetic glycoforms, it was found that many factors, including glycosylation site, glycan structure and/or size, and amino acid sequence, contributed to the effects of glycosylation. Small changes in either glycans or amino acids could lead to large differences in the ability of glycosylation to affect CBM folding. It was also revealed that glycosylation had different effects on the rates of each step in the folding process. Additionally, only the attachment of specific glycans, such as mono-mannose, to glycosylation sites nearest the N-terminus of CBM could significantly increase the overall folding rate. Together, the results pointed to the possibility that the interactions between glycan and local amino acids might be responsible for the increased folding rate.

3.2 Impact on stability

The term "protein stability" includes many meanings, including proteolytic stability, thermal stability, physical stability (aggregation), et al. Thermal stability is the stability of a molecule at elevated temperatures. Proteolytic stability refers to the resistance toward the action of proteolytic enzymes. Aggregation represents the most vexing physical instability of proteins, which is a process by which misfolded proteins form insoluble precipitates.

Through the characterization and comparison of the properties of a series of synthetic glycoforms, the Tan group was able to reveal how glycans, and their site occupancy, can affect the thermal and proteolytic stability, as well as resistance to aggregation. Data from these studies showed that the effects of glycosylation are strongly dependent on the glycosylation sites and the structures of glycans directly attached to proteins (Chen et al., 2014; Guan et al., 2015; Guan, Chaffey, Chen, et al., 2018; Prates et al., 2018). Increasing the sizes of glycans could further increase or maintain the stability conferred by original glycan structures. Again, the observations from these studies also suggested that the interactions between the α-linked glycans and the local amino acids might be the molecular basis underlying the increased stability.

3.3 Impact on structure

Glycosylation tends to significantly increase the solubility of proteins. This makes crystallization of glycoproteins difficult. Solution NMR was thus chosen by the Tan group to investigate the effect of glycosylation on protein structure and the molecular basis underlying the increase in protein folding rate and stability. By systematically comparing the gathered high-resolution NMR structures, the results confirmed that glycosylation did not significantly alter the structures of proteins. Its main role is to introduce direct interactions between glycan residues and side chains of local amino acids (Chaffey, Guan, Chen, et al., 2017; Chaffey, Guan, Wang, et al., 2017). If these interactions can be propagated throughout the rest of the backbone, they will be able to strengthen the global network of intramolecular interactions and lead to a decrease in the backbone dynamics of proteins. Lower backbone flexibility is most likely to be the reason for the increased folding rate and stability.

3.4 Impact on biological activity

In addition to examining the detailed effects of glycans on the properties related to folding, stability and structure, the Tan group also used the CBM as a model molecule to investigate the effect of glycosylation on the binding affinity (Chen et al., 2014; Guan et al., 2015; Payne et al., 2013). By systematically comparing the properties of glycosylated variants, the results showed that the effect of glycosylation on the binding affinity was not dependent on glycosylation sites nor glycan structures. Small glycan residues at each of the glycosylation sites may increase the binding affinity. The molecular mechanism by which glycans affect the binding of CBM to its substrates was investigated by NMR studies and molecular simulations (Happs et al., 2015). The results revealed that the interactions between glycans and substrates could be responsible for increased binding.

4. Summary

In summary, through many years of efforts and development, the Danishefsky group was able to pave the way for the synthesis of multiply glycosylated proteins. The broad applicability of the developed methodologies was demonstrated in the studies of the impacts of glycosylation on protein properties. Through systematic chemical glycobiology studies, it was revealed that the effects of glycosylation were determined by a combination of factors, including glycosylation site, glycan size, structure, and amino acid sequence. The combined observations led to a hypothesis that is important for simplifying protein glycoengineering: "glycoforms with better overall properties can be generated by collaboratively varying glycan structures and adjacent amino acids within unstructured regions crucial to biological function and/or stability". This hypothesis was confirmed by studying two molecules, human insulin and RANTES, otherwise known as CCL5 (Guan, Chaffey, Chen, et al., 2018; Guan, Chaffey, Wei, et al., 2018).

Acknowledgments

The authors are grateful to the National Science Foundation CAREER Award (Award number: CHE-1454925), the National Institutes of Health (Award number: R01EB025892) and the National Natural Science Foundation of China (Grant Number: 91853120) for their support.

References

Anthony, R. M., Wermeling, F., & Ravetch, J. V. (2012). Novel roles for the IgG Fc glycan. *Annals of the New York Academy of Sciences*, *1253*, 170–180.

Chaffey, P. K., Chi, L., & Tan, Z. (2017). *Chemical biology of protein O-glycosylation chemical biology of glycoproteins*. The Royal Society of Chemistry, pp. 48–93.

Chaffey, P. K., Guan, X., Chen, C., Ruan, Y., Wang, X., Tran, A. H., et al. (2017). Structural insight into the stabilizing effect of O-glycosylation. *Biochemistry*, *56*(23), 2897–2906.

Chaffey, P. K., Guan, X., Li, Y., & Tan, Z. (2018). Using chemical synthesis to study and apply protein glycosylation. *Biochemistry*, *57*(4), 413–428.

Chaffey, P. K., Guan, X., Wang, X., Ruan, Y., Li, Y., Miller, S. G., et al. (2017). Quantitative effects of O-linked glycans on protein folding. *Biochemistry*, *56*(34), 4539–4548.

Chen, L., Drake, M. R., Resch, M. G., Greene, E. R., Himmel, M. E., Chaffey, P. K., et al. (2014). Specificity of O-glycosylation in enhancing the stability and cellulose binding affinity of family 1 carbohydrate-binding modules. *Proceedings of the National Academy of Sciences of the United States of America*, *111*(21), 7612–7617.

Dwek, R. A. (1995). Glycobiology: More functions for oligosaccharides. *Science*, *269*(5228), 1234–1235.

Fernandez-Tejada, A., Brailsford, J., Zhang, Q., Shieh, J. H., Moore, M. A., & Danishefsky, S. J. (2015). Total synthesis of glycosylated proteins. In *Vol. 362. Topics in current chemistry* (pp. 1–26). New York: Academic Press.

Greene, E. R., Himmel, M. E., Beckham, G. T., & Tan, Z. (2015). Glycosylation of cellulases: Engineering better enzymes for biofuels. *Advances in Carbohydrate Chemistry and Biochemistry*, *72*, 63–112.

Guan, X., Chaffey, P. K., Chen, H., Feng, W., Wei, X., Yang, L. M., et al. (2018). O-GalNAcylation of RANTES improves its properties as a human immunodeficiency virus type 1 entry inhibitor. *Biochemistry*, *57*(1), 136–148.

Guan, X., Chaffey, P. K., Wei, X., Gulbranson, D. R., Ruan, Y., Wang, X., et al. (2018). Chemically precise glycoengineering improves human insulin. *ACS Chemical Biology*, *13*(1), 73–81.

Guan, X., Chaffey, P. K., Zeng, C., Greene, E. R., Chen, L., Drake, M. R., et al. (2015). Molecular-scale features that govern the effects of O-glycosylation on a carbohydrate-binding module. *Chemical Science*, *6*(12), 7185–7189.

Happs, R. M., Guan, X., Resch, M. G., Davis, M. F., Beckham, G. T., Tan, Z., et al. (2015). O-glycosylation effects on family 1 carbohydrate-binding module solution structures. *FEBS Journal*, *282*(22), 4341–4356.

Li, Y., Clark, K. A., & Tan, Z. (2018). Methods for engineering therapeutic peptides. *Chinese Chemical Letters*, *29*(7), 1074–1078.

Meany, D. L., & Chan, D. W. (2011). Aberrant glycosylation associated with enzymes as cancer biomarkers. *Clinical Proteomics*, *8*(1), 7.

Moremen, K. W., Tiemeyer, M., & Nairn, A. V. (2012). Vertebrate protein glycosylation: Diversity, synthesis and function. *Nature Reviews Molecular Cell Biology*, *13*(7), 448–462.

Nagorny, P., Fasching, B., Li, X., Chen, G., Aussedat, B., & Danishefsky, S. J. (2009). Toward fully synthetic homogeneous beta-human follicle-stimulating hormone (beta-hFSH) with a biantennary N-linked dodecasaccharide. Synthesis of beta-hFSH with chitobiose units at the natural linkage sites. *Journal of the American Chemical Society*, *131*(16), 5792–5799.

Ohtsubo, K., & Marth, J. D. (2006). Glycosylation in cellular mechanisms of health and disease. *Cell*, *126*(5), 855–867.

Payne, C. M., Resch, M. G., Chen, L., Crowley, M. F., Himmel, M. E., Taylor, L. E., 2nd, et al. (2013). Glycosylated linkers in multimodular lignocellulose-degrading enzymes dynamically bind to cellulose. *Proceedings of the National Academy of Sciences of the United States*, *110*(36), 14646–14651.

Prates, E. T., Guan, X., Li, Y., Wang, X., Chaffey, P. K., Skaf, M. S., et al. (2018). The impact of O-glycan chemistry on the stability of intrinsically disordered proteins. *Chemical Science*, *9*(15), 3710–3715.

Rich, J. R., & Withers, S. G. (2009). Emerging methods for the production of homogeneous human glycoproteins. *Nature Chemical Biology*, *5*(4), 206–215.

Rudd, P. M., & Dwek, R. A. (1997). Glycosylation: Heterogeneity and the 3D structure of proteins. *Critical Reviews in Biochemistry and Molecular Biology*, *32*(1), 1–100.

Sames, D., Chen, X. T., & Danishefsky, S. J. (1997). Convergent total synthesis of a tumour-associated mucin motif. *Nature*, *389*(6651), 587–591.

Shang, S., Tan, Z., & Danishefsky, S. J. (2011). Application of the logic of cysteine-free native chemical ligation to the synthesis of human parathyroid hormone (hPTH). *Proceedings of the National Academy of Sciences of the United States*, *108*(15), 5986–5989.

Shang, S., Tan, Z., Dong, S., & Danishefsky, S. J. (2011). An advance in proline ligation. *Journal of the American Chemical Society*, *133*(28), 10784–10786.

Spiro, R. G. (2002). Protein glycosylation: Nature, distribution, enzymatic formation, and disease implications of glycopeptide bonds. *Glycobiology*, *12*(4), 43R–56R.

Tan, Z., Shang, S., & Danishefsky, S. J. (2010). Insights into the finer issues of native chemical ligation: An approach to cascade ligations. *Angewandte Chemie International Edition*, *49*(49), 9500–9503.

Tan, Z., Shang, S., & Danishefsky, S. J. (2011). Rational development of a strategy for modifying the aggregatibility of proteins. *Proceedings of the National Academy of Sciences of the United States*, *108*(11), 4297–4302.

Tan, Z., Shang, S., Halkina, T., Yuan, Y., & Danishefsky, S. J. (2009). Toward homogeneous erythropoietin: Non-NCL-based chemical synthesis of the Gln78-Arg166 glycopeptide domain. *Journal of the American Chemical Society*, *131*(15), 5424–5431.

Varki, A. (1993). Biological roles of oligosaccharides: All of the theories are correct. *Glycobiology*, *3*(2), 97–130.

Varki, A. (2017). Biological roles of glycans. *Glycobiology*, *27*(1), 3–49.

Walczak, M. A., & Danishefsky, S. J. (2012). Solving the convergence problem in the synthesis of triantennary N-glycan relevant to prostate-specific membrane antigen (PSMA). *Journal of the American Chemical Society, 134*(39), 16430–16433.

Wan, Q., & Danishefsky, S. J. (2007). Free-radical-based, specific desulfurization of cysteine: A powerful advance in the synthesis of polypeptides and glycopolypeptides. *Angewandte Chemie International Edition, 46*(48), 9248–9252.

Wang, P., Dong, S., Brailsford, J. A., Iyer, K., Townsend, S. D., Zhang, Q., et al. (2012). At last: Erythropoietin as a single glycoform. *Angewandte Chemie International Edition, 51*(46), 11576–11584.

Wang, P., Dong, S., Shieh, J. H., Peguero, E., Hendrickson, R., Moore, M. A. S., et al. (2013). Erythropoietin derived by chemical synthesis. *Science, 342*(6164), 1357–1360.

Wang, Z. G., Zhang, X., Live, D., & Danishefsky, S. J. (2000). From glycals to glycopeptides: A convergent and stereoselective total synthesis of a high mannose N-linked glycopeptide. *Angewandte Chemie International Edition, 39*(20), 3652–3656.

Wilson, R. M., Dong, S. W., Wang, P., & Danishefsky, S. J. (2013). The winding pathway to erythropoietin along the chemistry-biology frontier: A success at last. *Angewandte Chemie International Edition, 52*(30), 7646–7665.

Wu, B., Hua, Z., Warren, J. D., Ranganathan, K., Wan, Q., Chen, G., et al. (2006). Synthesis of the fucosylated biantennary N-glycan of erythropoietin. *Tetrahedron Letters, 47*(31), 5577–5579.

CHAPTER THIRTEEN

In bulla functional channel expression systems that mimic bacterial synthetic membranes

Masayuki Iwamoto, Shigetoshi Oiki*
Department of Molecular Physiology and Biophysics, University of Fukui, Faculty of Medical Sciences, Fukui, Japan
*Corresponding author: e-mail address: oiki-fki@umin.ac.jp

Contents

Abstract

A functional characterization of channel proteins has been performed using planar lipid bilayers as the following procedure. For bacterial channels, such as the KcsA potassium channel, channel proteins were synthesized in *Escherichia coli*, followed by solubilization, purification, and incorporation into liposomes. Similarly, channel proteins were synthesized using an *in vitro* transcription/translation kit in the presence of liposomes. Then, these liposome-incorporated channels were served for electrophysiological recordings after liposome fusion into a preformed planar lipid bilayer. Here, we established a straightforward method for concurrent channel synthesis and functional

Methods in Enzymology, Volume 621
ISSN 0076-6879
https://doi.org/10.1016/bs.mie.2019.02.011

measurement using a water-in-oil bubble bilayer system. Channel proteins were synthesized *in vitro* within a water-in-oil bubble, having a lipid bilayer at the contact with another bubble (*in bulla* synthesis). The channels were spontaneously incorporated into the lipid bilayer under application of the membrane potential, and we successfully detected nascent channel activities. This way our experiment has mimicked bacterial synthetic membrane in the presence of a resting membrane potential. Technical details for establishing the *in bulla* expression system are described.

1. Introduction

Cell-free protein synthesis generates functional proteins *in vitro*, and various types of *in vitro* transcription/translation (IVTT) system provide opportunities for studying proteins under defined experimental conditions (Kim, Kigawa, Choi, & Yokoyama, 1996; Mathews & Korner, 1970; Nirenberg & Matthaei, 1961; Sawasaki, Ogasawara, Morishita, & Endo, 2002). Cell extracts from various cellular sources have been used for high-throughput protein synthesis. However, some components of extracts may deteriorate synthesized proteins (Kim & Swartz, 2000). Alternatively, a defined set of purified components relevant to transcription and translation can be reconstituted as a cocktail (Shimizu et al., 2001; Shimizu, Kanamori, & Ueda, 2005), which generates efficient protein synthesis under controlled conditions. In synthetic biology, this system yields combinations of protein species for generating metabolic networks (Jewett, Calhoun, Voloshin, Wuu, & Swartz, 2008). However, for membrane proteins, there is an additional problem in the cell-free synthesis. Membrane proteins have been synthesized *in vitro* in the presence of membrane fractions (Kalmbach et al., 2007; Shimono et al., 2009; Shinoda et al., 2016) and spontaneously incorporated into the membrane. However, they may be randomly oriented in the membrane (Sumino, Uchihashi, & Oiki, 2017), which is critical in examining functional features of membrane proteins, such as measuring flux of substance across the membrane.

Here, we present a novel platform for functional expression and examination of channel proteins in water-in-oil bubble system (*in bulla* [bubble in Latin] functional expression system) (Iwamoto, Elfaramawy, Yamatake, Matsuura, & Oiki, 2018). To understand structure-function relationships of ion channels, the KcsA potassium channel from *Streptomyces lividans* has been used as a prototypical channel (Oiki, 2015). Channel proteins have been synthesized using *Escherichia coli* (Iwamoto et al., 2006) or *in vitro* synthesis in the presence of liposomes and nanodiscs (Banerjee & Nimigean, 2011) for serving various experimental techniques (Fig. 1). Single-channel current recordings are a prerequisite method to examine the functional features of the channels (Iwamoto & Oiki, 2013) using the planar lipid bilayer (PLB) method (Oiki, 2012). In this method, only a single KcsA channel molecule is expected to be incorporated into PLB; hence, a large number of molecules are not necessary. Accordingly, we anticipated decades ago that *in vitro* synthesis of channel proteins inside the PLB chamber may allow detection of newly synthesized channel currents *via* spontaneous incorporation into the membrane. This idea was examined but was technically unpractical because a large amount of IVTT (~1 mL) was necessary to fill a compartment, and stable bilayers could not be formed. Now that advanced lipid bilayer techniques such as the DIB method (Bayley et al., 2008) using a drop-in-well chamber (see Chapter 13; Urakubo, Iwamoto & Oiki, 2019) as well as the contact bubble bilayer (CBB) method (Iwamoto & Oiki, 2015, 2019; Oiki & Iwamoto, 2018) are available, we considered that our old idea seemed to be achievable in two respects. The droplet size is only ~1 μL, which reduces the amount of the IVTT cocktail that is required. Moreover, rather than crude extracts, the reconstituted *E. coli*-based IVTT cocktail (PUREfrex) (Shimizu et al., 2001), in which relevant components are optimized, would minimize membrane perturbations. Accordingly, we constructed a concurrent IVTT synthesis and functional measurement system by circumventing several technical difficulties (Fig. 2). The KcsA channel was synthesized *in bulla* followed by the detection of nascent channel activities.

The *in bulla* expression system casts light on processes used to form membrane proteins. Experimentally, the membrane potential was applied throughout the protein synthesis to detect single-channel currents. We noted that this procedure was not simply for the recording purpose but was representative of mimicking bacterial synthetic membrane. The bacterial synthetic membrane has the steady resting membrane potential (~−200 mV), distinguishing eukaryotic counterparts with practically null membrane potential. Accordingly, we addressed the effects of membrane potential on the protein expression process in the membrane. Even though translocon-mediated membrane insertion has been examined in earlier studies (Hessa et al., 2005, 2007; Ojemalm et al., 2011), the processes following the translocation, such as folding and oligomerization, as well

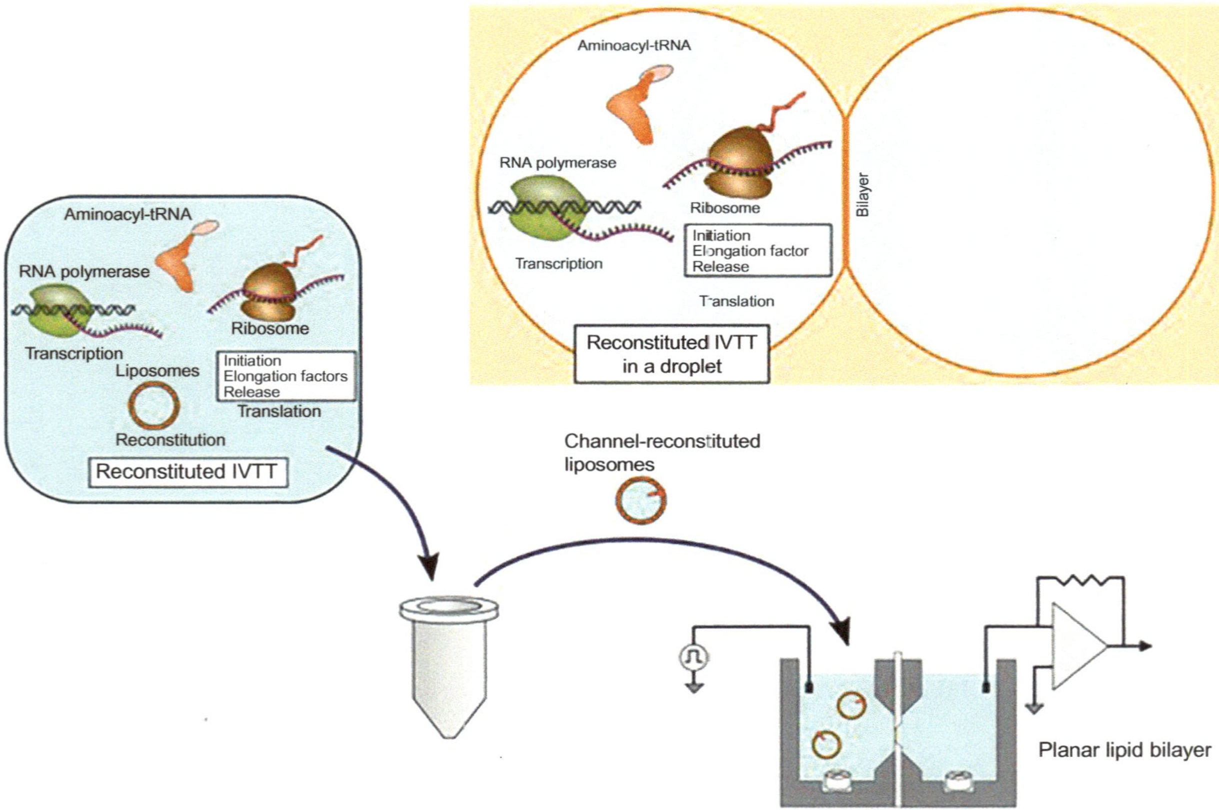

Fig. 1 *In vitro vs. in bulla* synthesis of the KcsA channel. Previously, two-step synthesis and reconstitution have been performed. Now the channel has been synthesized in a recording bubble, where the bilayer is available.

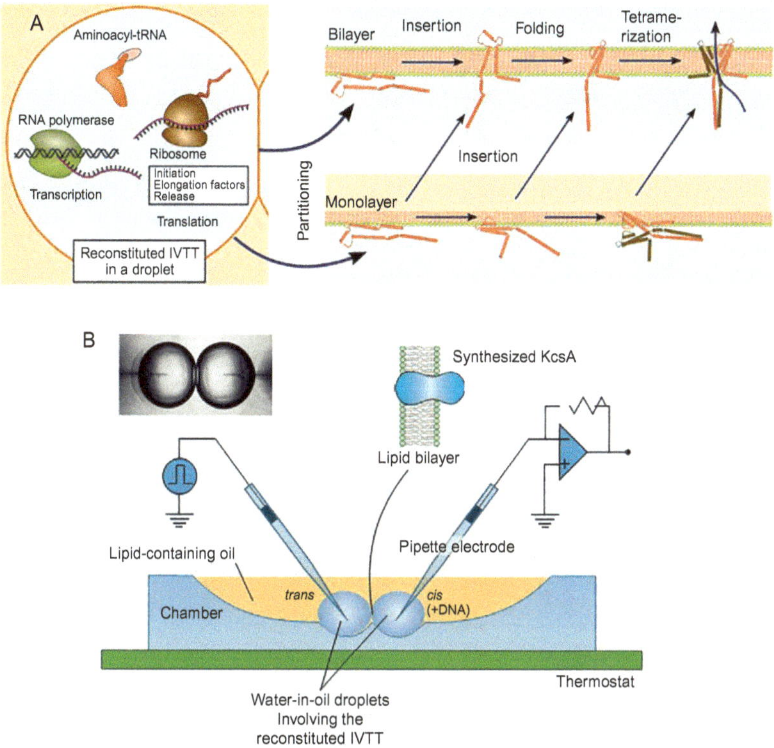

Fig. 2 The KcsA channel expression system. (A) Channel protein expression using the reconstituted *in vitro* transcription/translation system (reconstituted IVTT; PUREfrex1.0) in the droplet interface bilayer (DIB). The synthesis of the KcsA potassium channel was performed in a bubble (left panel). The right panel shows a membrane system of the DIB, such that the monolayer and bilayer coexist and are contiguous to each other. The synthesized protein was partitioned to the interface of either the monolayer or bilayer, followed by insertion, folding, and tetramerization. (B) The concurrent expression and functional detection system. One microliter of IVTT solution was dropped into a drop-in-well chamber filled with azolectin-containing hexadecane. Two droplets were placed in spontaneous contact. Two glass microelectrodes were inserted into the droplets, and electrophysiological measurements were performed under a two-electrode voltage clamp configuration. Under the continuously applied membrane potential, the first-born channel was detected as a single-channel current. Inset shows a photo of the DIB system. *Adapted from Iwamoto, M., Elfaramawy, M. A., Yamatake, M., Matsuura, T., & Oiki, S., (2018). Concurrent in vitro synthesis and functional detection of nascent activity of the KcsA channel under a membrane potential.* ACS Synthetic Biology, *7(4), 1003–1011.*

as the effects of the membrane potential, remain elusive (Cymer, von Heijne, & White, 2015). The *in bulla* functional channel expression system serves as a platform for such fundamental studies, and here, we present a simple protocol (Fig. 2).

2. Experimental tips for *in bulla* functional channel expression

To establish a bacterial membrane-mimicking system, the following tips are crucial for properly performing the experiments. We chose a reconstituted *E. coli*-based *in vitro* transcription/translation system, which is commercially available as PUREfrex®. The channel synthesis in the bubble was performed in the absence of liposomes such that the membrane environment for synthesized channels is served exclusively from the surface monolayer and bilayer. Otherwise, synthesized proteins are captured into liposomes before reaching the surface membrane, and the reconstituted channels in liposomes are transferred to the bilayer through membrane fusion. This is an aberrant route for the protein expression in the bacterial membrane. In the liposome-free IVTT bubble, the synthesized proteins diffused and reached the surface membrane, where they remained at the interface until they were spontaneously incorporated into the bilayer without any translocon machinery (unassisted pathway). This process mimics the bacterial membrane protein synthesis (Fig. 2).

2.1 The lipid-out method with high concentration of lipids for stable membranes

In the absence of liposomes inside the IVTT-bubble, phospholipids for bilayer formation are furnished by the lipid-out method such that the phospholipids were dispersed in the oil (hexadecane) phase. For the membrane phospholipids, we used azolectin, a mixture of phospholipids with a predominant component of phosphatidylcholine. To make a stable bilayer, the azolectin concentration in hexadecane was increased to 100 mg/mL. Pure phospholipids, such as palmitoyloleoylphosphatidylcholine cannot be dispersed at such high concentrations in hexadecane. In the lipid-out method, lipids were transferred slowly toward the oil-water interface (Venkatesan et al., 2016). Thus, two droplets were left undocked for 3 min (see Chapter 13).

2.2 Glass microelectrodes for electrophysiological measurements

In the conventional DIB, AgCl electrodes coated with hydrogel were used (Bayley et al., 2008). In the IVTT containing bubble, however, some constituents of the IVTT may interfere with the electrode potential. To circumvent this possible interference, microelectrodes were used (see Chapter 13 for drop-in-wells method).

2.3 A pH-independent mutant of KcsA

Membrane incorporated KcsA channels are anticipated to be orientated with their cytoplasmic domain remaining in the IVTT side of the bubble, similar to a native membrane orientation. If this is the case, the wild-type channel never opens because it is activated at an acidic pH on the intracellular side, while the PUREfrex is optimized at a neutral pH. To detect nascent channel activity, a KcsA mutant exhibiting channel activity even at neutral pH was used (Posson, Thompson, McCoy, & Nimigean, 2013). Furthermore, the open probability of the wild-type channel is low, and the gating kinetics is fast. Thus, the inactivation-free mutant E71A was used (Cordero-Morales, et al., 2006), which allowed readily counting of a number of functional channels with its nearly 100% open probability.

3. Materials

3.1 Equipment

1. Inverted microscope (IX73, Olympus, Tokyo, Japan)
2. Glass plate heater (TP-110NLR, Tokai Hit, Fujinomiya, Japan)
3. Digital camera (ORCA-Flash2.8, HAMAMATSU PHOTONICS, Hamamatsu, Japan)
4. Patch-clamp amplifier (EPC800USB, HEKA, Lambrecht/Pfalz, Germany)
5. AD converter (Digidata1550A, Molecular Devices, Sunnyvale, CA)
6. Motor-driven micromanipulator × 2 (EMM-3NV, Narishige, Tokyo, Japan)
7. Pneumatic microinjector × 2 (IM-11-2, Narishige)
8. Micropipette puller (P-87, Sutter Instrument, Novato, CA, USA)
9. Thin wall borosilicate capillary glass (length; 100 mm, OD/ID; 1.5/1.1 mm)
10. Pipette holder equipped with a pressure port × 2
11. AgCl wire electrode × 2
12. Rotary evaporator
13. N_2 gas cylinder
14. Vacuum pump
15. Desiccator
16. Eggplant flask (5–10 mL)
17. Tuberculine syringe (1 mL)
18. Injection needle (25G, 60 mm)
19. Acryl chamber with shallow holes at the bottom (handmade)

3.2 Chemicals

1. Azolectin (P3644, Sigma-Aldrich, St. Louis, MO)
2. Hexadecane
3. Potassium chloride
4. IVTT system (PUREfrex 1.0, Gene Frontier, Kashiwa, Japan)
5. DNA solution of the constitutive open mutant (H25R/E71A/E118A) of the KcsA channel
6. Distilled water

3.3 Drop-in-well method

The method of the drop-in-well method is described in Chapter 13 in this book (Urakubo et al., 2019).

4. Protocol

4.1 Preparation of lipid solution

1 Disperse azolectin (*e.g.*, 10 mg in powder) in chloroform at a desired concentration (*e.g.*, 50 mg/mL).
2 Place the azolectin solution in an eggplant flask and set it on a rotary evaporator connected to a N_2 gas cylinder. Rotate the flask under the N_2 flow at room temperature until a thin phospholipid film appears (20–40 min).
3 Place the flask into a desiccator connected to a vacuum pump. Aspirate the inside of the desiccator using the vacuum pump for several hours to evaporate the chloroform thoroughly.
4 Add an appropriate volume of hexadecane to the flask to obtain a 100 mg/mL azolectin solution.

4.2 Preparation of glass microelectrodes and their installation

1 Set a glass capillary to a micropipette puller. Pull the capillary by two steps to fabricate a pair of micropipettes with fine tapered tip (diameter; 1–3 μm).
2 Fill 3 M KCl solution into the micropipettes using the tuberculine syringe and an injection needle. Connect the AgCl wire electrodes to the pipette holders equipped with a pressure port. Mount the micropipette to the pipette holders. Confirm that the AgCl wire electrodes are soaked into the 3 M KCl solution. Connect one of the holders to the head stage of the patch-clamp amplifier, and the other one, to the electrical ground.

3 Adjust the electrode offset potential by inserting both electrodes in a bubble (Fig. 2).
4 Insert each electrode to a bubble.

4.3 IVTT reaction and concurrent electrophysiological recording

1 Mix the followings to prepare an IVTT reaction solution, and keep the solution on ice

PUREfrex 1.0 solution I	10 μL
Solution II	1 μL
Solution III	1 μL
Distilled water	7 μL
	Total 19 μL

2 Add 0.5 μL of the KcsA DNA solution (10 mM), or the distilled water, to 9.5 μL of the IVTT reaction solution, respectively, and preserve them on ice.
3 Set the glass plate heater on the stage of the inverted microscope, and place the acryl chamber on it. Add hexadecane (*ca.* 50 μm) into the chamber. At this time, the heater is off.
4 Add 1 μL of the IVTT reaction solutions (+/− DNA), respectively, into the hexadecane to form water-in-oil bubbles. Keep the bubbles separated and leave them at the bottom of the chamber for 3 min.
5 Make sure that the bubbles are in contact with each other. Insert the tip of the glass microelectrodes in the respective bubble.
6 Set the membrane voltage to the desired value by the patch-clamp amplifier. Set the glass plate heater to 37 °C and start current recording.

5. Nascent channel activity

The applied membrane potential was +25 mV, and current recordings were started. The channel synthesis was initiated by raising the temperature to 37 °C using the transparent glass plate underneath the chamber.

As shown in Fig. 3, a single-channel current of the KcsA channel appeared 28 min after the initiation of the synthesis (Fig. 3). This is the first

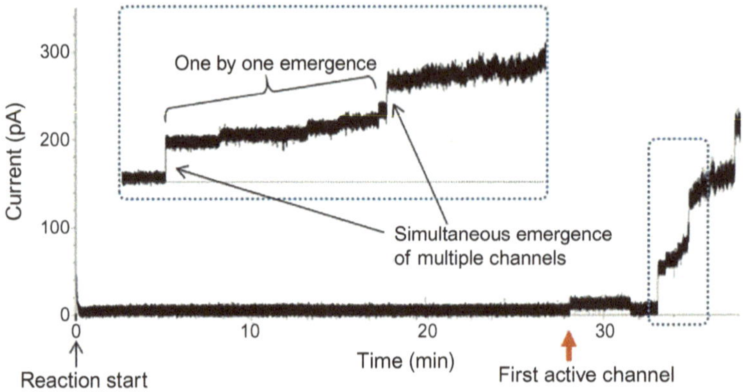

Fig. 3 Time course of the appearance of the KcsA channel. The IVTT reaction was started by raising the temperature to 37 °C at time 0 and immediately after, the membrane potential was set to +50 mV. In the time course presented, the first active channel emerged 28 min after the reaction was started, which was detected as a rectangular single-channel current (red arrow). Simultaneous emergence of multiple channels was observed in addition to one by one emergence as indicated in the inset. *Adapted from Iwamoto, M., Elfaramawy, M. A., Yamatake, M., Matsuura, T., & Oiki, S., (2018). Concurrent in vitro synthesis and functional detection of nascent activity of the KcsA channel under a membrane potential.* ACS Synthetic Biology, *7(4), 1003–1011.*

appearance time of the functional expression, involving the time course of transcription/translation, membrane insertion, folding and oligomerization. More channels were found to be functional successively. The membrane broke frequently when massive channel proteins were incorporated.

The membrane potential was set differently, such as at +100 mV from the beginning of the synthesis, and we found surprisingly that the channel current appeared only after 6 min. Furthermore, at negative membrane potentials, the first appearance time was similar to those at positive potentials (Fig. 4). Thus, the appearance time varied substantially with the membrane potentials, which polarity was instead found to be irrelevant. This unprecedented symmetrical voltage dependency of the first appearance time provides clues for understanding underlying mechanism of functional channel formation under the membrane potential.

6. Summary of additional experimental results

Orientation of the membrane-embedded functional channels were examined by exploiting the acid sensitivity of the wild-type KcsA channel with its cytoplasmic pH-sensor. For the wild-type channel, functional

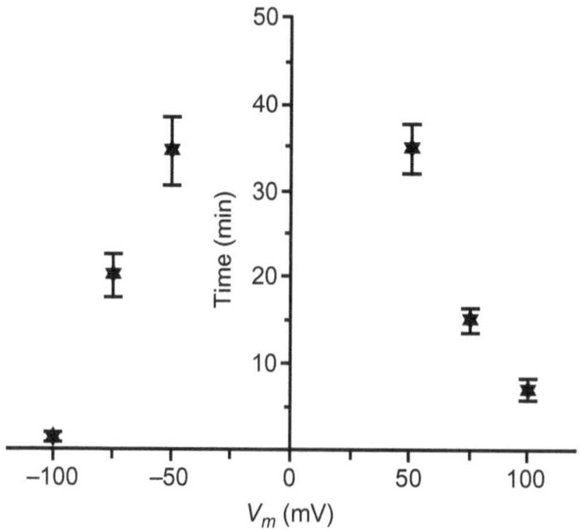

Fig. 4 First appearance time from the initiation of transcription/translation as a function of the membrane potential. The number of the observation was 3–11. *Adapted from Iwamoto, M., Elfaramawy, M. A., Yamatake, M., Matsuura, T., & Oiki, S., (2018). Concurrent in vitro synthesis and functional detection of nascent activity of the KcsA channel under a membrane potential.* ACS Synthetic Biology, *7(4), 1003–1011.*

expression was recognized *a posteriori* when the IVTT-containing bubble was made acidic by injecting acid solution. When the *trans* bubble was made acidic, no channel currents were observed, indicating that the channels were never oriented oppositely. Thus, the *in bulla* expression system served as an oriented channel expression system.

7. Summary

We established an *in bulla* channel expression system, where concurrent synthesis and functional expression of channel proteins under applied membrane potential were achieved. The goal of this method was to mimic bacterial synthetic membrane, so as to understand the formation process of functional channels, and we were able to successfully capture nascent channel activities. This system allowed oriented reconstitution of the channel, which is crucially important in membrane proteins expression. We found that the first appearance time was shortened dramatically as the amplitude of the membrane potential was increased in both polarities, which has never been anticipated (Iwamoto et al., 2018). This suggests that the first appearance time is not related to electric field-driven membrane insertion, but to physical features of the membrane, such as its thickness and bilayer tension

(Iwamoto & Oiki, 2018), facilitate the appearance of functional channels. There still remain questions about the process of channel formation after the partitioning to the membrane interface: Are the channels inserted into the membrane as monomers? Are the channels folded after membrane insertion? How and when are the channel proteins oligomerized?

There are several limitations in this system. Unlike the small cell size of bacteria, size of the bubble as large as several hundred micrometers in diameter must be taken into consideration. Some channel proteins synthesized in the center of the bubble were aggregated before they reached the membrane interface. While reaching the membrane surface, most channel proteins remained at the interface without trespassing the membrane. When channel proteins are inserted into the membrane, they hunt for other molecules for tetramerization through diffusion. Eventually, only a fraction of the synthesized proteins was functionally detected.

There remain some technical problems. (1) We are still unable to control the number of functional channels, even though protein synthesis can be stopped by lowering the temperature. A large number of synthesized proteins remained at the interface and were incorporated spontaneously into the membrane even after stopping the synthesis. (2) The membrane is not stable enough. The membrane breaks at higher membrane potential as the number of functional channels increase.

Apart from the goal of mimicking bacterial synthetic membrane, this technique is an efficient system for generating channel proteins *in bulla*. By adding liposomes into the bubble, more synthesized proteins are retrieved into liposomes without aggregations. Channel reconstituted liposomes diffuse to reach the membrane surface and fuse there. Even though the orientation of reconstituted channels may vary, more channel proteins should be functionally expressed.

We are now examining other channel proteins for functional expression. For more complex membrane proteins, having more transmembrane helices, a protein-transporting machinery might be necessary. Cell-free protein synthesis is a generic technology in synthetic biology, and metabolic networks are generated by synthesizing a combination of proteins. Thus, the *in bulla* functional expression system serves an elementary step toward the generation of artificial cells through functional expression of membrane proteins.

Acknowledgment

This work was supported in part by KAKENHI Grant numbers 25440067, 17K07360 (M.I.), and 16H00759, 17H04017 (S.O.).

References

Banerjee, S., & Nimigean, C. M. (2011). Non-vesicular transfer of membrane proteins from nanoparticles to lipid bilayers. *The Journal of General Physiology*, *137*(2), 217–223.

Bayley, H., Cronin, B., Heron, A., Holden, M. A., Hwang, W. L., Syeda, R., et al. (2008). Droplet interface bilayers. *Molecular BioSystems*, *4*(12), 1191–1208.

Cordero-Morales, J. F., Cuello, L. G., Zhao, Y., Jogini, V., Cortes, D. M., Roux, B., et al. (2006). Molecular determinants of gating at the potassium-channel selectivity filter. *Nature Structural and Molecular Biology*, *13*, 311–318.

Cymer, F., von Heijne, G., & White, S. H. (2015). Mechanisms of integral membrane protein insertion and folding. *Journal of Molecular Biology*, *427*(5), 999–1022.

Hessa, T., Kim, H., Bihlmaier, K., Lundin, C., Boekel, J., Andersson, H., et al. (2005). Recognition of transmembrane helices by the endoplasmic reticulum translocon. *Nature*, *433*, 377–381.

Hessa, T., Meindl-Beinker, N. M., Bernsel, A., Kim, H., Sato, Y., Lerch-Bader, M., et al. (2007). Molecular code for transmembrane-helix recognition by the Sec61 translocon. *Nature*, *450*, 1026–1030.

Iwamoto, M., Elfaramawy, M. A., Yamatake, M., Matsuura, T., & Oiki, S. (2018). Concurrent in vitro synthesis and functional detection of nascent activity of the KcsA channel under a membrane potential. *ACS Synthetic Biology*, 7(4), 1003–1011.

Iwamoto, M., & Oiki, S. (2013). Amphipathic antenna of an inward rectifier K^+ channel responds to changes in the inner membrane leaflet. *Proceedings of the National Academy of Sciences of the United States of America*, *110*(2), 749–754.

Iwamoto, M., & Oiki, S. (2015). Contact bubble bilayers with flush drainage. *Scientific Reports*, *5*, 9110.

Iwamoto, M., & Oiki, S. (2018). Constitutive boost of a K^+ channel via inherent bilayer tension and a unique tension-dependent modality. *Proceedings of the National Academy of Sciences of the United States of America*, *115*, 13117–13122.

Iwamoto, M., & Oiki, S. (2019). Lipid bilayer experiments with contact bubble bilayers for patch-clampers. *Journal of Visualized Experiments*, *143*, e58840.

Iwamoto, M., Shimizu, H., Inoue, F., Konno, T., Sasaki, Y. C., & Oiki, S. (2006). Surface structure and its dynamic rearrangements of the KcsA potassium channel upon gating and tetrabutylammonium blocking. *The Journal of Biological Chemistry*, *281*(38), 28379–28386.

Jewett, M. C., Calhoun, K. A., Voloshin, A., Wuu, J. J., & Swartz, J. R. (2008). An integrated cell-free metabolic platform for protein production and synthetic biology. *Molecular Systems Biology*, *4*, 220.

Kalmbach, R., Chizhov, I., Schumacher, M. C., Friedrich, T., Bamberg, E., & Engelhard, M. (2007). Functional cell-free synthesis of a seven helix membrane protein: In situ insertion of bacteriorhodopsin into liposomes. *Journal of Molecular Biology*, *371*(3), 639–648.

Kim, D. M., Kigawa, T., Choi, C. Y., & Yokoyama, S. (1996). A highly efficient cell-free, protein synthesis system from *Escherichia coli*. *European Journal of Biochemistry*, *239*(3), 881–886.

Kim, D. M., & Swartz, J. R. (2000). Prolonging cell-free protein synthesis by selective reagent additions. *Biotechnology Progress*, *16*(3), 385–390.

Mathews, M. B., & Korner, A. (1970). Mammalian cell-free protein synthesis directed by viral ribonucleic acid. *European Journal of Biochemistry*, *17*(2), 328–338.

Nirenberg, M. W., & Matthaei, J. H. (1961). The dependence of cell-free protein synthesis in *E. coli* upon naturally occurring or synthetic polyribonucleotides. *Proceedings of the National Academy of Sciences of the United States of America*, *48*, 104–109.

Oiki, S. (2012). Planar lipid bilayer method for studying channel molecules. In Y. Okada (Ed.), *Patch clamp techniques: From beginning to advanced protocols* (pp. 229–275). Tokyo: Springer-Verlag.

Oiki, S. (2015). Channel function reconstitution and re-animation: A single-channel strategy in the postcrystal age. *The Journal of Physiology, 593*(12), 2553–2573.

Oiki, S., & Iwamoto, M. (2018). Channel-membrane interplay in lipid bilayer membranes manipulated through monolayer technologies. *Biological & Pharmaceutical Bulletin, 41*, 303–311.

Ojemalm, K., Higuchi, T., Jiang, Y., Langel, U., Nilsson, I., White, S. H., et al. (2011). Apolar surface area determines the efficiency of translocon-mediated membrane-protein integration into the endoplasmic reticulum. *Proceedings of the National Academy of Sciences of the United States of America, 108*(31), E359–E364.

Posson, D. J., Thompson, A. N., McCoy, J. G., & Nimigean, C. M. (2013). Molecular interactions involved in proton-dependent gating in KcsA potassium channels. *The Journal of General Physiology, 142*, 613–624.

Sawasaki, T., Ogasawara, T., Morishita, R., & Endo, Y. (2002). A cell-free protein synthesis system for high-throughput proteomics. *Proceedings of the National Academy of Sciences of the United States of America, 99*(23), 14652–14657.

Shimizu, Y., Inoue, A., Tomari, Y., Suzuki, T., Yokogawa, T., Nishikawa, K., et al. (2001). Cell-free translation reconstituted with purified components. *Nature Biotechnology, 19*(8), 751–755.

Shimizu, Y., Kanamori, T., & Ueda, T. (2005). Protein synthesis by pure translation systems. *Methods, 36*(3), 299–304.

Shimono, K., Goto, M., Kikukawa, T., Miyauchi, S., Shirouzu, M., Kamo, N., et al. (2009). Production of functional bacteriorhodopsin by an *Escherichia coli* cell-free protein synthesis system supplemented with steroid detergent and lipid. *Protein Science, 18*(10), 2160–2171.

Shinoda, T., Shinya, N., Ito, K., Ishizuka-Katsura, Y., Ohsawa, N., Terada, T., et al. (2016). Cell-free methods to produce structurally intact mammalian membrane proteins. *Scientific Reports, 6*(1), 30442.

Sumino, A., Uchihashi, T., & Oiki, S. (2017). Oriented reconstitution of the full-length KcsA potassium channel in a lipid bilayer for AFM imaging. *The Journal of Physical Chemistry Letters, 8*, 785–793.

Urakubo, K., Iwamoto, M., & Oiki, S. (2019). Drop-in-well chamber for the droplet interface bilayer with built-in electrodes. *Methods in Enzymology, 621*.

Venkatesan, G. A., Lee, J., Farimani, A. B., Heiranian, M., Collier, C. P., Aluru, N. R., et al. (2016). Adsorption kinetics dictate monolayer self-assembly for both lipid-in and lipid-out approaches to droplet interface bilayer formation. *Langmuir, 31*, 12883–12893.

CHAPTER FOURTEEN

A mass spectrometry-based isotope-coded mass tag method to map thiol accessibility in biological systems

John E. Gadbery[a], Nicole S. Sampson[a,b,*]
[a]Biochemistry and Structural Biology Graduate Program, Stony Brook University, New York, NY, United States
[b]Department of Chemistry, Stony Brook University, New York, NY, United States
*Corresponding author: e-mail address: nicole.sampson@stonybrook.edu

Contents

Abstract

Interactions at biological membranes are important for many cellular functions, but because of the dynamic nature of these interactions, traditional methods of structure investigation are limited. Here, we describe a method that utilizes thiol–maleimide chemistry to monitor the solvent-accessible surface of membrane–protein complexes and membrane dynamics in vitro in real time. This method, called the isotope-coded mass tag (ICMT) method, has been used previously to locate thiols in transmembrane peptides, to elucidate lipid flipping kinetics in bilayers, to determine the oxidation state

Methods in Enzymology, Volume 621
ISSN 0076-6879
https://doi.org/10.1016/bs.mie.2019.02.025

of disulfide bonds, and to investigate the protein–lipid interface of the peripheral membrane protein cholesterol oxidase. The method requires only microgram quantities of protein or thiolated lipid, uses common laboratory equipment and routine instrumentation, and can be applied to many different systems to investigate macromolecular interactions in the presence or absence of biological membranes.

1. Introduction

1.1 Theory

One of the most common methods for studying protein–protein and protein–lipid topology and dynamics involves conjugation of a reporting probe to a nucleophilic amino acid. Such probes can be used for various applications, including determination of the distance between two amino acids or the solvent accessibility of an amino acid (Bogdanov, Zhang, Xie, & Dowhan, 2005; dos Remedios & Moens, 1995).

Cysteine is the most nucleophilic amino acid and is therefore widely used as a target for conjugation to probes (Bischoff & Schluter, 2012). Thiol-reactive probes typically have maleimide or iodoacetamide moieties, which form stable thioether linkages at physiological pH via Michael addition (Fig. 1). Cysteines have a pK_a of 8–9 in solution, which means that the reactive thiolate anion is the predominant species in alkaline aqueous environments. However, in hydrophobic environments, the pK_a of cysteine increases to 14, and cysteines in the hydrocarbon core of lipid membranes are therefore much less reactive than solvent-exposed cysteines (Bednar, 1990).

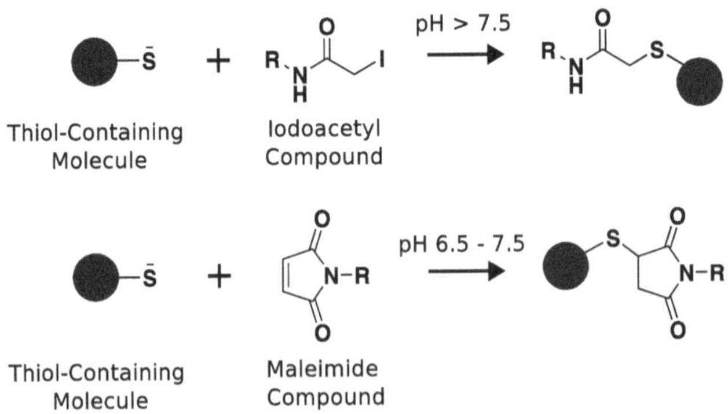

Fig. 1 Reactions of iodoacetyl- and maleimide-based probes with thiol-containing molecules.

Typically, probe-based methods rely on fluorescent molecules that are sensitive to their local environment. For example, in Förster resonance energy transfer (FRET) spectroscopy, two fluorophores, known as donor/acceptor FRET pairs, with overlapping emission/excitation spectra are chosen such that when the molecules are within a certain distance of each other, the acceptor is excited by emission from the donor. The intensity of the resulting acceptor fluorescence can then be used to measure the distance between the two fluorophores (dos Remedios & Moens, 1995; Kedziora & Jalink, 2015).

In addition, the λ_{max} and quantum yield of a single fluorescent molecule can be used to evaluate the hydrophobicity of its local environment. For example, the fluorophores acrylodan and monobromobimane, which can be covalently attached to cysteine residues, are sensitive to solvent polarity and can therefore be used to investigate protein–lipid and protein–protein interactions (Hibbs, Talley, & Taylor, 2004; Mansoor, McHaourab, & Farrens, 1999).

Although valuable information about the structures of protein–lipid complexes can be obtained by means of probe-based methods, they have several inherent disadvantages. Fluorescent probes typically contain long linker groups that separate the reporting group from the environment at the binding site, and this feature limits the resolution to the length of the linker group. Additionally, only a single site can be monitored at a time, so multiple proteins must be prepared if more than one site is to be investigated.

To elucidate macromolecular interactions, we have recently developed a mass-spectrometry-based method that involves isotope-coded thiol-reactive probes and overcomes many of the above-described limitations. This method, which we refer to as the isotope-coded mass tag (ICMT) method, uses two maleimide-based, membrane-impermeable probes that are synthesized in light and heavy (deuterated) forms (Figs. 2 and 3), and it takes advantage of the environment-dependent reactivity of cysteine residues. Like the substituted cysteine accessibility method (Bogdanov, 2017), the ICMT method can be used to monitor the solvent accessibility of cysteine residues in proteins and other biological membrane systems by means of a two-step labeling approach (Fig. 4). Cysteine mutations can be introduced into the protein, or native cysteines can be labeled.

In the first step, the protein in its native state is labeled with the heavy probe, and in the second step, the protein in its unfolded state is labeled with the light probe. The ratio of light probe labeling to heavy probe labeling

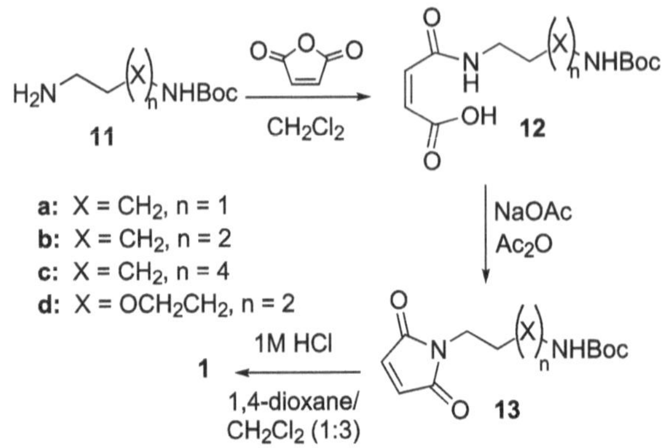

Fig. 2 Synthesis of ICMT probe precursors 1. Free amines 1 are prepared in three steps from mono-BOC-protected diamines and maleic anhydride. *Adapted with permission from Su, C. Y., London, E., & Sampson, N. S. (2013). Mapping peptide thiol accessibility in membranes using a quaternary ammonium isotope-coded mass tag (ICMT).* Bioconjugate Chemistry, 24, *1235–1247. Copyright 2013 American Chemical Society.*

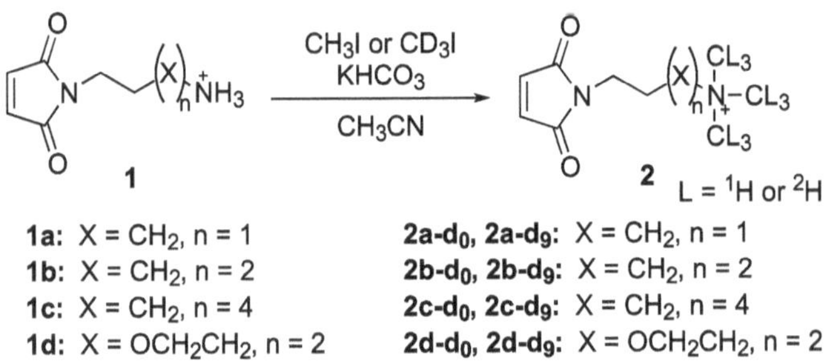

Fig. 3 Synthesis of ICMT maleimide probes. Permethylation of free amines 1 with iodomethane-d_0 (CH_3I) or iodomethane-d_3 (CD_3I) under mild basic conditions provides probes **2a-d_0–2d-d_0** or **2a-d_9–2d-d_9**, respectively. *Adapted with permission from Su, C. Y., London, E., & Sampson, N. S. (2013). Mapping peptide thiol accessibility in membranes using a quaternary ammonium isotope-coded mass tag (ICMT).* Bioconjugate Chemistry, 24, *1235–1247. Copyright 2013 American Chemical Society.*

can be followed as a function of time to determine the labeling rate, which corresponds to the solvent accessibility of each cysteine.

The labeled protein is proteolyzed into short peptides that are subsequently analyzed by matrix-assisted laser desorption/ionization time-of-flight (MALDI-TOF) mass spectrometry. In this way, multiple cysteines can be monitored in a single experiment, reducing the time spent constructing and assaying individual cysteine mutants or removing native cysteines.

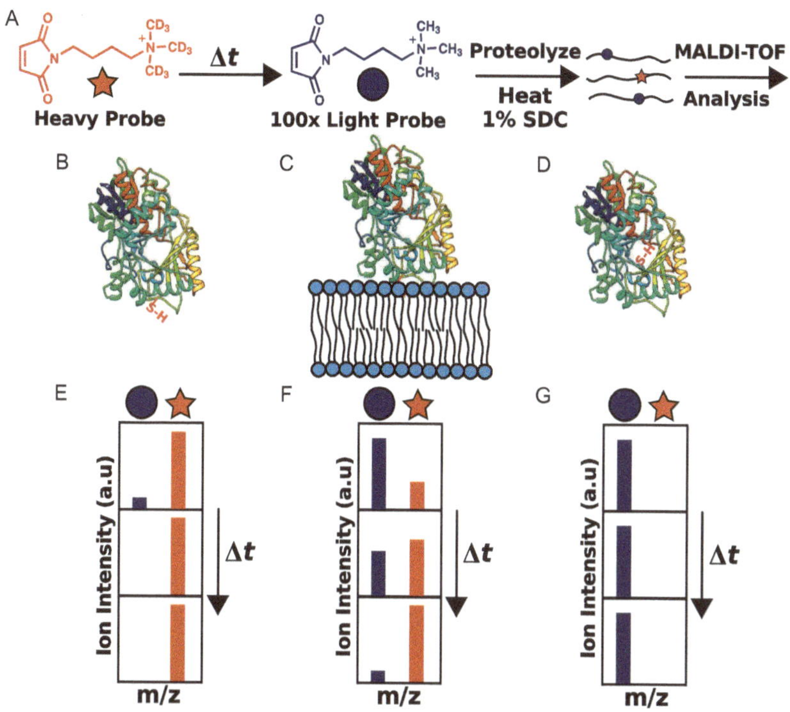

Fig. 4 Two-step ICMT labeling procedure. (A) Schematic of ICMT labeling procedure. (B–D) Cysteine-containing protein (PDB: 1MXT) is labeled with the heavy probe (red) in the presence of vesicles (C) or in buffer alone (B, D). At specified time points, an aliquot of the reaction mixture is added to a 100-fold molar excess of the light probe (blue). Then the reaction mixture is heated in the presence of a detergent to solubilize the membrane and unfold the protein, and the unfolded protein is digested into short peptides, which are analyzed via MALDI-TOF mass spectrometry. (E–G) The position of a cysteine residue with respect to the membrane can easily be determined from the MALDI-TOF spectra: (E) solvent-exposed residue in buffer alone, (F) the same residue at the protein–lipid interface in the presence of vesicles, and (G) a solvent-protected residue. *Adapted with permission from Gadbery, J. E., & Sampson, N. S. (2018). Use of an isotope-coded mass tag (ICMT) method to determine the orientation of cholesterol oxidase on model membranes.* Biochemistry, 57, *5370–5378. Copyright 2018 American Chemical Society.*

We selected a maleimide-based moiety as the thiol-reactive group because at pH 7.0, maleimides react with thiolate anions much faster and more specifically than do iodoacetamides, which are more reactive under slightly alkaline conditions (Rogers, Leinweber, & Smith, 2006). Additionally, the synthesis of maleimide derivatives is straightforward (Niwayama, Kassar, Zhao, Sutton, & Altenberg, 2011).

In this chapter we describe the procedures for using the ICMT method to map protein–lipid interfaces (Gadbery & Sampson, 2018), elucidate membrane dynamics (Su, London, & Sampson, 2013), and detect the redox state of disulfide bonds (Touchette et al., 2015).

1.2 Equipment and instrumentation

Dry heat block
Centrifuge
Micropipettes
Microcentrifuge tubes
Thermocycler or water bath
Timer
MALDI-TOF mass spectrometer

1.3 Materials

1.3.1 ICMT probe synthesis

Dichloromethane (CH_2Cl_2)
1,4-Dioxane
Diamines
Iodomethane-d_0 (CH_3I)
Iodomethane-d_3 (CD_3I)
Potassium bicarbonate ($KHCO_3$)
Acetonitrile (CH_3CN)
Sodium acetate (NaOAc)
Acetic anhydride (Ac_2O)
Hydrochloric acid (HCl)
Maleic anhydride ($C_4H_2O_3$)

1.3.2 ICMT labeling

HEPES
Protein to be labeled
Protein storage buffer
Lipid vesicles
Trifluoroacetic acid (TFA)
Sodium deoxycholate (SDC)
Triton X-100
Trypsin/chymotrypsin
2,5-Dihydroxybenzoic acid (DHB)
Acetonitrile
Tris(2-carboxyethyl)phosphine (TCEP)
ICMT probes

1.4 Solutions and reaction mixtures

1.4.1 ICMT probes (in ddH_2O)

	Stock concentration	Final concentration
First-Step Heavy ICMT Probe	0.5 m*M*	50 μ*M*
Second-Step Light ICMT Probe	50 m*M*	5 m*M*

1.4.2 Matrix solution (DHB)

DHB (2% w/v) in 7:3 0.1% TFA/acetonitrile

1.4.3 ICMT reaction mixtures

For protein–lipid interface mapping

	Stock concentration	Final concentration	Amount vesicle condition	Amount no-vesicle condition
Protein to be labeled[a]	As available	5 μ*M*	As available	As available
HEPES pH 7.0	500 m*M*	50 m*M*	45 μL	45 μL
Vesicles	6 m*M* (total lipid)	3 m*M* (total lipid)	225 μL	0 μL
ddH_2O	n/a	n/a	to 450 μL	to 450 μL
First-Step Heavy ICMT Probe	0.5 m*M*	50 μ*M*	45 μL	45 μL
Second-Step Light ICMT Probe[b]	50 m*M*	5 m*M*	5.5 μL	5.5 μL

[a]Amount depends on stock concentration of protein to be labeled, which may vary.
[b]Not present in the initial reaction.

For lipid flipping kinetics

	Stock concentration	Final concentration	Amount
1,2-Dipalmitoyl-*sn*-glycero-3-phosphatidylthioethanol/phosphatidylcholine vesicles (2/98)	8 m*M* (total lipid)	4 m*M* (total lipid)	225 μL
First-Step Heavy ICMT Probe	0.5 m*M*	50 μ*M*	45 μL
ddH_2O	n/a	n/a	to 450 μL
Second-Step Light ICMT Probe[a]	50 m*M*	5 m*M*	5.5 μL

[a]Not present in the initial reaction.

For redox state detection

	Stock concentration	Final concentration	Amount
Protein to be labeled	As available	5 μ*M*	45 μL
First-Step Heavy ICMT Probe	0.5 m*M*	50 μ*M*	5 μL
Second-Step Light ICMT Probe[a]	50 m*M*	5 m*M*	5 μL

[a]Not present in the initial reaction.

2. ICMT probe synthesis

2.1 Synthesis of ICMT probe precursors

tert-Butyl-aminoalkyl carbamates **11** were prepared with various diamines (Fig. 2) by means of the procedure reported by Guy et al. (2007), and then compounds **12** and **13** were synthesized from **11** using a modification of a procedure for preparing *N*-ethylmaleimides (Niwayama et al., 2011). Specifically, maleic anhydride was allowed to react with **11** to form the corresponding maleamic acid intermediates. Ring closure was achieved by dehydration with acetic anhydride/sodium acetate, and deprotection was performed with hydrochloric acid in 1:3 1,4-dioxane/CH_2Cl_2 to afford free terminal amines **1** (Su et al., 2013).

2.2 Synthesis of ICMT maleimide probes

Free amines **1** were permethylated with iodomethane or iodomethane-d_3 to obtained light and heavy ICMT probes **2**, respectively (Fig. 3).

3. ICMT labeling applications

With slight variations, the ICMT method can be used in a number of applications to map the accessibilities of thiols and cysteines.

3.1 ICMT labeling procedure to map protein–lipid interactions

For mapping of protein–lipid interactions, the protein of interest in its native state is first incubated with the heavy ICMT probe in the presence and absence of vesicles (Fig. 4A). Then, at specified time points, an aliquot of the initial reaction mixture is added to a 100-fold molar excess of the complementary light probe. Exposure to the light probe essentially quenches the heavy probe reaction, allowing for strict control of temporal resolution. Solvent-exposed cysteine(s) are rapidly labeled with the light probe (Fig. 4B and E), whereas cysteine(s) that are partially buried are labeled at a much slower rate (Fig. 4C and F). Completely buried cysteines are labeled only after the protein is unfolded and/or the membrane is solubilized with the aid of a detergent (Fig. 4D and G). SDC is the detergent of choice because it can be easily removed from solution by acidification with TFA, and SDC concentrations up to 1% are compatible with trypsin and chymotrypsin digestion. Upon addition of detergent, the samples are incubated at 70 °C to completely unfold the protein and ensure complete labeling of all cysteines by the light probe. The samples are then cooled to room temperature, and a protease is added to digest the protein into short peptides, which are then analyzed by MALDI-TOF mass spectrometry to obtain light probe/heavy probe labeling ratios at each time point.

Labeling protocol

(1) To prepare the vesicle and no-vesicle reaction mixtures, combine all the reagents listed in the ICMT Reaction Mixtures section except for the ICMT probes. For the no-vesicle reaction add an equivalent volume of ddH_2O in lieu of vesicles. Gently vortex the solutions and allow them to incubate for 15 min at 30 °C in the dry heat block.

Note: For vesicle preparation see (Akbarzadeh et al., *2013)*

(2) Add 45 μL 0.5 m*M* first-step heavy ICMT probe to the vesicle-containing reaction mixture while simultaneously starting the timer.

(3) When the timer reads exactly 10 s, add 45 μL 0.5 m*M* first-step heavy ICMT probe to the no-vesicle reaction mixture.

Note: By delaying the no-vesicle experiment by 10 s, the two reactions can be run in parallel, limiting the variability between the two experiments for accurate results.

(4) When the timer reads exactly 30 s, remove a 50 μL aliquot of the vesicle-containing reaction mixture and place it in a 600 μL microcentrifuge tube containing 5.5 μL 50 m*M* second-step light ICMT probe.

(5) When the timer reads exactly 40 s, remove a 50 μL aliquot from the no-vesicle reaction mixture and place it in a 600 μL microcentrifuge tube containing 5.5 μL 50 m*M* second-step light ICMT probe.

Tip: Especially at the early time points and at time points that are close together, it is often necessary to pre-pipette the second-step light ICMT probe into a series of tubes before the first-step heavy ICMT probe is added to the reaction mixtures (before step 2).

(6) At each additional time point (1 min, 2 min, 5 min, etc.), repeat step 4 and step 5 (each time point +10 s), removing a 50 μL aliquot and ensuring that each aliquot is pipetted exactly at the specified time point.

Vesicle membrane lysis and protein denaturation

(1) Add SDC (final concentration, 1% [w/v]) to each tube, and briefly vortex the contents.

(2) Incubate the tubes for 1 h at 70 °C in a thermocycler or water bath to fully solubilize the vesicle membranes and unfold the protein.

Protease digestion

(1) Remove the tubes from the thermocycler or water bath and allow them to cool to room temperature. To each tube, add protease (chymotrypsin or trypsin) to a final protein/protease ratio of 30:1–100:1 and gently mix the contents.

(2) Incubate the tubes for 12–16 h at 25 °C for chymotrypsin or at 37 °C for trypsin.

MALDI-TOF sample preparation

(1) Stop the protease digestion by adding TFA (final concentration, 3% [v/v]) to each tube. Addition of TFA also precipitates the SDC. Briefly vortex each tube and then incubate the tubes at 25 °C for 10 min.

(2) Centrifuge the tubes for 10 min at 18,000 × *g*.

(3) From each tube, carefully remove the supernatant and mix it with the DHB matrix solution in a 1:1 (v/v) ratio.

(4) Spot 1 μL of each sample/matrix mixture on a MALDI target plate and allow the spot to dry completely. Repeat up to four times for a total of up to five spots.

3.2 ICMT labeling procedure to measure lipid flipping kinetics

If 1,2-dipalmitoyl-*sn*-glycero-3-phosphatidylthioethanol, which contains a terminal thiol in the headgroup, is present as a constituent of synthetic lipid vesicles, the ICMT method can be used to obtain information about the fluid state of the vesicle membrane and to measure transverse diffusion rates of lipids.

For this application, the two-step ICMT labeling procedure is essentially the same as that described for analysis of protein–lipid interactions. However, a different detergent is selected, and the proteolysis step is omitted. Because there is no proteolysis step, Triton X-100 can be used as the detergent, and it can be added at a concentration low enough not to interfere with acquisition of the MALDI-TOF mass spectra.

Labeling protocol

(1) Prepare 100–200 nm unilamellar 1,2-dipalmitoyl-*sn*-glycero-3-phosphatidylthioethanol/phosphatidylcholine (2/98) vesicles (4 m*M* total lipid) and incubate them for 5 min at 30 °C in the dry heat block.

(2) Add 45 μL 0.5 m*M* first-step heavy ICMT probe to the vesicles while simultaneously starting the timer.

(3) At specified time points, remove a 50 μL aliquot of the reaction mixture and place it in a 600 μL microcentrifuge tube containing 5.5 μL 50 m*M* second-step light ICMT probe.

(4) Repeat step 3 at each time point, ensuring that each aliquot is pipetted exactly at the specified time point.

Vesicle membrane lysis

(1) Add Triton X-100 (final concentration, 0.03% [w/v]) to each tube, and briefly mix the contents of the tubes.

(2) Incubate the tubes at 25 °C for 1 h.

MALDI-TOF sample preparation

(1) Mix each sample with the DHB matrix solution in a 1:1 ratio.

(2) Spot 1 μL of each sample/matrix solution on a MALDI target plate and allow the spot to dry completely. Repeat up to four times for a total of up to five spots.

3.3 ICMT labeling procedure to determine redox state of disulfide bonds

The ICMT method can be used to determine the redox state of disulfide bonds; whether a given cysteine is expected to participate in a disulfide bond can be determined by evaluating whether it can be readily labeled by the ICMT probe. Compared with the above-described applications of the ICMT method, this application requires more substantial modification of the basic procedure.

First, the protein is incubated with the first-step heavy ICMT probe. Next, a 100-fold excess of the complementary second-step light ICMT probe is added, along with a 100-fold molar excess (with respect to protein concentration) of TCEP to reduce all the disulfide bonds. If the target cysteines exist as disulfides, they will be labeled only by the second-step light ICMT probe. Conversely, cysteines not involved in a disulfide bond will be labeled with the first-step heavy ICMT probe. TCEP is used as a reducing agent because it contains no thiols and is therefore compatible with downstream labeling of the protein with the ICMT probes. However, to minimize the potential for interference with ICMT probe labeling, excess TCEP is removed via buffer exchange before the final labeling step. The subsequent denaturation and unfolding steps are carried out by a procedure identical to that in the applications described previously.

For disulfide bonds that are slow to reduce, an alternative method can be employed. Proteins containing such bonds can be incubated with TCEP for a long time to ensure reduction of the disulfide bonds prior to labeling. The TCEP-treated and untreated proteins can then be labeled with either the light or the heavy ICMT probe. Upon MALDI-TOF mass spectrometry, the TCEP-treated protein should show a single peak corresponding to light or heavy probe labeling. The untreated protein will be unlabeled, and thus no peak will be observed.

Labeling protocol

(1) Incubate the protein (5 μ*M* in 45 μL protein storage buffer) for 15 min at 30 °C in a dry heat block.

(2) Add 5 μL 0.5 m*M* first-step heavy ICMT probe (50 μ*M* final concentration) and incubate for 1 h at 30 °C.

Note: The incubation time and temperature can be altered depending on the stability of the protein.

(3) Add 5.5 μL 50 m*M* second-step light ICMT probe and briefly mix the contents.

(4) Add a 100-fold molar excess of TCEP and incubate for 1 h at 30 °C.

(5) Using a 500 μL microfuge spin concentrator with an appropriate molecular weight cutoff, concentrate the sample 10-fold and then dilute the residue 10-fold to exchange the buffer and remove excess TCEP.

(6) Repeat step 5 at least three to five times to ensure sufficient reduction of the TCEP concentration.

(7) Repeat the second labeling step (step 3) by adding 5 μL 50 m*M* second-step light ICMT probe.

(8) Add SDC (final concentration, 1% [w/v]) and incubate for 1 h at 70 °C in a thermocycler or water bath.

Protease digestion

(1) Remove from the thermocycler or water bath and allow them to cool to room temperature.

(2) Add protease to a final protein/protease ratio of 30:1–100:1 to each tube and gently mix the contents.

(3) Incubate the tubes for 12–16 h at 25 °C for chymotrypsin or at 37 °C for trypsin.

MALDI-TOF sample preparation

(1) Stop the protease digestion by adding TFA (final concentration, 3% [v/v]) to each tube. Addition of TFA also precipitates the SDC. Briefly vortex each tube and incubate at 25 °C for 10 min.

(2) Centrifuge the tubes for 10 min at 18,000 × *g*.

(3) From each tube, carefully remove the supernatant and mix it with the DHB matrix solution in a 1:1 ratio.

(4) Spot 1 μL of each sample/matrix mixture on a MALDI target plate and allow the spot to dry completely. Repeat up to four times for a total of five spots.

4. Acquisition and analysis of MALDI-TOF mass spectra

MALDI-TOF mass spectrometry is the most effective method for analyzing ICMT-labeled peptides or lipids, and the ICMT probes have been optimized accordingly. Specifically, the positive charge on the quaternary ammonium group aids in the ionization of the labeled species, and there are no isotope effects inherent to MALDI-TOF mass spectrometry, which makes integration and comparison of light and heavy labeled peaks more straightforward than is the case for other methods, such as liquid chromatography/mass spectrometry. Additionally, acquisition of MALDI-TOF mass spectra is a high-throughput process, and the data can be rapidly acquired and analyzed.

For acquisition of mass spectrometry data, the MALDI-TOF instrument should be operated in positive mode, because the labeled species carry a positive charge, and in reflectron mode to enhance the resolution of each peak. The theoretical m/z value of a peptide containing a labeled cysteine of interest can be determined by using a mass-spectrometry-based sequence editing program. Once the spectra are acquired, they can be analyzed using software such as Bruker flexAnalysis.

4.1 Redox state detection

The analysis of labeled peptides resulting from redox state detection experiments is straightforward. The peak that corresponds to a peptide containing a cysteine of interest will be labeled either with the second-step light probe or with the first-step heavy probe. Because the results are based purely upon which probe labels the cysteines, no further analysis is required.

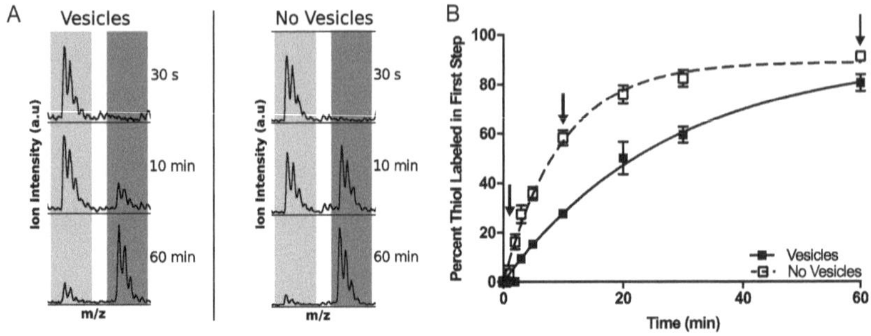

Fig. 5 Analysis of ICMT Labeling in Protein–Lipid Dynamics Experiments. (A) MALDI-TOF spectra of light and heavy labeled cysteine-containing peptides (light and dark gray, respectively) in the presence (left) and absence of vesicles (right). (B) ICMT labeling as a function of time. Arrows represent time points shown in panel A. *Adapted with permission from Gadbery, J. E., & Sampson, N. S. (2018). Use of an isotope-coded mass tag (ICMT) method to determine the orientation of cholesterol oxidase on model membranes.* Biochemistry, 57, *5370–5378. Copyright 2018 American Chemical Society.*

4.2 Protein–lipid interactions and membrane dynamics

Determination of the rate of ICMT labeling in kinetics experiments requires that the isotopic envelope of each light and heavy labeled peptide (Fig. 5A) be calculated. This calculation can be done using the flexAnalysis software. The following equation can be used to determine the percentage of first-step heavy-labeled peptide (%H) at each time point:

$$\text{Percentage heavy labeled }(\%\text{H}) = \frac{\text{Area}_{\text{heavy labeled}}}{\left(\text{Area}_{\text{heavy labeled}} + \text{Area}_{\text{light labeled}}\right)}$$

The %H at each time point can then be plotted vs time to generate a one-phase exponential association curve (Fig. 5B) from which the observed rate of labeling, $k_{\text{obs}}^{\text{ICMT}}$, for each cysteine can be acquired by fitting the data to the following equation:

$$Y = Y_{\max}\left(1 - e^{-kt}\right)$$

where Y is the %H at each time point, $Y_{\max}$ is the maximum %H, k is the observed rate constant, and t is time.

From the $k_{\text{obs}}^{\text{ICMT}}$ for the vesicle and no-vesicle conditions, one can then determine, in the case of protein–lipid interactions, which residues are at the protein–lipid interface. In the case of membrane dynamics $k_{\text{obs}}^{\text{ICMT}}$ can determine the relative fluid states of the membrane and resulting transverse diffusion rates of lipids between leaflets.

An additional consideration when determining k_{obs}^{ICMT} is ensuring that pseudo-first-order reaction conditions are maintained. In systems containing multiple cysteines, this is challenging because determining the amount of available ICMT probe for reaction with each cysteine during the experiment is difficult. In these systems, it is best, therefore, to first utilize the ICMT method to qualitatively determine differences in labeling rates to identify the region where interactions occur, that is, where there is simply a difference in labeling rates between experiments conducted in the presence and absence of vesicles. Single cysteine mutations can then be targeted to this region if more-accurate quantitative labeling rates are desired.

5. Conclusions and future perspectives

The ICMT method has several advantages over other labeling methods and can be easily adapted for use in many different systems to determine thiol accessibilities. The synthesis of the ICMT probes is straightforward, and analysis by means of MALDI-TOF mass spectrometry enables a high-throughput workflow. Because the mass of the labeled peptides is being detected, there is no loss of spatial resolution, which can be a limiting factor with other fluorescence-based labeling methods. Additionally, only microgram quantities of protein are required, and the method does not require specialized equipment or complicated experimental setups. A routine MALDI-TOF mass spectrometer is utilized for detection and this instrumentation is typically available to non-specialist laboratory researchers.

We expect that the ICMT method will find applications beyond those described herein. For example, this method has the potential to be useful for mapping thiol accessibilities of membrane proteins and membrane–protein complexes in living cells, information that is otherwise difficult to obtain.

References

Akbarzadeh, A., Rezaei-Sadabady, R., Davaran, S., Joo, S. W., Zarghami, N., Hanifehpour, Y., et al. (2013). Liposome: Classification, preparation, and applications. *Nanoscale Research Letters*, *8*(1), 102. https://doi.org/10.1186/1556-276X-8-102.

Bednar, R. A. (1990). Reactivity and pH dependence of thiol conjugation to N-ethylmaleimide: Detection of a conformational change in chalcone isomerase. *Biochemistry*, *29*(15), 3684–3690. https://doi.org/10.1021/bi00467a014.

Bischoff, R., & Schluter, H. (2012). Amino acids: Chemistry, functionality and selected non-enzymatic post-translational modifications. *Journal of Proteomics*, *75*(8), 2275–2296. https://doi.org/10.1016/j.jprot.2012.01.041.

Bogdanov, M. (2017). Mapping of membrane protein topology by substituted cysteine accessibility method (SCAM). *Methods in Molecular Biology*, *1615*, 105–128. https://doi.org/10.1007/978-1-4939-7033-9_9.

Bogdanov, M., Zhang, W., Xie, J., & Dowhan, W. (2005). Transmembrane protein topology mapping by the substituted cysteine accessibility method (SCAM(TM)): Application to lipid-specific membrane protein topogenesis. *Methods*, *36*(2), 148–171. https://doi.org/10.1016/j.ymeth.2004.11.002.

dos Remedios, C. G., & Moens, P. D. (1995). Fluorescence resonance energy transfer spectroscopy is a reliable "ruler" for measuring structural changes in proteins. Dispelling the problem of the unknown orientation factor. *Journal of Structural Biology*, *115*(2), 175–185. https://doi.org/10.1006/jsbi.1995.1042.

Gadbery, J. E., & Sampson, N. S. (2018). Use of an isotope-coded mass tag (ICMT) method to determine the orientation of cholesterol oxidase on model membranes. *Biochemistry*, *57*, 5370–5378. https://doi.org/10.1021/acs.biochem.8b00788.

Guy, J., Caron, K., Dufresne, S., Michnick, S. W., Skene, W. G., & Keillor, J. W. (2007). Convergent preparation and photophysical characterization of dimaleimide dansyl fluorogens: Elucidation of the maleimide fluorescence quenching mechanism. *Journal of the American Chemical Society*, *129*(39), 11969–11977. https://doi.org/10.1021/ja0738125.

Hibbs, R. E., Talley, T. T., & Taylor, P. (2004). Acrylodan-conjugated cysteine side chains reveal conformational state and ligand site locations of the acetylcholine-binding protein. *The Journal of Biological Chemistry*, *279*(27), 28483–28491. https://doi.org/10.1074/jbc.M403713200.

Kedziora, K. M., & Jalink, K. (2015). Fluorescence resonance energy transfer microscopy (FRET). *Methods in Molecular Biology*, *1251*, 67–82. https://doi.org/10.1007/978-1-4939-2080-8_5.

Mansoor, S. E., McHaourab, H. S., & Farrens, D. L. (1999). Determination of protein secondary structure and solvent accessibility using site-directed fluorescence labeling. Studies of T4 lysozyme using the fluorescent probe monobromobimane. *Biochemistry*, *38*(49), 16383–16393. https://doi.org/10.1021/bi991331v.

Niwayama, S., Kassar, A. S., Zhao, T., Sutton, R. B., & Altenberg, G. A. (2011). A pyrene maleimide with a flexible linker for sampling of longer inter-thiol distances by excimer formation. *PLoS One*, *6*(10), e26691. https://doi.org/10.1371/journal.pone.0026691.

Rogers, L. K., Leinweber, B. L., & Smith, C. V. (2006). Detection of reversible protein thiol modifications in tissues. *Analytical Biochemistry*, *358*(2), 171–184. https://doi.org/10.1016/j.ab.2006.08.020.

Su, C. Y., London, E., & Sampson, N. S. (2013). Mapping peptide thiol accessibility in membranes using a quaternary ammonium isotope-coded mass tag (ICMT). *Bioconjugate Chemistry*, *24*(7), 1235–1247. https://doi.org/10.1021/bc400171j.

Touchette, M. H., Bommineni, G. R., Delle Bovi, R. J., Gadbery, J. E., Nicora, C. D., Shukla, A. K., et al. (2015). Diacyltransferase activity and chain length specificity of mycobacterium tuberculosis PapA5 in the synthesis of alkyl beta-diol lipids. *Biochemistry*, *54*, 5457–5468. https://doi.org/10.1021/acs.biochem.5b00455.

CHAPTER FIFTEEN

Quick-soaking of crystals reveals unprecedented insights into the catalytic mechanism of glycosyltransferases

David Albesa-Jové[a,b], Javier O. Cifuente[a], Beatriz Trastoy[a], Marcelo E. Guerin[a,b,*]

[a]Structural Biology Unit, CIC bioGUNE, Derio, Spain
[b]IKERBASQUE, Basque Foundation for Science, Bilbao, Spain
*Corresponding author: e-mail address: mrcguerin@cicbiogune.es

Contents

Abstract

Glycosyltransferases (GTs) catalyze the transfer of a sugar moiety from nucleotide-sugar or lipid-phospho-sugar donors to a wide range of acceptor substrates, generating a remarkable amount of structural diversity in biological systems. Glycosyl transfer reactions can proceed with either inversion or retention of the anomeric configuration with respect to the sugar donor substrate. In this chapter, we discuss the application of a quick soaking method of substrates and products into protein crystals to visualize native ternary complexes of retaining glycosyltransferases. The crystal structures provide different snapshots of the catalytic cycle, including the Michaelis complex. During this sequence of events, we visualize how the enzyme guides the substrates into the reaction center where the

Methods in Enzymology, Volume 621
ISSN 0076-6879
https://doi.org/10.1016/bs.mie.2019.02.034

glycosyl transfer reaction takes place, and unveil the mechanism of product release, involving multiple conformational changes not only in the substrates and products but also in the enzyme. The methodology described here provides unprecedented insights into the catalytic mechanism of glycosyltransferases at the molecular level, and can be applied to the study a myriad of enzymatic mediated reactions.

1. Introduction

Glycosyltransferases (GTs) catalyze the transfer of a sugar residue from nucleotide-sugar or lipid-phospho-sugar donors to a variety of acceptor molecules, including mono-, oligo- and polysaccharides, lipids, proteins, small organic molecules and nucleic acids (Albesa-Jové & Guerin, 2016; Lairson, Henrissat, Davies, & Withers, 2008). As a consequence, GTs generate substantial structural diversity, particularly important for biological processes, such as cell signaling, cell-pathogen interactions, and the maintenance of the cellular architecture and metabolism (Rini & Esko, 2017). GTs are highly selective in nature, allowing the recognition of subtle structural differences in the sequences and stereochemistry of their sugar substrates. Based on amino acid sequence similarities GTs have been distinguished in 106 different families (CAZy database, http://www.cazy.org/; Lombard, Golaconda Ramulu, Drula, Coutinho, & Henrissat, 2014). Interestingly, only three structural folds have been described among the families for which three-dimensional structures have been reported. The GT-A and GT-B topologies are variations of Rossmann fold domains, and employ nucleoside diphosphate activated sugars as donors (Albesa-Jové & Guerin, 2016; Breton, Fournel-Gigleux, & Palcic, 2012). In addition, many of the uncharacterized families are also predicted to adopt one of these two folds, suggesting that the majority of GTs has evolved from a small number of progenitor sequences (Lairson et al., 2008). In contrast, GT-C protein topology displays an integral polytopic membrane domain that recognize the lipid-phospho-sugar donor, and a variable soluble domain mainly involved in acceptor binding, with the reaction center found at the interface between these two domains (Rini & Esko, 2017; Wild et al., 2018).

2. Reaction mechanisms of inverting and retaining glycosyltransferases

Glycosyl transfer reactions can proceed with retention or inversion of the anomeric configuration (Albesa-Jové & Guerin, 2016; Ardèvol & Rovira, 2015; Lairson et al., 2008; Sinnott, 1990). Importantly, both inverting and retaining glycosyltransferases have been identified within both the GT-A and GT-B fold classes (Lairson et al., 2008). Analogous to inverting glycoside hydrolases, inverting GTs proceed via a SN_2 reaction in a single displacement mechanism that involves a residue acting as a general base catalyst that activates the acceptor substrate and the formation of an oxocarbenium ion-like transition state (Fig. 1A; (Ardèvol & Rovira, 2015). In contrast, the reaction mechanism of retaining GTs is a matter of debate. By analogy with glycoside hydrolases, a double-displacement mechanism involving a covalently bound glycosyl-enzyme intermediate

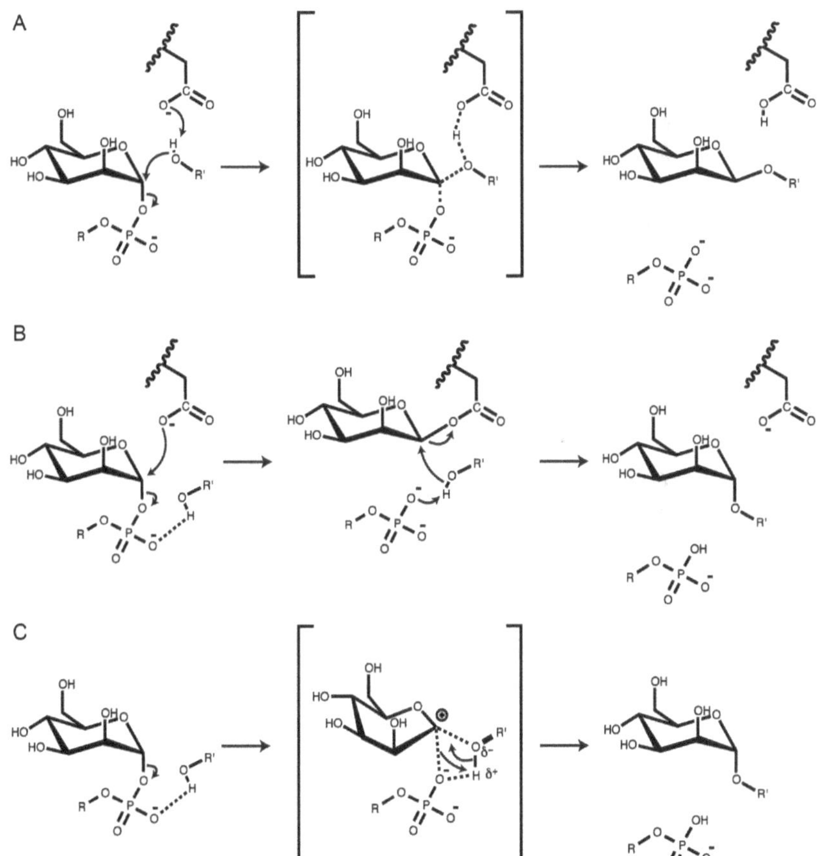

Fig. 1 Proposed reaction mechanisms for enzymatic glycosyl transfer. (A) Single displacement mechanism proposed for inverting GTs. (B) Double displacement mechanism proposed for retaining GTs containing a negatively charged residue on the β-face of the sugar donor. (C) Concerted front-face mechanism proposed for retaining GTs without a negatively charged residue on the β-face of the sugar donor.

was first proposed (Fig. 1B; Koshland, 1953). This mechanism requires the existence of a correctly positioned nucleophile (Asp or Glu residue) in the catalytic center that could mediate the formation of a glycosyl-enzyme intermediate during the first part of the reaction, resulting in a first anomeric inversion. In the second part of the reaction, the activated acceptor substrate attacks the glycosyl-enzyme intermediate, resulting in the product formation with total retention of the anomeric configuration (Lairson et al., 2008). However, to date, only members of family-6 GT have been identified to display such a putative nucleophile. In the absence of a residue near the reaction center that can act as a nucleophile to form the glycosyl-enzyme intermediate, an alternative mechanism known as the SNi "internal return," also called the SNi-like mechanism, has been proposed for retaining GTs (Fig. 1B and C; Albesa-Jové & Guerin, 2016; Ardèvol & Rovira, 2015; Lairson et al., 2008; Sinnott, 1990). In this mechanism, leaving group departure and nucleophilic attack occur on the same face of the sugar (Lee et al., 2011) and involve either a short-lived oxocarbenium ion intermediate (SNi like; Ardèvol & Rovira, 2011; Bobovská, Tvaroška, & Kóňa, 2014; Gómez, Lluch, & Masgrau, 2013; Lira-Navarrete et al., 2014) or an oxocarbenium ion transition state (SNi; Gómez, Polyak, Thiel, Lluch, & Masgrau, 2012).

Understanding the molecular basis of enzyme catalysis has been a major goal of biochemistry and structural biology. The study of enzyme structures at the molecular level of detail has been possible thanks to the implementation of X-ray crystallography (Blow & Steitz, 1970) The visualization of a native GT in complex with natural substrate(s) before the reaction takes place, including the Michaelis complex, would contribute to understand the substrate binding and catalytic mechanism. Similarly, the observation of structural snapshots of a GT in complex with natural products would provide unique information on the release mechanism along the catalytic cycle. Obtaining such native enzyme-substrate(s) and/or enzyme-product(s) complexes remain a challenging task. To further advance the understanding of the catalytic mechanism of glycosyl transfer reactions, we investigated (i) the glucosyl-3-phosphoglycerate synthase (GpgS) from *Mycobacterium tuberculosis* (Albesa-Jové et al., 2015; Albesa-Jové, Romero-García, et al., 2017; Urresti et al., 2012) and (ii) the α1, 3-galactosyltransferase (α3GalT) from *Bos taurus,* as models of retaining GTs lacking or containing a putative nucleophile in the active site, respectively (Albesa-Jové, Sainz-Polo, Marina, & Guerin, 2017).

3. GpgS, a retaining glycosyltransferase lacking a putative catalytic nucleophile

Mycobacteria assemble the intracellular amphiphilic 6-*O*-methylglucose lipopolysaccarides (MGLPs), a group of molecules postulated to be involved in the regulation of fatty acid metabolism, due to their ability to form stable 1:1 complexes with fatty acyl chains (Jackson & Brennan, 2009; Nunes-Costa, Maranha, Costa, Alarico, & Empadinhas, 2017). Glucosyl-3-phosphoglycerate Synthase (GpgS, EC 2.4.1.266) catalyzes the first step of the pathway, by transferring a glucosyl residue from UDP-Glc to the 2-position of 3-phospho-D-glycerate (3PGA) to bring forth the precursor 2-*O*-(α-D-glucosyl)-3-phospho-D-glycerate (GPG) and UDP (Empadinhas, Albuquerque, Mendes, Macedo-Ribeiro, & da Costa, 2008; Kaur et al., 2009). GpgS requires the presence of a divalent metal cation to facilitate catalysis, with Mg^{2+} displaying maximum activity *in vitro* (Urresti et al., 2012). GpgS was proposed to use a sequential ordered mechanism, with the acceptor PGA binding first, followed by UDP-glucose (Kumar, Guan, & Frantom, 2014).

GpgS is a homodimer, with each protomer displaying a classic GT-A architecture composed of two side-by-side tightly bound Rossmann-fold domains (Fig. 2A; Albesa-Jové et al., 2015; Albesa-Jové, Romero-García, et al., 2017; Fulton et al., 2008; Urresti et al., 2012). The N-terminal domain contains the UDP-glucose binding site, whereas the C-terminal domain holds the 3PGA acceptor binding site (Fig. 2A). The C-terminal extension participates into the stabilization of the dimeric architecture. The active site contains four loops that play a prominent role in catalysis including: (i) β5-α6 loop (residues 133–142) contains the Asp-Xaa-Asp motif (residues 134–136) involved in metal coordination and the UDP-Glc binding, (ii) β6-α7 loop (residues 165–183) located in the neighborhood of the 3PGA binding site, and absent in the density maps, (iii) α7-β7 loop (residues 204–212), located at the base of the active site, provides important interactions with both substrates, and (iv) the "L" Loop (residues 253–262) located at the interface between the UDP-Glc and 3-PGA binding sites. Importantly, different crystal structures disclosed that the L-loop can undergo drastic conformational changes along the catalytic cycle (Albesa-Jové, Romero-García, et al., 2017). GpgS does not display a putative nucleophilic residue positioned within the active site to act as receptor of a covalent intermediate for a double displacement mechanism (Albesa-Jové et al., 2015).

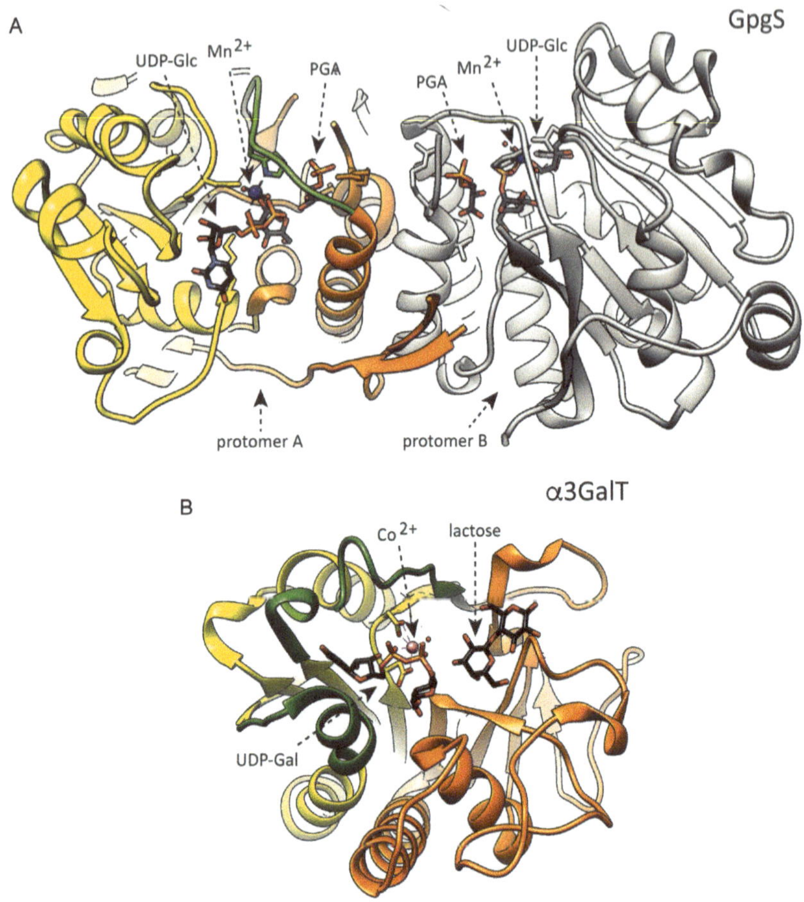

Fig. 2 Overall fold of GpgS and α3GalT. (A) Dimeric GpgS with UDP-Glc, Mn^{2+} and PGA. (B) Monomeric α3GalT with UDP-Gal, Co^{2+} and lactose (LAT).

4. α3GalT, a retaining glycosyltransferase containing a putative catalytic nucleophile

The crystal structures of only four GTs has been reported to contain a putative nucleophile in the active site, all classified into the GT6 family: α3GalT (Boix et al., 2001; Gastinel et al., 2001), the two ABO(H) blood group A and B glycosyltransferases, the *N*-acetylgalactosaminyltransferase (GTA) and galactosyltransferase (GTB; (Alfaro et al., 2008; Patenaude et al., 2002), and the UDP-GalNAc:2′-fucosyl lactose α-*N*-acetylgalactosaminyltransferase (BoGT6a; (Pham et al., 2014). Several lines of experimental data suggest a role of this residue in the formation of a covalent glycosyl-enzyme intermediate. Mutation of the putative nucleophile in several GT6 family members resulted in a significant reduction of the catalytic activity but also in the acceptor binding efficiency (Blackler et al., 2017; Zhang et al., 2003). Chemical rescue of α3GalT Glu317Ala mutant by using azide as an exogenous nucleophile results in the cleavage of UDP-α-Gal to give β-D-galactosyl azide (Monegal & Planas, 2006). A covalent adduct has been identified by mass spectrometry in GTA/GTB cysteine mutants of the putative nucleophile Glu303, structurally equivalent to Glu317 in α3GalT, but not in the wild type enzyme (Soya, Fang, Palcic, & Klassen, 2011). However, they do not provide definitive evidence of the role of this residue during catalysis in the native enzymes (Lairson et al., 2008). Experimental observation of the glycosyl-enzyme intermediate by X-ray crystallography has proven elusive (Gastinel et al., 2001; Lairson et al., 2008). Computational studies available for α3GalT, support the occurrence of a covalent glycosyl-enzyme intermediate and/or a stepwise front-face reaction with the formation of an ion-pair intermediate (SNi-like mechanism; Bobovská, Tvaroška, & Kóňa, 2015; Gómez et al., 2013; Rojas-Cervellera, Ardèvol, Boero, Planas, & Rovira, 2013).

α3GalT from *Bos taurus* catalyzes the synthesis of galactosyl-α-1,3-β-galactosyl structures in mammalian glycoconjugates. Steady state kinetics (Zhang, Wang, & Brew, 2001) indicates that α3GalT transfers a galactosyl moiety from UDP-Gal to β-linked galactosides, such as lactose, by a sequential mechanism (Zhang et al., 2003), with an expected obligatory ordered binding of donor and acceptor substrates (Boix, Zhang, Jawahar Swaminathan, Brew, & Ravi Acharya, 2002). The catalytic core of α3GalT consists of two tightly bound Rossmann fold domains typical of GT-A GTs, and contains a conserved Asp-Xaa-Asp motif that participates in divalent metal cation and UDP-Gal binding (Fig. 2B; Lairson et al., 2008). The enzyme shows maximum enzymatic activity with Mn^{2+}, followed by Co^{2+}, and Zn^{2+} (Zhang et al., 2001). Conformational changes in α3GalT are essential for substrate binding and catalysis, in particular three regions close to the active site allows assembly of the catalytic center: two internal loops (residues 136–139 and 190–202) and a C-terminal extension (residues 356–368; Gagnon et al., 2018; Zhang et al., 2001).

5. Quick-soaking protocol

Soaking techniques have been pivotal to the development of macromolecular crystallography. Since early days, soaking has been used for classic heavy-atom derivatization to phase the first macromolecular crystal structures (Kendrew et al., 1958; Perutz et al., 1960). Soaking has been also employed for changing the crystal's mother liquor to a stabilization solution containing cryo-protecting reagents prior to cooling them with liquid nitrogen (Haas, 1968; Teng, 1990). Later, quick-soaking techniques have been introduced for fast heavy-atom derivatization to minimize non-isomorphism or crystal damage (Sun, Radaev, & Kattah, 2002), which has further been refined to very short soakings for phasing with halides that bind to the ordered solvent region around protein molecules in the crystal (Dauter, Dauter, & Rajashankar, 2000). With the development of structure-based drug design, quick-soaking techniques have also been extensively utilized for identification, structural assessment and optimization of lead compounds (Blundell, Jhoti, & Abell, 2002; Nienaber et al., 2000). Structural enzymology has made extensive use of soaking techniques to study the molecular bases for enzyme-substrate recognition, in which the diffusion of ligands/ions into a protein crystal can be very fast (Roberts & Davies, 2012). Nonetheless, most of the work in this area has avoided the use of native enzymes and/or unmodified substrates. Instead, work has focused on the use of inactivated enzymes and/or substrate/transition state analogues, resulting in dead-end enzyme-substrate complexes. Motivated by the above results, we decided to attempt trapping native Michaelis complexes and structural snapshots in crystals using active GpgS and α3GalT and unmodified substrates.

The quick-soaking protocol we employ to study enzyme catalysis in a crystal can be divided into 3 steps (Fig. 3): (i) Crystallization of the unliganded forms of GpgS and α3GalT; (ii) Quick-soaking of GpgS and α3GalT with substrates or products and cofactors; and (iii) flash-cooling soaked crystals by plunging into liquid nitrogen.

5.1 Crystallization of the unbound GpgS and α3GalT

In this step, typically hundreds of different crystallization conditions are tested in order to obtain initial crystallization conditions that can be further optimized if necessary. The main objective is to identify as many polymorphs as possible (i.e., identify crystallization conditions that yield different crystal packing). Following structural solution of the different polymorphs, structures are carefully inspected. One needs to search for a crystal form that is catalytically competent. That is, the enzyme's catalytic center is accessible to the solvent and any dynamic region involved in catalysis is also free to move (i.e., not involved in crystal contacts). The above requirements are restrictive, since many enzymes will not yield the required crystal form.

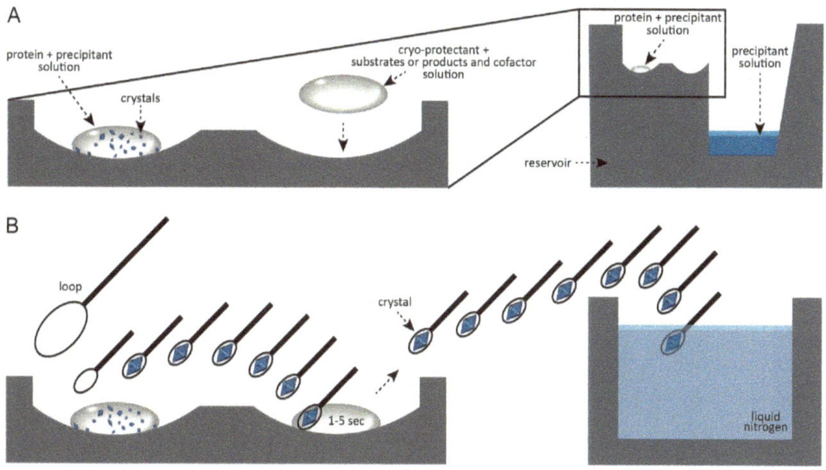

Fig. 3 Schematic representation of the quick-soaking protocol. (A) schematic view of a well in a crystallization plate containing two sub-wells to perform a sitting-drop vapor-diffusion crystallization screen. After opening the well, the cryo-protecting solution containing the enzyme substrates is dispensed into the available sub-wells. (B) A cryo-loop of similar size to the crystal is used for picking up a crystal and sweep it through the cryo-protecting solution before plunging in liquid nitrogen.

Crystals of the unliganded form of GpgS ($I4_1$ space group) were obtained by mixing 0.25 μL of GpgS at 5 mg mL^{-1}, in 30 mM Tris-HCl pH 7.5 with 0.25 μL of a mother liquor containing 14% PEG 8000 and 0.3–0.5 M Li_2SO_4 using the sitting drop vapor diffusion method. Crystals of the unliganded form of α3GalT (C 2 space group) were obtained by mixing 0.25 μL of α3GalT at 5.8 mg mL^{-1} in 20 mM MES pH 6.0 and 10% glycerol with 0.25 μL of a mother liquor containing: 0.02 M L-Na-glutamate, 0.02 M alanine (racemic), 0.02 M glycine, 0.02 M L-lysine (racemic), 0.02 M serine (racemic), 0.1 M sodium HEPES-MOPS (acid) mix pH 7.5, 20% (v/v) glycerol and 10% (w/v) PEG 4000.

5.2 Quick-soaking of GpgS and α3GalT with substrates or products and cofactors

To capture a native Michaelis complex, crystals of unbound enzyme are then soaked for a very short period of time in a cryo-protectant solution containing a high concentration (5–10 mM) of substrates and required cofactors. To determine the soaking time, there are two important variables to consider: the crystal size and the enzyme's turnover number (k_{cat}). The crystal size determines the substrate's diffusion time; that is, how long it takes for the substrate to disperse into the crystal lattice. Ideally, in order to trap a Michaelis complex, diffusion time needs to be faster than the reaction time. In this situation, we would expect substrate binding occurs almost simultaneously throughout the crystal. This requirement will significantly reduce the size of crystals that have better chance to yield a Michaelis complex. The diffusion time (τ_D) of a soluble substrate into a crystal has been previously approximated using Fick's 2nd law of diffusion (Eq. 1), which describes the rate of accumulation (or depletion) of concentration within a volume as proportional to the local curvature of the concentration gradient,

$$D\nabla^2 c = \frac{\partial c}{\partial t} \tag{1}$$

where D is the diffusion coefficient, ∇^2 is the Laplace operator and c is the time-dependent concentration of the diffusing species. Eq. (1) is a second-order partial differential equation that can be solved by Fourier methods, and if appropriate boundary conditions are applied it can have an analytic solution (Eq. 2; Schmidt, 2013),

$$\begin{aligned} C(x, y, z, t) = C_{\text{out}} \Bigg[1 - \frac{64}{\pi^3} \sum_{l=0}^{\infty} \sum_{m=0}^{\infty} \sum_{n=0}^{\infty} \frac{(-1)^{l+m+n}}{(2l+1)(2m+1)(2n+1)} \\ x \cos \frac{(2l+1)\pi x}{2a} x \cos \frac{(2m+1)\pi y}{2b} x \cos \frac{(2n+1)\pi z}{2c} \\ x \exp^{-\left(\frac{D\pi^2}{4}\right)\left[\left((2l+1)^2/a^2\right) + \left((2m+1)^2/b^2\right) + \left((2n+1)^2/c^2\right)\right] \cdot t} \Bigg] \end{aligned} \tag{2}$$

where the concentration (C) at a given point within the crystal (x, y, z) and time (t) is a function of the concentration outside the crystal (C_{out}) and a time-dependent exponential term that describes the diffusion time (τ_D) (Eq. 3),

$$\tau_D = \frac{4}{D\pi^2\left[\left((2l+1)^2/a^2\right) + \left((2m+1)^2/b^2\right) + \left((2n+1)^2/c^2\right)\right]} \tag{3}$$

where τ_D contains the diffusion coefficient D and (a, b, c) are half the sides of a model orthogonal crystal. Solving Eq. (3), when integers l, m, $n = 0$, approximates the slowest diffusion time to reach 69% of the outside concentration at the crystal center (x, y, $z = 0$). The diffusion coefficient (D) depends on the temperature, molecular weight, shape, flexibility and solvation of a solute in a given solvent and can either be experimentally determined (Fate & Lynn, 1990; Wall, Grieger, & Childers, 1952) or approximated (see, for example, Evans et al., 2013). In the case of uridine-diphosphate-glucose (UDP-Glc), the diffusion coefficient in water at room temperature using the above approximation is *ca.* 2.9×10^{-6} cm^2 s^{-1} and therefore, the calculated diffusion time (τ_D) for a crystal of dimensions $200 \times 300 \times 400$ μm^3 approximates to 8 s, reducing the crystal size to $50 \times 100 \times 150$ μm^3 reduces the diffusion time (τ_D) to 0.6 s, and down to 8.3 ms for a crystal with dimensions $5 \times 25 \times 50$ μm^3. The crystal size should be chosen so $\tau_D \ll 1/k_{cat}$. Thus, in the case of GpgS with a k_{cat} value of 0.8 s^{-1} (Kumar et al., 2014), crystals up $10 \times 50 \times 100$ μm^3 in size should be preferentially used. An important consideration is the possibility to reduce the k_{cat} to increase the crystal size. For example, in α3GalT the K_{cat} values range between 6.4 and 0.72 s^{-1} depending on the metal cofactor (Zhang et al., 2001) (please see Evans et al., 2013) on how to approximate diffusion coefficients and Supplementary Information—containing a spreadsheet to calculate diffusion times (τ_D) using the above approximations).

5.3 Flash-cooling soaked GpgS and α3GalT into liquid nitrogen

Following quick-soaking, crystals are cooled by quickly plunging them into liquid nitrogen for later diffraction analysis. Soaking times should be similar or bellow reaction times. For example, soaking times for GpgS and α3GalT were *ca.* 1 s. In practice, unbound crystals are fished out from their mother liquor using a loop of similar size and swept through a cryo-protecting solution containing substrate(s) and cofactor(s), followed by plunging them into a near-by liquid nitrogen Dewar (Fig. 3B). In our hands, the success rate of this approach is between 3% and 5%, which means 3–5 crystals in 100 contain a ternary complex. Importantly, it is anticipated that enzyme-substrate(s) complexes obtained with this approach might differ, providing valuable insight into the role of dynamics and conformational changes in catalysis.

6. A model for the reaction mechanism of retaining glycosyltransferases

Using our crystal quick-soaking protocol, we obtained a number of native ternary complexes of GpgS in the presence of substrates or products and using Mn^{2+} as the divalent metal cation cofactor instead of Mg^{2+} (Fig. 4). We first reported three snapshots of the reaction center at resolutions of 2.3, 2.3, and 2.6 Å, thereby providing for the first time the atomic coordinates of a native Michaelis complex for a GT. Specifically, in the Michaelis, the anomeric carbon C1′ of the sugar approaches only 2.6 Å from the acceptor oxygen atom O_3 of PGA, which in turn hydrogen bonds with

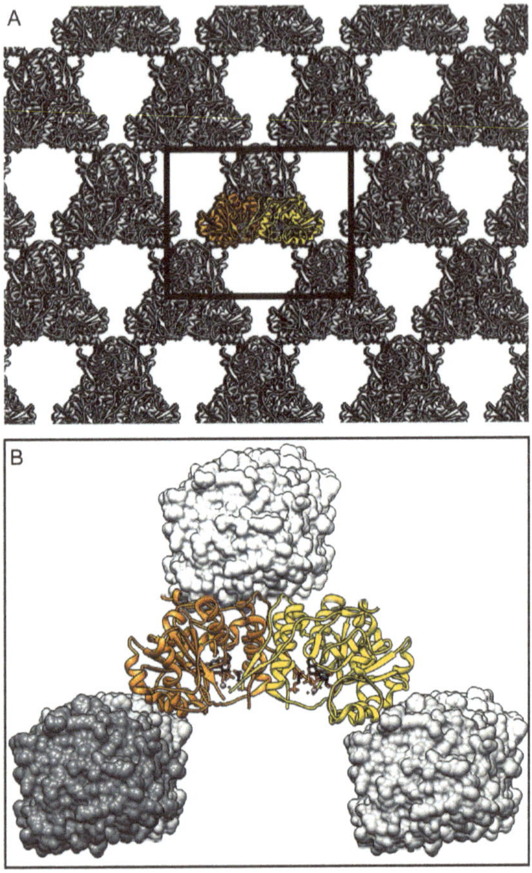

Fig. 4 Crystal packing GpgS. (A) Packing of GpgS dimers into the crystal (B) close view of crystal contacts within GpgS crystal.

the O1B atom of the β-phosphate (Fig. 5). The configuration of the active site in the native Michaelis complex of GpgS for the wild type enzyme and with the natural substrates complemented with quantum-mechanics/molecular-mechanics (QM/MM) calculations, provide a strong experimental support for a front-side substrate-assisted SNi-type reaction (Albesa-Jové et al., 2015; Lairson et al., 2008). Furthermore, we decided to apply this technique to capture different snapshots pre- and post-reaction reflective of the catalytic cycle (Albesa-Jové, Romero-García, et al., 2017; Kumar et al., 2014). During this sequence of events, we visualize how the GpgS guides the sugar acceptor and donor substrates, PGA and UDP-Glc, and the metal cofactor into the reaction center where the glycosyl transfer

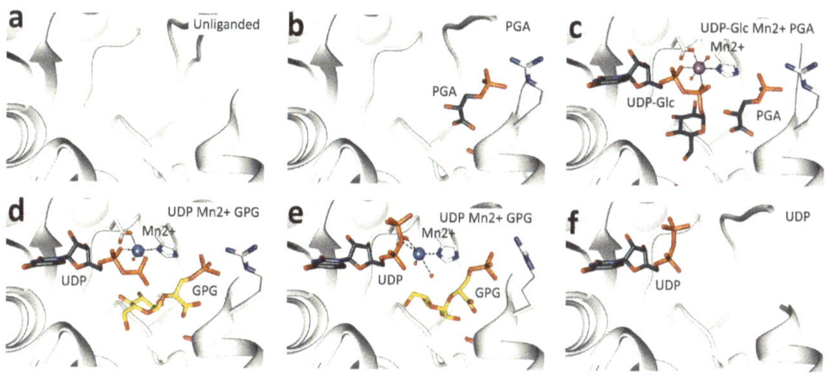

Fig. 5 Snapshots of GpgS catalytic cycle obtained by quick-soaking. (A) Unliganded GpgS crystal structure (B) GpgS in complex with PGA. (C) GpgS Michaelis complex (D) GpgS in complex with the products UDP, glucosyl-3-phosphoglycerate (GPG) and the metal cofactor Mn^{2+} (E) a second complex of GpgS with both products and the metal cofactor, highlighting the important conformational changes. (F) GpgS in complex with UDP product.

reaction takes place, and unveil the mechanism of product release. This process involves multiple conformational changes not only in the substrates/products but also in the GT (Fig. 5)

More recently, we obtained the two first ternary complexes of α3GalT in the presence of its sugar donor UDP-Gal, the acceptor substrate lactose (LAT), and the divalent cation Co^{2+} instead of the most active Mn^{2+} (Fig. 6; Albesa-Jové, Sainz-Polo, et al., 2017). These structures provide insight into the catalytic mechanism of α3GalT, supporting the occurrence of a SNi-type mechanism instead of the double displacement previously described for this type of enzymes with a putative nucleophile residue in the active site. The first step of the reaction consists in breaking the glycosidic O_{3P}-$C1'_{\beta\text{-GAL}}$ bond, stabilized by the interaction of the acceptor hydroxyl group and the β-phosphate of the nucleotide-sugar donor. In our two ternary complexes structures, UDP-Gal shows two conformations. The first conformation represents a pre-Michaelis complex state, and the O_{3P}-$HO_{\beta\text{-GAL}}$ hydrogen bond between donor and acceptor substrate is essential to activate the conformational change in UDP-Gal. In the second conformation, the Michaelis complex state, the UDP-Gal approaches the α-face of the glycosyl moiety and exposes the anomeric carbon C1-α-GAL to the incoming acceptor OH_3 group for a front-side attack (Fig. 7). Our ternary complexes structures suggest that Glu317 stabilizes the transition state by an ion pair intermediate in a substrate-assisted SNi-like mechanism.

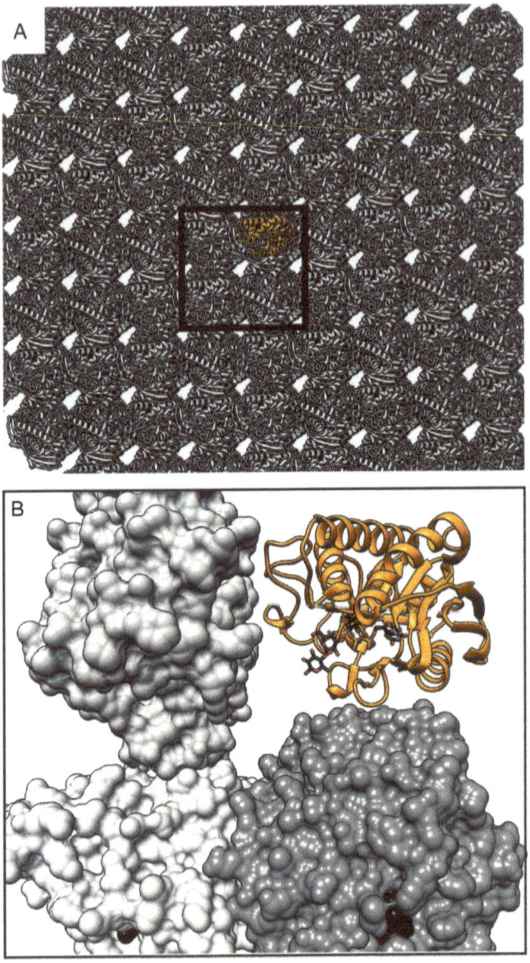

Fig. 6 Crystal packing of α3GalT. (A) Packing of α3GalT protomers into the crystal (B) close view of crystal contacts within α3GalT crystal.

The OE2 atom of the carboxylate group in the Glu317 residue ($OE2_{E317}$) remains at 4.4 Å from the donor $C1'_{\alpha\text{-GAL}}$ atom, while its oxygen $OE1_{E317}$ atom maintains a strong hydrogen bond with acceptor β-Gal O_4 atom. Furthermore, the observed angle between OE2(Glu317)-C1′-OP_{β}(GDU) is significantly distant to the optimal 180° for the nucleophilic attack (Fig. 7). Altogether, the observed position of Glu317 in the Michaelis complex structure does not support its nucleophilic role in a double displacement mechanism

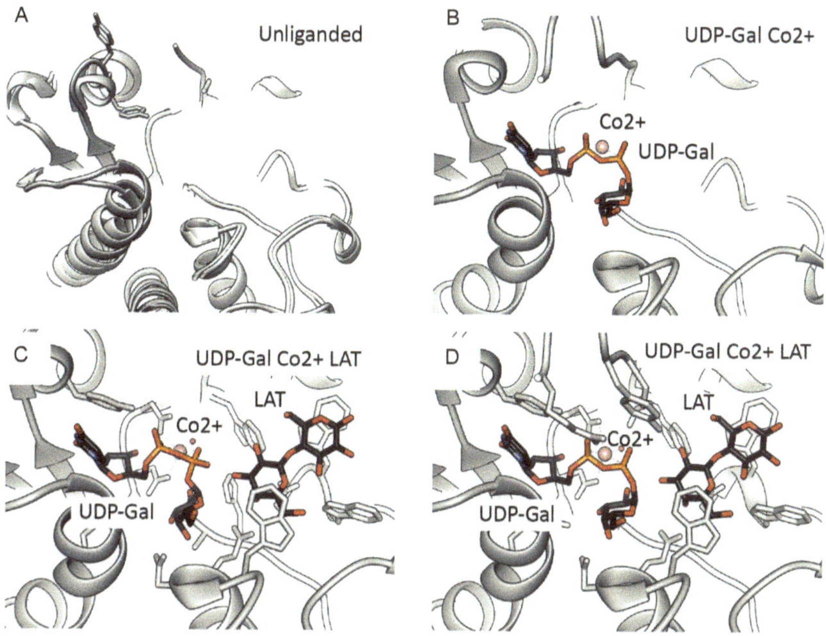

Fig. 7 Snapshots of α3GalT catalytic cycle obtained by quick-soaking. (A) Unliganded form of α3GalT crystal structure (B) α3GalT in complex with UDP-Gal and the metal cofactor Co^{2+}. (C) α3GalT in complex with substrates (D) α3GalT Michaelis complex.

7. Summary

To obtain catalytically competent ternary complexes of GpgS and α3GalT GTs, a precise spatial coordination of native substrates into the reaction center is a critical requirement. Thus, it was essential to work with a crystalline packing that not only tolerates the diffusion of natural substrates toward the active site of the enzyme model, but also accept the free movement of secondary structural elements involved in the protein dynamics and catalysis. The quick soaking protocols described here can certainly be applied to other enzymatic models providing an important contribution to understand the molecular basis of catalysis

Acknowledgments

This work was supported by MINECO/FEDER EU contracts BFU2016- 77427-C2-2-R, BFU2017-92223-EXP, and Severo Ochoa Excellence Accreditation SEV-2016-0644 (to M.E.G.); and CTQ2016-76941-R (to D.A.J.). B.T. acknowledges the "Juan de la Cierva" post-doctoral Contract IJCI-2014-19206 from the MINECO/FEDER EU.

Competing financial interest

The authors declare no competing financial interests.

References

Albesa-Jové, D., & Guerin, M. E. (2016). The conformational plasticity of glycosyltransferases. *Current Opinion in Structural Biology*, *40*, 23–32.

Albesa-Jové, D., Mendoza, F., Rodrigo-Unzueta, A., Gomollón-Bel, F., Cifuente, J. O., Urresti, S., et al. (2015). A native ternary complex trapped in a crystal reveals the catalytic mechanism of a retaining glycosyltransferase. *Angewandte Chemie, International Edition*, *54*, 9898–9902.

Albesa-Jové, D., Romero-García, J., Sancho-Vaello, E., Contreras, F.-X., Rodrigo-Unzueta, A., Comino, N., et al. (2017). Structural snapshots and loop dynamics along the catalytic cycle of glycosyltransferase GpgS. *Structure*, *25*, 1034–1044.e3.

Albesa-Jové, D., Sainz-Polo, M. Á., Marina, A., & Guerin, M. E. (2017). Structural snapshots of α-1,3-galactosyltransferase with native substrates: Insight into the catalytic mechanism of retaining glycosyltransferases. *Angewandte Chemie, International Edition*, *56*, 14853–14857.

Alfaro, J. A., Zheng, R. B., Persson, M., Letts, J. A., Polakowski, R., Bai, Y., et al. (2008). ABO(H) blood group A and B glycosyltransferases recognize substrate via specific conformational changes. *The Journal of Biological Chemistry*, *283*, 10097–10108.

Ardèvol, A., & Rovira, C. (2011). The molecular mechanism of enzymatic glycosyl transfer with retention of configuration: Evidence for a short-lived oxocarbenium-like species. *Angewandte Chemie, International Edition*, *50*, 10897–10901.

Ardèvol, A., & Rovira, C. (2015). Reaction mechanisms in carbohydrate-active enzymes: Glycoside hydrolases and glycosyltransferases. Insights from ab initio quantum mechanics/molecular mechanics dynamic simulations. *Journal of the American Chemical Society*, *137*, 7528–7547.

Blackler, R. J., Gagnon, S. M. L., Polakowski, R., Rose, N. L., Zheng, R. B., Letts, J. A., et al. (2017). Glycosyltransfer in mutants of putative catalytic residue Glu303 of the human ABO(H) A and B blood group glycosyltransferases GTA and GTB proceeds through a labile active site. *Glycobiology*, *27*, 370–380.

Blow, D. M., & Steitz, T. A. (1970). X-ray diffraction studies of enzymes. *Annual Review of Biochemistry*, *39*, 63–100.

Blundell, T. L., Jhoti, H., & Abell, C. (2002). High-throughput crystallography for lead discovery in drug design. *Nature Reviews. Drug Discovery*, *1*, 45–54.

Bobovská, A., Tvaroška, I., & Kóňa, J. (2014). A theoretical study on the catalytic mechanism of the retaining α-1,2-mannosyltransferase Kre2p/Mnt1p: The impact of different metal ions on catalysis. *Organic & Biomolecular Chemistry*, *12*, 4201.

Bobovská, A., Tvaroška, I., & Kóňa, J. (2015). Theoretical study of enzymatic catalysis explains why the trapped covalent intermediate in the E303C mutant of glycosyltransferase GTB was not detected in the wild-type enzyme. *Glycobiology*, *25*, 3–7.

Boix, E., Swaminathan, G. J., Zhang, Y., Natesh, R., Brew, K., & Acharya, K. R. (2001). Structure of UDP complex of UDP-galactose:β-Galactoside-α-1,3-galactosyltransferase at 1.53-Å resolution reveals a conformational change in the catalytically important C terminus. *The Journal of Biological Chemistry*, *276*, 48608–48614.

Boix, E., Zhang, Y., Jawahar Swaminathan, G., Brew, K., & Ravi Acharya, K. (2002). Structural basis of ordered binding of donor and acceptor substrates to the retaining glycosyltransferase, α-1,3-galactosyltransferase. *The Journal of Biological Chemistry*, *277*, 28310–28318.

Breton, C., Fournel-Gigleux, S., & Palcic, M. M. (2012). Recent structures, evolution and mechanisms of glycosyltransferases. *Current Opinion in Structural Biology*, *22*, 540–549.

Dauter, Z., Dauter, M., & Rajashankar, K. R. (2000). Novel approach to phasing proteins: Derivatization by short cryo-soaking with halides. *Acta Crystallographica, Section D: Biological Crystallography*, *56*, 232–237.

Empadinhas, N., Albuquerque, L., Mendes, V., Macedo-Ribeiro, S., & da Costa, M. S. (2008). Identification of the mycobacterial glucosyl-3-phosphoglycerate synthase. *FEMS Microbiology Letters*, *280*, 195–202.

Evans, R., Deng, Z., Rogerson, A. K., McLachlan, A. S., Richards, J. J., Nilsson, M., et al. (2013). Quantitative interpretation of diffusion-ordered NMR spectra: Can we rationalize small molecule diffusion coefficients? *Angewandte Chemie, International Edition*, *52*, 3199–3202.

Fate, G., & Lynn, D. G. (1990). Molecular diffusion coefficients: Experimental determination and demonstration. *Journal of Chemical Education*, *67*, 536.

Fulton, Z., McAlister, A., Wilce, M. C. J., Brammananth, R., Zaker-Tabrizi, L., Perugini, M. A., et al. (2008). Crystal structure of a UDP-glucose-specific glycosyltransferase from a mycobacterium species. *The Journal of Biological Chemistry*, *283*, 27881–27890.

Gagnon, S. M. L., Legg, M. S. G., Polakowski, R., Letts, J. A., Persson, M., Lin, S., et al. (2018). Conserved residues Arg188 and Asp302 are critical for active site organization and catalysis in human ABO(H) blood group A and B glycosyltransferases. *Glycobiology*, *28*, 624–636.

Gastinel, L. N., Bignon, C., Misra, A. K., Hindsgaul, O., Shaper, J. H., & Joziasse, D. H. (2001). Bovine α1,3-galactosyltransferase catalytic domain structure and its relationship with ABO histo-blood group and glycosphingolipid glycosyltransferases. *The EMBO Journal*, *20*, 638–649.

Gómez, H., Lluch, J. M., & Masgrau, L. (2013). Substrate-assisted and Nucleophilically assisted catalysis in bovine α1,3-galactosyltransferase. Mechanistic implications for retaining glycosyltransferases. *Journal of the American Chemical Society*, *135*, 7053–7063.

Gómez, H., Polyak, I., Thiel, W., Lluch, J. M., & Masgrau, L. (2012). Retaining glycosyltransferase mechanism studied by QM/MM methods: Lipopolysaccharyl-α-1,4-galactosyltransferase C transfers α-galactose via an oxocarbenium ion-like transition state. *Journal of the American Chemical Society*, *134*, 4743–4752.

Haas, D. J. (1968). X-ray studies on lysozyme crystals at −50°C. *Acta Crystallographica, Section B: Structural Crystallography and Crystal Chemistry*, *24*, 604.

Jackson, M., & Brennan, P. J. (2009). Polymethylated polysaccharides from mycobacterium species revisited. *The Journal of Biological Chemistry*, *284*, 1949–1953.

Kaur, D., Pham, H., Larrouy-Maumus, G., Rivière, M., Vissa, V., Guerin, M. E., et al. (2009). Initiation of methylglucose lipopolysaccharide biosynthesis in mycobacteria. *PLoS One*, *4*, e5447.

Kendrew, J. C., Bodo, G., Dintzis, H. M., Parrish, R. G., Wyckoff, H., & Phillips, D. C. (1958). A three-dimensional model of the myoglobin molecule obtained by X-ray analysis. *Nature*, *181*, 662–666.

Koshland, D. E. (1953). Sterechemistry and the mechanism of enzymatic reactions. *Biological Reviews*, *28*, 416–436.

Kumar, G., Guan, S., & Frantom, P. A. (2014). Biochemical characterization of the retaining glycosyltransferase glucosyl-3-phosphoglycerate synthase from mycobacterium tuberculosis. *Archives of Biochemistry and Biophysics*, *564*, 120–127.

Lairson, L. L., Henrissat, B., Davies, G. J., & Withers, S. G. (2008). Glycosyltransferases: Structures, functions, and mechanisms. *Annual Review of Biochemistry*, *77*, 521–555.

Lee, S. S., Hong, S. Y., Errey, J. C., Izumi, A., Davies, G. J., & Davis, B. G. (2011). Mechanistic evidence for a front-side, SNi-type reaction in a retaining glycosyltransferase. *Nature Chemical Biology*, *7*, 631–638.

Lira-Navarrete, E., Iglesias-Fernández, J., Zandberg, W. F., Compañón, I., Kong, Y., Corzana, F., et al. (2014). Substrate-guided front-face reaction revealed by combined structural snapshots and metadynamics for the polypeptide N-acetylgalactosaminyltransferase 2. *Angewandte Chemie (International Ed. in English)*, *53*, 8206–8210.

Lombard, V., Golaconda Ramulu, H., Drula, E., Coutinho, P. M., & Henrissat, B. (2014). The carbohydrate-active enzymes database (CAZy) in 2013. *Nucleic Acids Research*, *42*, D490–D495.

Monegal, A., & Planas, A. (2006). Chemical rescue of α3-galactosyltransferase. Implications in the mechanism of retaining glycosyltransferases. *Journal of the American Chemical Society*, *128*, 16030–16031.

Nienaber, V. L., Richardson, P. L., Klighofer, V., Bouska, J. J., Giranda, V. L., & Greer, J. (2000). Discovering novel ligands for macromolecules using X-ray crystallographic screening. *Nature Biotechnology*, *18*, 1105–1108.

Nunes-Costa, D., Maranha, A., Costa, M., Alarico, S., & Empadinhas, N. (2017). Glucosylglycerate metabolism, bioversatility and mycobacterial survival. *Glycobiology*, *27*, 213–227.

Patenaude, S. I., Seto, N. O. L., Borisova, S. N., Szpacenko, A., Marcus, S. L., Palcic, M. M., et al. (2002). The structural basis for specificity in human ABO(H) blood group biosynthesis. *Nature Structural Biology*, *9*, 685–690.

Perutz, M. F., Rossmann, M. G., Cullis, A. F., Muirhead, H., Will, G., & North, A. C. T. (1960). Structure of hæmoglobin: A three-dimensional Fourier synthesis at 5.5-Å. resolution, obtained by X-ray analysis. *Nature*, *185*, 416–422.

Pham, T. T. K., Stinson, B., Thiyagarajan, N., Lizotte-Waniewski, M., Brew, K., & Acharya, K. R. (2014). Structures of complexes of a metal-independent glycosyltransferase GT6 from Bacteroides ovatus with UDP- N -acetylgalactosamine (UDP-GalNAc) and its hydrolysis products. *The Journal of Biological Chemistry*, *289*, 8041–8050.

Rini, J. M., & Esko, J. D. (2017). Glycosyltransferases and glycan-processing enzymes. In *Essentials of glycobiology* (3rd ed.). Cold Spring Harbor Laboratory Press [Chapter 6].

Roberts, S. M., & Davies, G. J. (2012). The crystallization and structural analysis of cellulases (and other glycoside hydrolases). In *Methods in enzymology* (pp 141–168): Academic Press.

Rojas-Cervellera, V., Ardèvol, A., Boero, M., Planas, A., & Rovira, C. (2013). Formation of a covalent glycosyl-enzyme species in a retaining glycosyltransferase. *Chemistry—A European Journal*, *19*, 14018–14023.

Schmidt, M. (2013). Mix and inject: Reaction initiation by diffusion for time-resolved macromolecular crystallography. *Advances in Condensed Matter Physics*, *2013*, 1–10.

Sinnott, M. L. (1990). Catalytic mechanism of enzymic glycosyl transfer. *Chemical Reviews*, *90*, 1171–1202.

Soya, N., Fang, Y., Palcic, M. M., & Klassen, J. S. (2011). Trapping and characterization of covalent intermediates of mutant retaining glycosyltransferases. *Glycobiology*, *21*, 547–552.

Sun, P. D., Radaev, S., & Kattah, M. (2002). Generating isomorphous heavy-atom derivatives by a quick-soak method. Part I: Test cases. *Acta Crystallographica, Section D: Biological Crystallography*, *58*, 1092–1098.

Teng, T.-Y. (1990). Mounting of crystals for macromolecular crystallography in a free-standing thin film. *Journal of Applied Crystallography*, *23*, 387–391.

Urresti, S., Albesa-Jové, D., Schaeffer, F., Pham, H. T., Kaur, D., Gest, P., et al. (2012). Mechanistic insights into the retaining glucosyl-3-phosphoglycerate synthase from mycobacteria. *The Journal of Biological Chemistry*, *287*, 24649–24661.

Wall, F. T., Grieger, P. F., & Childers, C. W. (1952). A rapid method for measuring diffusion coefficients for solutions. *Journal of the American Chemical Society*, *74*, 3562–3567.

Wild, R., Kowal, J., Eyring, J., Ngwa, E. M., Aebi, M., & Locher, K. P. (2018). Structure of the yeast oligosaccharyltransferase complex gives insight into eukaryotic N-glycosylation. *Science*, *359*, 545–550.

Zhang, Y., Swaminathan, G. J., Deshpande, A., Boix, E., Natesh, R., Xie, Z., et al. (2003). Roles of individual enzyme – substrate interactions by α-1,3-galactosyltransferase in catalysis and specificity. *Biochemistry*, *42*, 13512–13521.

Zhang, Y., Wang, P. G., & Brew, K. (2001). Specificity and mechanism of metal ion activation in UDP-galactose:β-galactoside-α-1,3-galactosyltransferase. *The Journal of Biological Chemistry*, *276*, 11567–11574.

CHAPTER SIXTEEN

Cysteine-ethylation of tissue-extracted membrane proteins as a tool to detect conformational states by solid-state NMR spectroscopy

Daniel K. Weber[a], Taysir Bader[b], Erik K. Larsen[b], Songlin Wang[a], Tata Gopinath[a], Mark Distefano[b], Gianluigi Veglia[a,b,*]
[a]Department of Biochemistry, Molecular Biology, and Biophysics, University of Minnesota, Minneapolis, MN, United States
[b]Department of Chemistry, University of Minnesota, Minneapolis, MN, United States
*Corresponding author: e-mail address: vegli001@umn.edu

Contents

Abstract

Solid-state NMR (ssNMR) is an ideal tool to study structure and dynamics of membrane proteins in their native lipid environment. In principle, ssNMR has no size limitations. However, this feature is rarely exploited as large membrane proteins display severe resonance overlap. In addition, dismal yields from recombinant bacterial expression systems limit severely spectroscopic characterization of membrane proteins. For very large mammalian membrane proteins, extraction from the original organism remains the most viable approach. In this case, NMR-observable nuclei must be introduced

Methods in Enzymology, Volume 621
ISSN 0076-6879
https://doi.org/10.1016/bs.mie.2019.02.001

post-translationally, but the approaches developed so far are rather scarce. Here, we detail the synthesis and engineering of a reactive ^{13}C-ethylmethanethiosulfonate (^{13}C-EMTS) reagent for the post-translational alkylation of cysteine sidechains of a 110kDa sarcoplasmic reticulum Ca^{2+}-ATPase (SERCA) extracted from rabbit skeletal muscle tissue. When reconstituted into liposomes, it is possible to resolve the resonances of the engineered ethyl groups by magic-angle spinning (MAS) 2D [^{13}C,^{13}C]-DARR experiments. Notably, the ethyl-group modification does not perturb the function of SERCA, yielding well-resolved ^{13}C-^{13}C fingerprints that are used to image its structural states in the catalytic cycle and filtering out overwhelming naturally-abundant ^{13}C nuclei signals arising from the enzyme and lipids. We anticipate that this approach will be used together with ^{19}F NMR to monitor conformational transitions of enzymes and proteins that are difficult to express recombinantly.

1. Introduction

Solid-state NMR (ssNMR) is a powerful technique for investigating structure and dynamics of membrane proteins, which unlike solution NMR, does not rely on perturbing detergents for their functional reconstitution (Chipot et al., 2018). With no upper size limit, ssNMR is accessible to the largest membrane proteins embedded in their native lipid-bilayer environment, but poorer resolution, combined with unique signal arising from hundreds to thousands of amino acids, produce spectra far too overlapped for meaningful interpretation. Furthermore, large membrane proteins may be extraordinarily difficult to obtain by heterologous expression from *Escherichia coli*, or other common hosts, in a correctly folded state and in milligram quantities required for ssNMR. Although a few successful attempts have been reported for cell-free expression and yeast expression systems, the extraction from the host organism often remains the only viable option. In such cases, sparse, selective and segmental labeling methods built on recombinant expression to reduce spectral overlap (Göbl, Madl, Simon, & Sattler, 2014; Iwai & Züger, 2007; Larsen et al., 2018; Prestegard et al., 2014; Verardi, Traaseth, Masterson, Vostrikov, & Veglia, 2012; Walters et al., 2001; Xu, Ayers, Cowburn, & Muir, 1999) are inapplicable and NMR-observable nuclei must be introduced post-translationally. Fortunately, there is an extensive toolbox for post-translational modification of proteins, exploiting the chemistry of residues such as cysteine, lysine, tyrosine, arginine, glutamate, aspartate, serine, threonine, methionine, histidine, tryptophan as well as N- and C-termini (Boutureira & Bernardes, 2015; Debelouchina & Muir, 2017). However, only a few to date have been utilized by the ssNMR community.

In this chapter, we detail protocols for the synthesis of a ^{13}C-ethylmethanethiosulfonate (^{13}C-EMTS) reagent (Vostrikov et al., 2016) and its use for ethylating cysteines of a 110kDa sarcoplasmic reticulum Ca^{2+}-ATPase (SERCA) transmembrane enzyme isolated from rabbit skeletal muscle tissue. The ^{13}C-EMTS reagent is analogous to the methylating reagent methylmethanethiosulfonate (^{13}C-MMTS) introduced by Kay and co-workers (Religa, Ruschak, Rosenzweig, & Kay, 2011), but does not require deuterated recombinant proteins to increase signal intensity. Importantly, this approach eliminates the background from lipids and naturally-abundant ^{13}C-nuclei that plague ^{1}H-^{13}C correlation spectra. Instead, the ^{13}C-^{13}C dipolar coupling of ethyl-tagged cysteines (Cys-S-$^{13}CH_2$-$^{13}CH_3$) provides a double-quantum filter previously demonstrated to yield background-free and well-resolved solid-state ^{13}C-^{13}C Dipolar-Assisted Rotational Resonance (DARR) experiments (Takegoshi, Nakamura, & Terao, 2001; Vostrikov et al., 2016). This reagent reacts with cysteine (Fig. 1A) by well-known chemistry between nucleophilic thiols and alkyl alkanethiosulfonates, which is appealing for its applicability under mild reaction conditions, i.e., aqueous solutions and physiological pH. EMTS reagent enables fast reaction times, stoichiometric conversion of cysteine without requiring excess reagent, and reversibility upon addition of common thiol reagents such as dithiothreitol and β-mercaptoethanol (Kenyon & Bruice, 1977). More importantly, it does not perturb the enzyme's function.

SERCA has 24 cysteines, of which 6 solvent-exposed residues are preferentially labeled throughout both the cytoplasmic nucleotide-binding (N) and phosphorylation (P) domains (Fig. 1B). When using a fivefold molar excess of ^{13}C-EMTS, the ATPase activity is fully retained (Vostrikov et al., 2016) and the ^{13}C-^{13}C DARR spectra give compelling fingerprints of the different structural states throughout the enzymatic cycle (Toyoshima, 2008).

2. Protocols for using ^{13}C-EMTS

2.1 Synthesis and purification of the ^{13}C-EMTS reagent

The synthesis of ^{13}C-EMTS is described at 9.2mmol scale, which accounting for a total yield of 75%, produces 35mL of 200mM stock and enough for over 18,000 NMR sample preparations using the protocol described in the following section. The scale was chosen to match the quantities of $^{13}C_2$-bromoethane available commercially, but can be significantly reduced. In our hands, we have

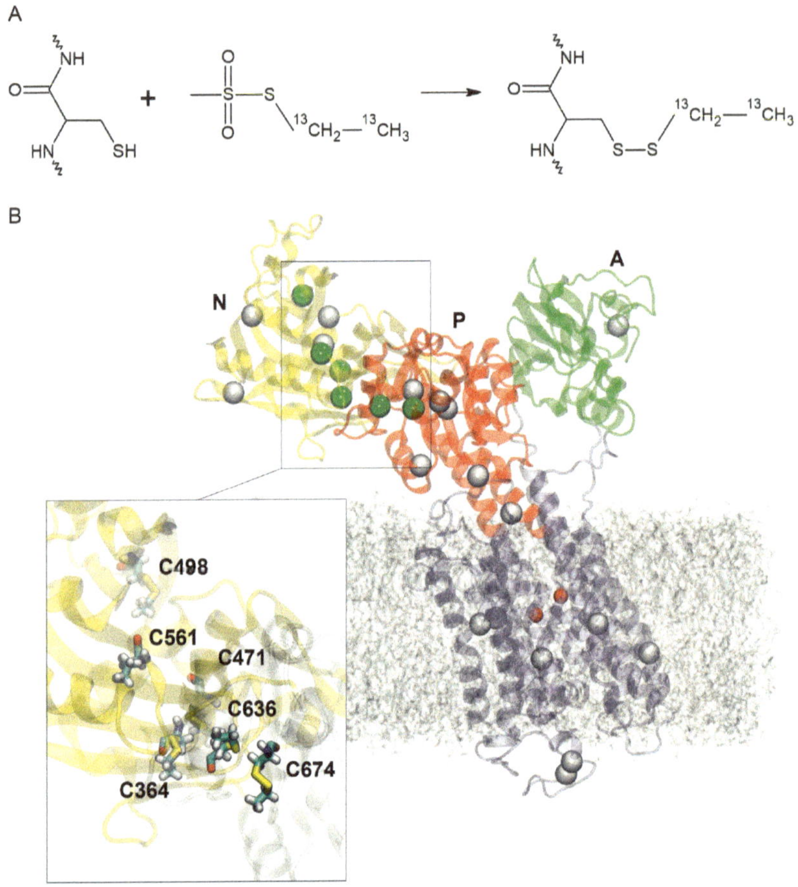

Fig. 1 (A) Reaction scheme for ethylating cysteine with ^{13}C-EMTS reagent. (B) Structural representation of ethylated cysteines (green spheres) and non-ethylated cysteines (white spheres) projected onto SERCA in the calcium-bound E1 state (E1·2Ca^{2+}, PDB 1SU4) (Toyoshima, Nakasako, Nomura, & Ogawa, 2000). Nucleotide-binding (N, yellow), phosphorylation (P, red) and actuator (A, green) domains are shown as well as Ca^{2+} sites (red spheres). Expanded region shows ethyl-modified cysteines (patch to convert cysteine to ethyl-cysteine for CHARMM molecular dynamics simulation are available on request). System was built using the CHARMM-GUI webserver (Jo, Kim, Iyer, & Im, 2008; Wu et al., 2014) and includes 400 POPC lipids.

successfully used this scheme at 0.7, 2, and 3.5 mmol scales to produce sufficient amounts for ssNMR studies. ^{13}C-EMTS is obtained by a nucleophilic SN_2 reaction between sodium methanethiosulfonate and $^{13}C_2$-bromoethane (Fig. 2A) as shown previously (Vostrikov et al., 2016), but with several improvements described here to make the protocol more efficient.

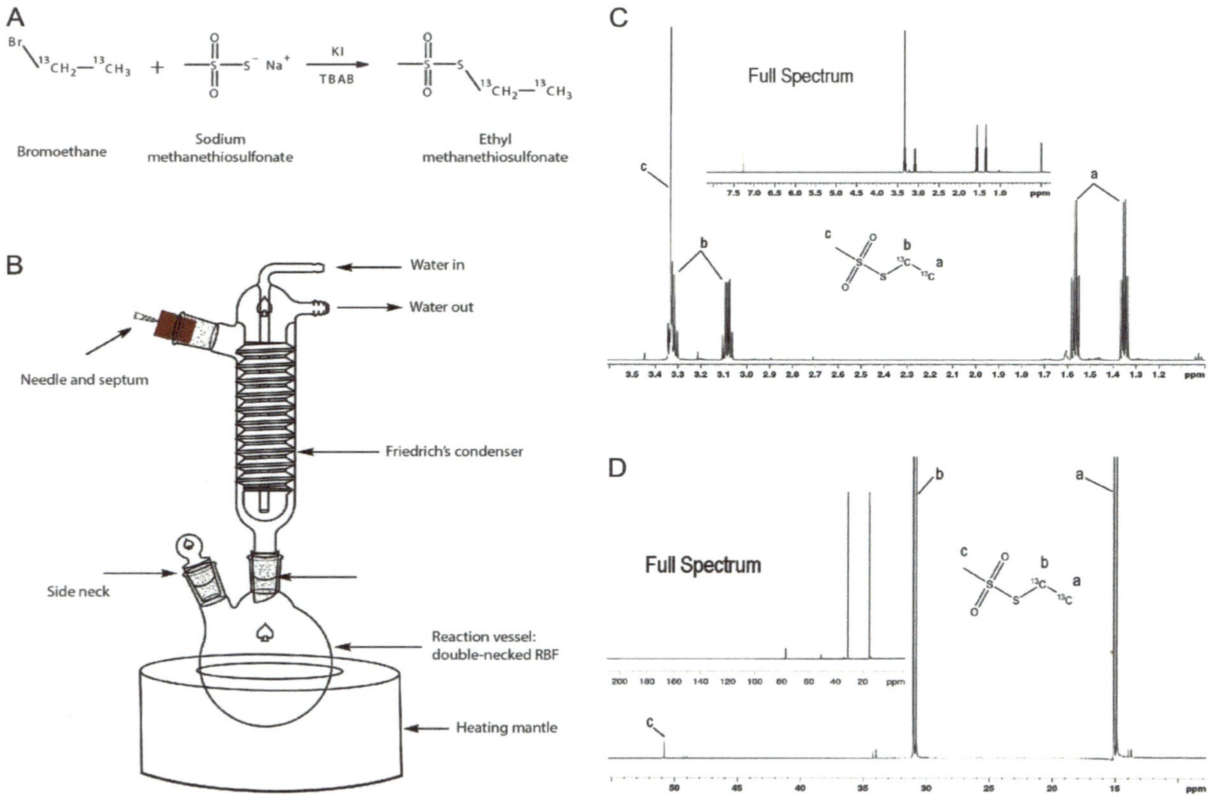

Fig. 2 (A) Reaction scheme and (B) apparatus for synthesis of ^{13}C-EMTS synthesis. (C) ^{1}H and (D) ^{13}C NMR spectrum of ^{13}C-EMTS in $CDCl_3$ acquired in a Varian 600 MHz spectrometer at 298 K. ^{1}H spectrum was acquired using 8 scans and 30° pulses. ^{13}C spectrum was collected using 512 scans with power-gated decoupling.

First, additional care must be taken to ensure the reaction is carried out under anhydrous conditions as bromoethane's C—Br bond is hydrolysable to ethanol and HBr in the presence water. Therefore, better yields were obtained using dry solvent and avoiding atmospheric exposure. Older reports of this reaction, however, compromise yields by using 5% water for dissolving sodium methanethiosulfonate (Hayashi et al., 1964; Pachter & Scheinbaum, 1974; Weidner & Block, 1964). To circumvent this issue, a phase transfer catalyst that shuttles ionic sodium methanethiosulfonate into the organic phase can be used. In our protocol, we use tetra-*n*-butyl ammonium bromide (TBAB). Second, the solvent of choice has a dramatic effect on the final yield of the reaction. Highest yields were obtained using CH_3CN as a solvent. CH_3CN is a polar aprotic solvent and is favorable for SN_2 reactions as it increases the strength of the nucleophile due to weak solvation. Higher yields were also obtained when a catalytic amount of KI was added to the reaction. Iodide generated from KI displaces the bromine in bromoethane, which results in a better leaving group. Lastly, several phase-separated extractions were used to purify the final product from ionic by-products. Using the apparatus depicted in Fig. 2B as a guide, the reaction protocol is as follows:

1. Combine dry reagents sodium methanethiosulfonate (1354 mg, 10.1 mmol), TBAB (10 mg, 0.03 mmol), and potassium iodide (10 mg, 0.06 mmol) in a two-necked 50-mL round-bottomed flask (RBF).
2. Attach a condenser column to the RBF. A standard Liebig condenser would be sufficient, but a higher efficiency condenser, such as a Friedrichs or a Jacketed Coil condenser, is recommended. Cheaper and more-common Dimroth or Davies condensers may also be acceptable.
3. Seal joints with Glindemann polytetrafluoroethylene (PTFE) rings to prevent volatile ^{13}C-bromomethane from escaping the reaction apparatus. Vacuum grease may also be used, but risks contamination especially after prolonged heating.
4. Plug the top of the condenser with a rubber septum and insert a syringe needle through it (Fig. 2B). The septum will serve to minimize the amount of volatile $^{13}C_2$-bromoethane escaping and keep the reaction vessel mostly under nitrogen, while the needle will prevent pressure build-up. Attach a tubing adapter to the side neck and fasten with a keck clip. Flush the apparatus with dry nitrogen for 5 min. Notice that there's no need to continuously flush nitrogen during the reaction. The reaction can also be run using a nitrogen reflux apparatus, but sufficient results were obtained with the described method.

5. De-gas anhydrous CH_3CN (15 mL) either by bath-sonication for 20 min or sparging with dry nitrogen for at least 5 min, then use it to dilute $^{13}C_2$-bromoethane (680 μL, 9.18 mmol) and add to reaction vessel via the side neck. The reaction mixture should appear as a cloudy white suspension.
6. Cap the side neck with a ground-glass stopper, PTFE joint seal and Keck clip.
7. Start a slow flow of water through the condenser (just faster than dripping) and turn on the heating mantle with a low setting. Reflux for at least 20 h. The suspension will remain cloudy due to unreacted starting materials and formation of inorganic salts.
8. NMR may be used to confirm completion of the reaction. This may be done by briefly opening the side neck and drawing 100 μL of the reaction with a glass syringe and transferring to an NMR tube. Dilute to 700 μL with $CDCl_3$ containing TMS standard and acquire 1D ^1H and ^{13}C NMR spectra. Example ^1H and ^{13}C-spectra of pure ^{13}C-EMTS are depicted in Fig. 2C and D, respectively. Chemical shifts of unreacted $^{13}C2$-bromoethane and ^{13}C-EMTS product are summarized in Table 1.
9. Filter the solution using a medium ground glass filter, then wash the solids in filter with 20 mL of $CHCl_3$. Collect the filtrate and transfer to a separatory funnel.
10. Add 75 mL of 1 M $MgSO_4$ (in water) to convert any $CHCl_3$-soluble I_2 formed to water-soluble I_3^-. The organic layer stays on the bottom and

Table 1 Chemical shifts of unreacted $^{13}C_2$-bromoethane and ^{13}C-EMTS product.

Nuclei	^1H δ (ppm)	Multiplicity	^{13}C δ (ppm)	Multiplicity
$^{13}C_2$-bromoethane				
$^{13}CH_2$	3.33 and 3.58	Doublet of quartets	19.32	Doublet
$^{13}CH_3$	1.59 and 1.80	Doublet of triplets	28.02	Doublet
^{13}C-EMTS				
$^{13}CH_2$	3.08 and 3.32	Doublet of quartets	14.99	Doublet
$^{13}CH_3$	1.35 and 1.57	Doublet of triplets	30.93	Doublet
CH_3	3.33	Singlet	50.86 (small)	Singlet

Chemical shifts were referenced to a tetramethylsilane (TMS) standard. Note that doublet ^1H signals arise from $^1J_{CH}$ coupling with ^{13}C carbons.

the aqueous rises to the top. Collect the organic layer and repeat the extraction a total of three times using fresh 75 mL 1 M $MgSO_4$ each time. Combine the aqueous layers and back extract with 20 mL $CHCl_3$.

11. Combine organic extractions and dry by adding anhydrous $MgSO_4$ powder until no clumping is observed. No more than 10 g should be required. Filter through a medium-glass filter directly into a pre-weighed 100 mL RBF. Rinse the remaining product from solid powder with 5 mL of anhydrous $CHCl_3$.
12. Remove organic solvent using a rotary evaporator with low heat (~30 °C). High heat may risk decomposition of ^{13}C-EMTS. The remaining product will be a colorless oil but may also be slightly yellow due to trace amounts of I_2 remaining. Trace I_2 impurity should not impede the ethylation reaction but could affect the activity of some enzymes. If this is an issue, then these traces may be removed by more phase-separated extractions in step 10.
13. Once all solvent is removed acquire 1D ^{1}H and ^{13}C NMR spectra in 0.7 mL $CDCl_3$. No CH_3CN solvent should remain, which will produce a singlet at 2.10 ppm in ^{1}H spectra; and singlets at 116.43 and 1.89 ppm in ^{13}C spectra (Gottlieb, Kotlyar, & Nudelman, 1997). The lack of CH_3CN peaks will also indicate that $CHCl_3$ was removed as it has a lower boiling point.
14. Weigh RBF flask and calculate the net mass of ^{13}C-EMTS and yield (MW of 142.003 g/mol). At this scale, the theoretical yield is 1303.59 mg, but expect an actual yield of around 75%.
15. Dissolve ^{13}C-EMTS into DMSO, quantitatively transfer to a 50-mL conical tube and dilute to 200 mM with DMSO using the graduations on the tube. Store at −20 °C. Aliquoting into smaller (polypropylene) tubes is recommended to limit repeated freeze-thawing of the main stock. The lifetime of the reagent has not been determined but has been stable for months in storage in our experience.

2.2 Ethylation and reconstitution of membrane proteins into lipid bilayers

The details regarding the purification of SERCA are highly specific to this system and outside the scope of this review, but may be briefly described in two stages. First, skeletal muscle was dissected from a freshly euthanized rabbit and crude sarcoplasmic reticulum (SR) extracted by several rounds of blending and centrifugation (Fernandez, Rosemblatt, & Hidalgo, 1980), then stored at −80 °C in sucrose buffer (20 mM MOPS, 0.3 M sucrose, 1 mM

NaN_3, pH 7.0). Second, aliquots of crude SR were thawed, SERCA solubilized from the SR membrane into $C_{12}E_8$ detergent and purified in a single step by affinity to a Reactive Red resin (Stokes & Michael Green, 1990). Purified SERCA was then stored in aliquots of elution buffer (0.5 mg/mL SERCA, 0.1% $C_{12}E_8$, 1 mM $CaCl_2$, 1 mM $MgCl_2$, 20 mM MOPS, 20% glycerol, 8 mM ADP, pH 7.00, 0.25 mM) at −80 °C and thawed again at 4 °C prior to use. Activity was confirmed by coupled-enzyme assay as a function of calcium concentration (Reddy, Cornea, Winters, McKenna, & Thomas, 2003). The following steps may be more generalizable to any membrane protein and scaled here to pack into a thin-walled 3.2 mm Varian MAS rotor (see Fig. 3A):

1. Dialyze 7.5 mg of SERCA (15 mL) twice into 1 L of dialysis buffer (20 mM HEPES, 100 mM KCl, 5 mM $CaCl_2$, 1 mM $MgCl_2$, 1 mM NaN_3, 5% glycerol, 0.25 mM $C_{12}E_8$, pH 7.0). The first dialysis is done over 3 h and the second overnight, both at 4 °C using 6–8 kDa MWCO dialysis tubing (Fisher Scientific). DTT must be removed as much as possible to avoid consumption of the ^{13}C-EMTS reagent. $C_{12}E_8$ detergent is included above its critical micelle concentration (CMC) of 0.1 mM (Le Maire, Kwee, Andersen, & Møller, 1983) to prevent SERCA denaturation during dialysis. Ca^{2+} and Mg^{2+} are included to stabilize SERCA in the Ca^{2+}-bound E1 state.
2. Transfer dialyzed SERCA to a 50-mL conical centrifuge tube with a small magnetic stirrer, dilute to 30 mL with dialysis buffer and bring to room temperature.
3. Add ^{13}C-EMTS reagent (200 mM) at a 5:1 M excess to SERCA and incubate for 2 h at room temperature. The molar ratio of the reagent used must by optimized to the tolerance of the protein of interest.

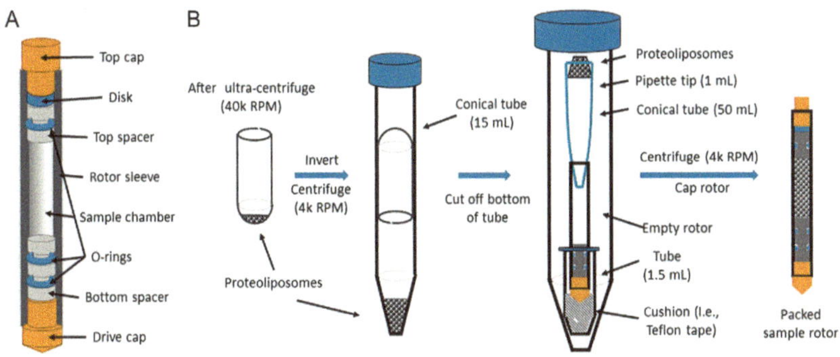

Fig. 3 (A) Rotor schematic. (B) Sample packing scheme for MAS-ssNMR.

For SERCA, this ratio has been previously determined at small scale by analyzing activity over various molar excesses of ^{13}C-EMTS reagent by enzyme-coupled ATPase assays (Reddy et al., 2003). At 5:1, there is no loss in SERCA activity with solvent-exposed cysteines 364, 471, 498, 561, 636 and 674 the most susceptible to ethylation. At this stage, a trypsin digest followed with liquid-chromatography electrospray ionization tandem mass spectrometry (LC-ESI-MS/MS) may be used to identify labeled residues.

4. Solubilize 9mg d_{54}-DMPC (1,2-dimyristoyl-*sn*-glycero-3-phosphocholine) into 1mL of dialysis buffer and 20mg of $C_{12}E_8$ detergent (i.e., 80μL of aqueous 25% w/v stock purchased from Anatrace), add to ethylated SERCA and incubate 30min at room temperature with stirring. Vortex and bath-sonicate to fully solubilize lipids and cool on ice if mixture becomes too viscous to transfer by pipette. Deuterated lipids are recommended to suppress long-relaxing signal from dynamic methyl groups that will create noise along the indirect dimension of DARR spectra.
5. Remove detergent by adding 1.5g of Bio-Beads SM-2 (Bio-Rad) over amounts of 0.25, 0.25, 0.5 and 0.5g, with 30-min intervals of stirring at room temperate after each addition. Bio-Beads have an absorptive capacity for $C_{12}E_8$ of 190mg/g, for which it is recommended that they are added in successive amounts to allow smooth passage through the micelle to vesicle transition and minimize losses of lipid (Rigaud, Levy, Mosser, & Lambert, 1998). The sample will become turbid due to formation of proteoliposomes.

2.3 Rotor packing

The following describes a simple method to pack sample into a thin-walled 3.2mm Varian MAS rotor. This protocol may be substituted for one that utilizes specialized rotor packing equipment from commercial suppliers.

1. Aspirate proteoliposomes (from Section 2.2, step 5) from Bio-Beads by syringe with a 25G needle and transfer to a 40-mL centrifuge tube. Centrifuge at 12,000rpm for 30min at 4°C using a Beckman JA-25.50 rotor.
2. Discard supernatant and resuspend pellet into 2mL of sample buffer (20mM HEPES, 100mM KCl, 5mM $CaCl_2$, 1mM $MgCl_2$, 1mM NaN_3, 2.5% glycerol, pH 7.0).
3. Transfer to a centrifuge tube (compatible for a Beckman TLA-100.3 rotor) and centrifuge at 40,000rpm for 30min at 4°C.

4. Discard supernatant and remove as much residual buffer as possible by tapping the tube inverted over a paper towel.
5. Nest the inverted tube inside a 15-mL conical centrifuge tube (Fig. 3B) and centrifuge 4000 RPM for 20 min at room temperature using a Beckman SX4400 swinging-bucket rotor.
6. Very carefully, cut off the bottom of the tube as close to possible to the pellet using a sharp razor blade.
7. Fix the pellet inverted inside a 1 mL pipette tip inserted into the top of the rotor. Place the assembly inside a 50-mL conical tube and screw on lid (Fig. 3B). If the 1 mL pipette tip does not fit the entrance of the rotor, then nesting the 1 mL tip inside a smaller 200 μL will work, but the top of the 1 mL pipette tip may have to be cut shorter to fit inside the 50-mL tube. The rotor should have a drive cap and bottom spacer already inserted.
8. Centrifuge 4000 RPM for 20 min at room temperature using a Beckman SX4400 swinging-bucket rotor.
9. Insert top spacer and seal rotor with top cap

3. Solid-state NMR spectroscopy

3.1 Hardware requirements

To further demonstrate the use of ^{13}C-EMTS labeling, we acquired MAS ssNMR spectra of tissue-extracted SERCA in the calcium-bound E1 state (E1·2Ca^{2+}) prepared according to Section 2. Detection of ^{13}C-ethylated protein is achievable without the need of highly-specialized ssNMR equipment. For example, all spectra presented in this review were acquired on a simple setup including a 700 MHz Varian spectrometer equipped with a low-E Black Fox 3.2 mm HXY MAS probe (designed by Peter Gor'kov, National High Magnetic Field Laboratory, Tallahassee, FL) operating at a MAS rate of 12 kHz for all experiments. All spectra were externally referenced to adamantane (Morcombe & Zilm, 2003) and processed using NMRPipe (Delaglio et al., 1995).

3.2 Cross-polarization and general considerations

Cross-polarization (CP) is a basic element in solid-state NMR pulse sequences and optimized here for detection of ^{13}C-ethyl-tagged cysteines by using standard 1D [^{13}C-^{1}H]-CP sequences (Fig. 4A). The ethyl tag will produce distinguishable signals in the regions of 31–35 ppm (CH_2) and

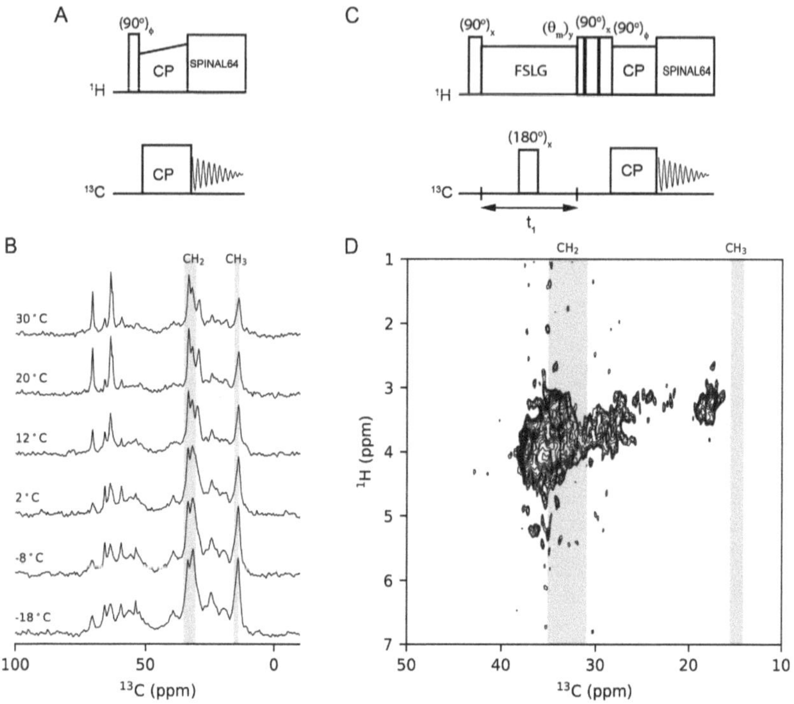

Fig. 4 (A) Schematic of a standard 1D [^{13}C,^{1}H]-CP experiment. (B) 1D [^{13}C,^{1}H]-CP spectra of ^{13}C-cysteine-ethylated SERCA at variable temperatures. Spectra were acquired with 1000 scans (51 min per experiment); 3 s recycle delays; 2.5 μs 90° pulse on ^{1}H; 2 ms CP contact time; CP field strength of 59 kHz on ^{1}H (with 80–100% linear ramp) and 35 kHz on ^{15}N; 100 kHz spectral width for ^{13}C; 100 kHz heteronuclear SPINAL decoupling applied over an acquisition time of 20 ms; and processed with 100 Hz exponential line broadening. (C) Schematic of a 2D [^{13}C,^{1}H]-FSLG-HETCOR pulse sequence. (D) 2D [^{13}C,^{1}H]-FSLG-HETCOR spectrum acquired at 2 °C and with parameters from 1D [^{13}C,^{1}H]-CP experiments, except using 480 scans (26 h experiment time); 100 μs CP contact time; 5207.66 Hz spectral width and 32 points for the indirect ^{1}H dimension; and FSLG ^{1}H homonuclear decoupling. Shaded regions in (B) and (D) specify expected signal from ^{13}C-ethylated cysteines.

14.5–15.5 ppm (CH_3) in CP spectra (Fig. 4B). During the CP block, magnetization is transferred via dipolar couplings from the high-γ and highly-abundant ^{1}H nuclei to the low-γ ^{13}C nuclei during a Hartman–Hahn-matched spin-lock period on both nuclei with the net result being an enhancement in signal on the directly-detected ^{13}C nuclei in proportion to the ratio of gyromagnetic ratios (γ) γ_H/γ_C and faster recycle delays allowed as T_1 relaxation is driven by the faster-relaxing ^{1}H spin (Hartmann & Hahn, 1962). The Hartmann–Hahn condition is optimized by varying the radio-frequency power levels (B_1) until nutation frequencies (ω) on ^{1}H and ^{13}C are equivalent or matched to a spinning sideband to achieve maximal ^{13}C signal. The optimal contact time must also be determined and depends on the rate of magnetization transfer, which is proportional to the strength of ^{13}C-^{1}H dipolar coupling in the sample; and decay of magnetization due to $T_1\rho$ relaxion, which is accelerated by molecular motions on timescale of ω (Michaeli, Sorce, Idiyatullin, Ugurbil, & Garwood, 2004). The spin-lock field on one of the nuclei is generally ramped for better efficiency of magnetization transfer, where linear ramped-amplitude CP (RAMP-CP) (Metz, Wu, & Smith, 1994) and adiabatic passage Hartmann–Hahn CP (APHH-CP) (Hediger, Meier, & Ernst, 1995; Hediger, Meier, Kurur, Bodenhausen, & Ernst, 1994) methods are common. High-power ^{1}H-decoupling is also applied to increase resolution during the acquisition period, and t_1 evolution periods for 2D experiments, as MAS at moderate spinning rates (<25 kHz) only partially eliminate line-broadening from dipolar coupling and chemical shift anisotropy (CSA). For this, Small Phase Incremental Alternation (SPINAL-64) (Fung, Khitrin, & Ermolaev, 2000) and Two Pulse Phase Modulated (TPPM) (Bennett, Rienstra, Auger, Lakshmi, & Griffin, 1995) methods are usually employed. Due to the high-power applied for decoupling, acquisition times are shortened as much as possible to reduce unnecessary wear on the probe and sample heating. Recycle delays between acquisitions are typically set on the order of 3 s and are limited to T_1 relaxation. Recycle delays of membrane-protein systems maybe reduced substantially with paramagnetic relaxation enhancement (PRE) by doping samples with paramagnetic ions (Wickramasinghe, Kotecha, Samoson, Past, & Ishii, 2007) and Cu^{2+}-chelated lipids (Wang, Gopinath, & Veglia, 2018). Note that high-power ^{1}H decoupling may cause probe damage and excessive sample heating. Faster MAS frequencies, however, will allow the use oflow-power decoupling and recycle delays to be reduced as low as100 ms for super-fast acquisition (Ishii et al., 2018).

3.3 Sample temperature

Sample temperature must also be considered in early stages of optimization. Low temperatures preserve the sample integrity, enhancing magnetization and rigidifying the system. This will enhance the efficiency of the CP transfer via stronger dipolar coupling interactions, but may also reduce resolution due to inhomogeneous exchange broadening from lipid and protein dynamics, particularly at temperatures below the gel-fluid phase transition of lipids.

Higher temperatures may provide better resolution, however, the flexible nature of the ^{13}C-ethyl tag, especially when attached to dynamic domains, may risk complete loss of signal. When CP-based experiments become insensitive, INEPT-based (Insensitive Nuclei Enhanced by Polarization Transfer) (Morris & Freeman, 1979) experiments may be used to detect highly dynamic regions (Andronesi et al., 2005; Gopinath, Nelson, Soller, & Veglia, 2017). An array of 1D [^{13}C,^{1}H]-CP experiments (Fig. 4B) show that signals arising from the ^{13}C-ethyl tag are retained over a temperature range of −18 °C to 30 °C, but with no observable change in resolution. Therefore, ^{13}C-EMTS can be used to study enzyme dynamics across a broad temperature range.

3.4 Optimization and acquisition of [^{13}C,^{13}C]-DARR spectra

By visual inspection of Fig. 4A, it is clear that the ^{13}C-ethyl tag is not the only signal observed in a simple 1D [^{13}C,^{1}H]-CP experiment. This is expected as the ethyl tag only introduces 10 to 12 ^{13}C-nuclei buried in the background of 4876 naturally-abundant (1.1% for ^{13}C) carbons of SERCA (110 kDa and 994 residues) plus carbons from endogenous lipids bound to the enzyme. This is especially problematic for the [^{13}C,^{1}H]-Frequency-Switched Lee–Goldburg (FSLG) HETeronuclear CORrelation (HETCOR) experiment (Lee & Goldburg, 1965; van Rossum, Förster, & de Groot, 1997) displayed in Fig. 4C. In Fig. 4D, background noise entirely swamps the signals expected from the $^{13}CH_2$ groups of the ethyl tag, and the signal-to-noise ratio is too low to detect $^{13}CH_3$ groups in the background-free region of the spectrum. Subsequently, $^{13}CH_2$-$^{13}CH_3$ dipolar coupling within the ethyl tag may be exploited in a 2D [^{13}C,^{13}C]-DARR experiment (Fig. 5A) to reduce this background by 100-fold. In this experiment, ^{13}C-^{13}C spins are recoupled under MAS conditions during a mixing time with a ^{1}H RF field (DARR) at a multiple of the spinning rate applied to recover ^{13}C-^{1}H dipolar interactions driving polarization transfer (Takegoshi et al., 2001). Note that we found that [^{13}C,^{13}C]-DARR experiments were suitable under our moderate MAS rates, although, other homonuclear recoupling methods may also be applicable, including Proton-Driven Spin Diffusion (PDSD) (Szeverenyi, Sullivan, & Maciel, 1982), Phase-Altered Recoupling Irradiation Scheme (PARIS) (Weingarth, Demco, Bodenhausen, & Tekely, 2009) and Second-order Hamiltonian among Analogous Nuclei Generated by Heteronuclear Assistance Irradiation (SHANGHAI) (Hu, Lafon, Trébosc, Chen, & Amoureux, 2011). Furthermore, Constant-Time-DARR

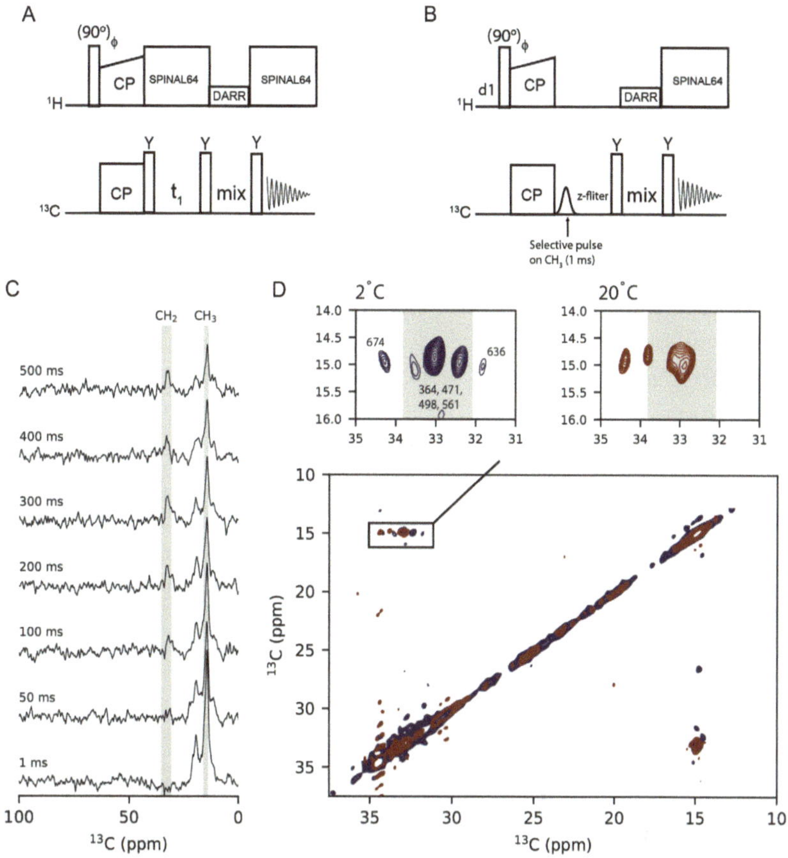

Fig. 5 (A) Schematic of 2D [^{13}C,^{13}C]-DARR pulse and (B) 1D [^{13}C,^{13}C]-DARR pulse sequence highlighting a selective 1 ms Gaussian pulse on CH_3 carbons at 15 ppm. (C) Array of selective 1D [^{13}C,^{13}C]-DARR spectra of ^{13}C-ethylated SERCA acquired at 20 °C with variable DARR mixing times show build-up of intensity of CH_2 signal. Each 1D spectrum was acquired with parameters from 1D [^{13}C,^{1}H]-CP experiments detailed in the caption of Fig. 4, but with a 12 kHz field on ^{1}H during DARR mixing and 5000 scans (4 h each experiment). Spectra were processed with 100 Hz exponential line-broadening. (D) 2D [^{13}C,^{13}C]-DARR spectra acquired at 2 °C (blue) and 20 °C (red) with 300 ms DARR; 60 t_1 increments and 5 kHz spectral width on the second ^{13}C dimension; and 1000 scans (52 h experiment time for each spectrum). 2D spectra were processed with 20 Hz Lorentzian line-sharpening and 20 Hz Gaussian line-broadening. Shaded region highlights residues of the P-domain as previously assigned (Vostrikov et al., 2016).

(CT-DARR) experiments (Chen et al., 2006) may also be used to improve resolution in the directly-detected ^{13}C dimension, but in our experience this experiment requires at least four times the acquisition time for sufficient signal-to-noise ratio.

In addition to optimizing [^{13}C,^{1}H]-CP parameters, the length of mixing times must also be optimized for 2D [^{13}C,^{13}C]-DARR experiments. Obviously, optimization by an array 2D experiments is expensive from an NMR time viewpoint. Therefore, similar to initial reports (Takegoshi et al., 2001), we took advantage of the large separation of $^{13}CH_2$ and $^{13}CH_3$ shifts to optimize parameters using selective 1D [^{13}C,^{13}C]-DARR experiments (Fig. 5B). The sequence differs to the 2D sequence (Fig. 5A) in that the magnetization of $^{13}CH_3$ is selectively inverted after CP, the t_1 chemical shift evolution period is removed and build-up of $^{13}CH_2$ signal is observed over an array of DARR-mixing times. For our sample at 20 °C, we found that 300 ms provided optimal cross-peak intensity using an array of 4 h long 1D experiments (Fig. 5C).

At the scale described in Section 2.2, with 7.5 mg of SERCA reconstituted into 9 mg of d_{54}-DMPC, a 2D [^{13}C,^{13}C]-DARR spectrum can be acquired overnight with good signal-to-noise ratio. In both spectra acquired at 2 °C and 20 °C over 2 days (Fig. 5D), the filtering ability of the ^{13}C-ethyl tag is exemplified by the observation that only labeled carbons are observed in the off-diagonal regions and that separation of $^{13}CH_2$ and $^{13}CH_3$ resonance places cross-peaks well away from the intense diagonal. Furthermore, resolution is afforded mostly by the greater chemical shift dispersion of the $^{13}CH_2$ nuclei. Five resolved signals are observed in spectra acquired at 2 °C, while only four at 20 °C, probably owing to faster conformational dynamics. In our case, higher resolution is not observed at higher temperature, likely due to the need to apply line-broadening to increase signal-to-noise ratio and time-constraints to acquiring additional points along the indirect dimension.

3.5 Assignment of labeled cysteines

Since site-specific mutations are unfeasible, the assignment ^{13}C-ethylated cysteine resonances in DARR spectra is a non-trivial task. The approach to be pursued will likely take advantage of chemical, structural and functional features unique to the protein of interest. At the very least, tryptic digests of ^{13}C-ethylated protein will help identify which cysteines have been labeled. Furthermore, as done previously (Vostrikov et al., 2016), a large excess of iodoacetamide may be added following incubation with ^{13}C-EMTS to identify partially-reacted and non-reacted cysteines through detection of carbamidomethylated (CAM) peptides. Specific assignments may be made by well-informed PRE-experiments. For example, the ethylated-cysteines in the N-domain (nucleotide-binding domain) of SERCA were unambiguously assigned previously (Vostrikov et al., 2016) by selective quenching upon addition of a paramagnetic ADP-TEMPO ligand (ADP = adenosine diphosphate; TEMPO = 2,2,6,6-tetramethyl-1-piperidinyloxy) and selective spin-labeling of C674 with iodoacetamide TEMPO (Wawrzynow, Collins, & Coan, 1993); while ^{13}C-ethylated P-domain (phosphate-binding domain) cysteines closest the membrane were assigned by quenching by a DOPE (1,2-dioleoyl-*sn*-glycero-3-phosphoethanolamine) lipid engineered with a TEMPO headgroup.

4. Conclusions and perspectives

In this chapter, we reported the protocols for utilizing ^{13}C-EMTS to post-translationally introduce ^{13}C-ethyl tags to the 110 kDa SERCA Ca^{2+}-pump, extracted from the skeletal muscle of rabbits, for ssNMR characterization in lipid membranes. Similar post-translational methods based on ^{13}C-methyl labeling have enabled solution NMR characterization of large systems including the 180 kDa α7 ring of the 20S proteasome (Religa et al., 2011), the 300 kDa ClpP protease (Religa et al., 2011), the 100 kDa ClC-ec1 chloride channel (Abraham et al., 2015) and a 240 kDa NADPH-cytochrome P450 oxidoreductase-nanodisc complex (Galiakhmetov, Kovrigina, Xia, Kim, & Kovrigin, 2018). ^{13}C-methyl labeling, however, requires extensive deuteration of the surrounding protein to enhance signal intensity of the label and eliminate background in 2D [^{1}H,^{13}C]-spectra (Religa et al., 2011), or deuteration of the label to filter out background using ^{2}H-^{13}C scalar coupling-edited experiments (Larda, Bokoch, Evanics, & Prosser, 2012) that cannot be applied to ssNMR due to short coherence lifetimes. Furthermore, good resolution is required to resolve poorly-dispersed methyl peaks. For MAS-ssNMR, the use of ^{13}C-EMTS overcomes several major practical issues, including: (1) the use of ^{13}C-^{13}C dipolar coupling to eliminate background from naturally abundant ^{13}C nuclei in protein and lipid carried through from extraction; (2) wider chemical shift dispersion of $^{13}CH_2$ to provide much-needed additional resolution; and (3) not at any stage being limited to obtaining protein by heterologous expression. This labeling strategy, therefore, enables the structural states of large membrane proteins, obtained directly from their host organism, to be fingerprinted by ssNMR. Furthermore, this ^{13}C-ethyl tag may also be adapted to other MAS experiments to extract additional structural and topological information (Cady, Goodman, Tatko, DeGrado, & Hong, 2007).

The synthetic protocol described in Section 2.1 may also be modified to introduce other useful nuclei, such as ^{2}H and ^{19}F that have no issues with background in biological systems. These reaction schemes can be carried out using commercially available bromoalkane precursors. For example, order parameters extracted from ^{2}H quadrupolar couplings can report structure and dynamics of the target proteins (Strandberg, Esteban-Martín, Salgado, & Ulrich, 2009; Tugarinov, Ollerenshaw, & Kay, 2005) and may also be introduced by post-translational site-selective protein α-deuteration (Galan et al., 2018). ^{19}F also has a large γ providing greater sensitivity and ability to measure long-range distances between probes, but requires specialized probes either equipped high-power ^{1}H and ^{19}F channels, or fast-MAS to remove ^{1}H-^{19}F-dipolar interactions to improve resolution (Wang, Lu, et al., 2018). In solution NMR, post-translational modification with CF_3 tags has enabled the conformational dynamics of G protein-coupled receptors (GPCRs) to be studied (Prosser & Kim, 2015). Subsequently, a number reagents have been utilized in solution NMR to post-translationally modify cysteines with ^{19}F nuclei (Kalbitzer et al., 1992; Liu, Horst, Katritch, Stevens, & Wüthrich, 2012; Manglik et al., 2015; Shekhawat, Pham, Prabakaran, & Strieter, 2014; Ye, Larda, Frank Li, Manglik, & Prosser, 2015). ^{19}F may also be introduced to tyrosine either electrochemically (Jeuken, 2009; Richards, Coles, Heptinstall, & Walton, 1994) or chemically using acetyl hypofluorite (Hebel, Kirk, Cohen, & Labroo, 1990). Furthermore, glutamines may also be enzymatically modified with ^{19}F (Hattori et al., 2017).

Another residue prone to post-translational methylation is lysine, which undergoes reductive methylation in the presence of ^{13}C-formaldehyde yielding a tertiary ammine. Lysine has also been modified with ^{19}F-tags with reagents designed to preserve charge and hydrophilicity of the protein (Larda, Pichugin, & Prosser, 2015; Mehta, Kulkarni, Mason, Constantinescu, & Antich, 1994). Partial (monomethylation) or complete methylation (di-methylation) of lysines is possible by titrating ^{13}C-formaldehyde (Abraham, Hoheisel, & Gaponenko, 2008; Xie, Fulton, & Andreotti, 2015). Lysine methylation can yield tremendous insights into protein structure and dynamics (Larda et al., 2012; Yokoyama, 2003). Lysines have high occurrence (6–7%) in soluble and solvent-exposed membrane proteins. Often, these residues are located nearby or within active sites and can report on protein function. In general, lysine methylation is structurally non-perturbing. However, when these residues participate to the chemistry of enzymes, then capping their charged side chain may result in a significant reduction or loss of activity as we have found for SERCA (data not shown). Nonetheless, there are several reports on the successful application of reductive methylation of lysines (Moore et al., 1998; Rayment, 1997; Rypniewski, Holden, & Rayment, 1993; White & Rayment, 1993; Zhang, Thulin, & Vogel, 1994) and efforts toward developing protocols for mono-ethylation have provided enhanced spectral quality and narrower line-widths (Larda et al., 2012). This method, however, has had its largest utilization for liquid-state NMR spectroscopy, where the fast tumbling rates of globular proteins and their internal flexibility of solvent-exposed lysine enhance the sensitivity of the methyl groups (Abraham et al., 2015). For ssNMR, however, lysine methyl labeling is still limited as the resonances overlap with the ^{13}C natural abundance background.

Acknowledgment

This work was supported by the National Institute of Health (GM 64742 and HL 144130).

References

Abraham, S. J., Cheng, R. C., Chew, T. A., Khantwal, C. M., Liu, C. W., Gong, S., et al. (2015). ^{13}C NMR detects conformational change in the 100-kDa membrane transporter ClC-ec1. *Journal of Biomolecular NMR*, *61*(3–4), 209–226. https://doi.org/10.1007/s10858-015-9898-7.

Abraham, S. J., Hoheisel, S., & Gaponenko, V. (2008). Detection of protein–ligand interactions by NMR using reductive methylation of lysine residues. *Journal of Biomolecular NMR*, *42*(2), 143–148. https://doi.org/10.1007/s10858-008-9274-y.

Andronesi, O. C., Becker, S., Seidel, K., Heise, H., Young, H. S., & Baldus, M. (2005). Determination of membrane protein structure and dynamics by magic-angle-spinning solid-state NMR spectroscopy. *Journal of the American Chemical Society*, *127*(37), 12965–12974. https://doi.org/10.1021/ja0530164.

Bennett, A. E., Rienstra, C. M., Auger, M., Lakshmi, K. V., & Griffin, R. G. (1995). Heteronuclear decoupling in rotating solids. *The Journal of Chemical Physics*, *103*(16), 6951–6958. https://doi.org/10.1063/1.470372.

Boutureira, O., & Bernardes, G. J. L. (2015). Advances in chemical protein modification. *Chemical Reviews*, *115*(5), 2174–2195. https://doi.org/10.1021/cr500399p.

Cady, S. D., Goodman, C., Tatko, C. D., DeGrado, W. F., & Hong, M. (2007). Determining the orientation of uniaxially rotating membrane proteins using unoriented samples: A 2H, 13C, and 15N solid-state NMR investigation of the dynamics and orientation of a transmembrane helical bundle. *Journal of the American Chemical Society*, *129*(17), 5719–5729. https://doi.org/10.1021/ja070305e.

Chen, L., Olsen, R. A., Elliott, D. W., Boettcher, J. M., Zhou, D. H., Rienstra, C. M., et al. (2006). Constant-time through-bond ^{13}C correlation spectroscopy for assigning protein resonances with solid-state NMR spectroscopy. *Journal of the American Chemical Society*, *128*(31), 9992–9993. https://doi.org/10.1021/ja062347t.

Chipot, C., Dehez, F., Schnell, J. R., Zitzmann, N., Pebay-Peyroula, E., Catoire, L. J., et al. (2018). Perturbations of native membrane protein structure in alkyl phosphocholine detergents: A critical assessment of NMR and biophysical studies. *Chemical Reviews*, *118*(7), 3559–3607. https://doi.org/10.1021/acs.chemrev.7b00570.

Debelouchina, G. T., & Muir, T. W. (2017). A molecular engineering toolbox for the structural biologist. *Quarterly Reviews of Biophysics*, *50*, e7. https://doi.org/10.1017/S0033583517000051.

Delaglio, F., Grzesiek, S., Vuister, G. W., Zhu, G., Pfeifer, J., & Bax, A. (1995). NMRPipe: A multidimensional spectral processing system based on UNIX pipes. *Journal of Biomolecular NMR*, *6*(3), 277–293. https://doi.org/10.1007/BF00197809.

Fernandez, J., Rosemblatt, M., & Hidalgo, C. (1980). Highly purified sarcoplasmic reticulum vesicles are devoid of $^{Ca2+}$-independent ('basal') ATPase activity. *Biochimica et Biophysica Acta (BBA)—Biomembranes*, *599*(2), 552–568. https://doi.org/10.1016/0005-2736(80)90199-6.

Fung, B. M., Khitrin, A. K., & Ermolaev, K. (2000). An improved broadband decoupling sequence for liquid crystals and solids. *Journal of Magnetic Resonance*, *142*(1), 97–101. https://doi.org/10.1006/jmre.1999.1896.

Galan, S. R. G., Wickens, J. R., Dadova, J., Ng, W.-l., Zhang, X., Simion, R. A., et al. (2018). Post-translational site-selective protein backbone α-deuteration. *Nature Chemical Biology*, *14*(10), 955–963. https://doi.org/10.1038/s41589-018-0128-y.

Galiakhmetov, A. R., Kovrigina, E. A., Xia, C., Kim, J. J. P., & Kovrigin, E. L. (2018). Application of methyl-TROSY to a large paramagnetic membrane protein without perdeuteration: ^{13}C-MMTS-labeled NADPH-cytochrome P450 oxidoreductase. *Journal of Biomolecular NMR*, *70*(1), 21–31. https://doi.org/10.1007/s10858-017-0152-3.

Göbl, C., Madl, T., Simon, B., & Sattler, M. (2014). NMR approaches for structural analysis of multidomain proteins and complexes in solution. *Progress in Nuclear Magnetic Resonance Spectroscopy*, *80*, 26–63. https://doi.org/10.1016/j.pnmrs.2014.05.003.

Gopinath, T., Nelson, S. E. D., Soller, K. J., & Veglia, G. (2017). Probing the conformationally excited states of membrane proteins via ^{1}H-detected MAS solid-state NMR spectroscopy. *The Journal of Physical Chemistry B*, *121*(17), 4456–4465. https://doi.org/10.1021/acs.jpcb.7b03268.

Gottlieb, H. E., Kotlyar, V., & Nudelman, A. (1997). NMR chemical shifts of common laboratory solvents as trace impurities. *Journal of Organic Chemistry*, *62*(21), 7512–7515. https://doi.org/10.1021/jo971176v.

Hartmann, S. R., & Hahn, E. L. (1962). Nuclear double resonance in the rotating frame. *Physical Review*, *128*(5), 2042–2053. https://doi.org/10.1103/PhysRev.128.2042.

Hattori, Y., Heidenreich, D., Ono, Y., Sugiki, T., Yokoyama, K.-i., Suzuki, E.-i., et al. (2017). Protein ^{19}F-labeling using transglutaminase for the NMR study of intermolecular interactions. *Journal of Biomolecular NMR*, *68*(4), 271–279. https://doi.org/10.1007/s10858-017-0125-6.

Hayashi, S., Ueki, H., Harano, S., Komiya, J., Iyama, S., Harano, K., et al. (1964). Studies on antitumor substances. III. Syntheses of bis(methanesulphonylthio)alkanes. *Chemical & Pharmaceutical Bulletin*, *12*(11), 1271–1276. https://doi.org/10.1248/cpb.12.1271.

Hebel, D., Kirk, K. L., Cohen, L. A., & Labroo, V. M. (1990). First direct fluorination of tyrosine-containing biologically active peptides. *Tetrahedron Letters*, *31*(5), 619–622. https://doi.org/10.1016/S0040-4039(00)94583-0.

Hediger, S., Meier, B. H., & Ernst, R. R. (1995). Adiabatic passage Hartmann-Hahn cross polarization in NMR under magic angle sample spinning. *Chemical Physics Letters*, *240*(5–6), 449–456. https://doi.org/10.1016/0009-2614(95)00505-X.

Hediger, S., Meier, B. H., Kurur, N. D., Bodenhausen, G., & Ernst, R. R. (1994). NMR cross polarization by adiabatic passage through the Hartmann-Hahn condition (APHH). *Chemical Physics Letters*, *223*(4), 283–288. https://doi.org/10.1016/0009-2614(94)00470-6.

Hu, B., Lafon, O., Trébosc, J., Chen, Q., & Amoureux, J.-P. (2011). Broad-band homonuclear correlations assisted by ^{1}H irradiation for bio-molecules in very high magnetic field at fast and ultra-fast MAS frequencies. *Journal of Magnetic Resonance*, *212*(2), 320–329. https://doi.org/10.1016/j.jmr.2011.07.011.

Ishii, Y., Wickramasinghe, A., Matsuda, I., Endo, Y., Ishii, Y., Nishiyama, Y., et al. (2018). Progress in proton-detected solid-state NMR (SSNMR): Super-fast 2D SSNMR collection for nano-mole-scale proteins. *Journal of Magnetic Resonance*, *286*, 99–109. https://doi.org/10.1016/j.jmr.2017.11.011.

Iwai, H., & Züger, S. (2007). Protein ligation: Applications in NMR studies of proteins. *Biotechnology and Genetic Engineering Reviews*, *24*(1), 129–146. https://doi.org/10.1080/02648725.2007.10648096.

Jeuken, L. J. C. (2009). Electrodes for integral membrane enzymes. *Natural Product Reports*, *26*(10), 1234–1240. https://doi.org/10.1039/b903252e.

Jo, S., Kim, T., Iyer, V. G., & Im, W. (2008). CHARMM-GUI: A web-based graphical user interface for CHARMM. *Journal of Computational Chemistry*, *29*(11), 1859–1865. https://doi.org/10.1002/jcc.20945.

Kalbitzer, H. R., Rohr, G., Nowak, E., Goody, R. S., Kuhn, W., & Zimmermann, E. (1992). A new high sensitivity ^{19}F probe for labeling cysteine groups of proteins. *NMR in Biomedicine*, *5*(6), 347–350. https://doi.org/10.1002/nbm.1940050605.

Kenyon, G. L., & Bruice, T. W. (1977). Novel sulfhydryl reagents. *Methods in Enzymology*, *47*, 407–430.

Larda, S. T., Bokoch, M. P., Evanics, F., & Prosser, R. S. (2012). Lysine methylation strategies for characterizing protein conformations by NMR. *Journal of Biomolecular NMR*, *54*(2), 199–209. https://doi.org/10.1007/s10858-012-9664-z.

Larda, S. T., Pichugin, D., & Prosser, R. S. (2015). Site-specific labeling of protein lysine residues and N-terminal amino groups with indoles and indole-derivatives. *Bioconjugate Chemistry*, *26*(12), 2376–2383. https://doi.org/10.1021/acs.bioconjchem.5b00457.

Larsen, E., Olivieri, C., Walker, C., V.S, M., Gao, J., Bernlohr, D., et al. (2018). Probing protein-protein interactions using asymmetric labeling and carbonyl-carbon selective heteronuclear NMR spectroscopy. *Molecules*, *23*(8), 1937. https://doi.org/10.3390/molecules23081937.

Le Maire, M., Kwee, S., Andersen, J. P., & Møller, J. V. (1983). Mode of interaction of polyoxyethyleneglycol detergents with membrane proteins. *European Journal of Biochemistry*, *129*(3), 525–532. https://doi.org/10.1111/j.1432-1033.1983.tb07080.x.

Lee, M., & Goldburg, W. I. (1965). Nuclear-magnetic-resonance line narrowing by a rotating rf field. *Physical Review*, *140*(4A), A1261–A1271. https://doi.org/10.1103/PhysRev.140.A1261.

Liu, J. J., Horst, R., Katritch, V., Stevens, R. C., & Wüthrich, K. (2012). Biased signaling pathways in β2-adrenergic receptor characterized by ^{19}F-NMR. *Science*, *335*(6072), 1106–1110. https://doi.org/10.1126/science.1215802.

Manglik, A., Kim, T. H., Masureel, M., Altenbach, C., Yang, Z., Hilger, D., et al. (2015). Structural insights into the dynamic process of β2-adrenergic receptor signaling. *Cell*, *161*(5), 1101–1111. https://doi.org/10.1016/j.cell.2015.04.043.

Mehta, V. D., Kulkarni, P. V., Mason, R. P., Constantinescu, A., & Antich, P. P. (1994). Fluorinated proteins as potential ^{19}F magnetic resonance imaging and spectroscopy agents. *Bioconjugate Chemistry*, *5*(3), 257–261. https://doi.org/10.1021/bc00027a011.

Metz, G., Wu, X. L., & Smith, S. O. (1994). Ramped-amplitude cross polarization in magic-angle-spinning NMR. *Journal of Magnetic Resonance, Series A*, *110*(2), 219–227. https://doi.org/10.1006/jmra.1994.1208.

Michaeli, S., Sorce, D. J., Idiyatullin, D., Ugurbil, K., & Garwood, M. (2004). Transverse relaxation in the rotating frame induced by chemical exchange. *Journal of Magnetic Resonance*, *169*(2), 293–299. https://doi.org/10.1016/j.jmr.2004.05.010.

Moore, G. R., Cox, M. C., Crowe, D., Osborne, M. J., Rosell, F. I., Bujons, J., et al. (1998). Nε,Nε-dimethyl-lysine cytochrome c as an NMR probe for lysine involvement in protein-protein complex formation. *Biochemical Journal*, *332*(Pt 2), 439–449. https://doi.org/10.1042/bj3320439.

Morcombe, C. R., & Zilm, K. W. (2003). Chemical shift referencing in MAS solid state NMR. *Journal of Magnetic Resonance*, *162*(2), 479–486. https://doi.org/10.1016/S1090-7807(03)00082-X.

Morris, G. A., & Freeman, R. (1979). Enhancement of nuclear magnetic resonance signals by polarization transfer. *Journal of the American Chemical Society*, *101*(3), 760–762. https://doi.org/10.1021/ja00497a058.

Pachter, I. J., & Scheinbaum, M. L. (1974). Trimethylene dithiotosylate and ethylene dithiotosylate. *Organic Syntheses*, *54*(September), 33. https://doi.org/10.15227/orgsyn.054.0033.

Prestegard, J. H., Agard, D. A., Moremen, K. W., Lavery, L. A., Morris, L. C., & Pederson, K. (2014). Sparse labeling of proteins: Structural characterization from long range constraints. *Journal of Magnetic Resonance*, *241*(1), 32–40. https://doi.org/10.1016/j.jmr.2013.12.012.

Prosser, R. S., & Kim, T. H. (2015). Nuts and bolts of CF_3 and CH_3 NMR toward the understanding of conformational exchange of GPCRs. In M. Filizola (Ed.), *Methods in molecular biology*: *Vol. 1335. G protein-coupled receptors in drug discovery* (pp. 39–51). Clifton, N.J: Humana Press. https://doi.org/10.1007/978-1-4939-2914-6_4.

Rayment, I. (1997). Reductive alkylation of lysine residues to alter crystallization properties of proteins. *Methods in Enzymology*, *276*, 171–179.

Reddy, L. G., Cornea, R. L., Winters, D. L., McKenna, E., & Thomas, D. D. (2003). Defining the molecular components of calcium transport regulation in a reconstituted membrane system. *Biochemistry*, *42*(15), 4585–4592. https://doi.org/10.1021/bi026995a.

Religa, T. L., Ruschak, A. M., Rosenzweig, R., & Kay, L. E. (2011). Site-directed methyl group labeling as an NMR probe of structure and dynamics in supramolecular protein systems: Applications to the proteasome and to the ClpP protease. *Journal of the American Chemical Society*, *133*(23), 9063–9068. https://doi.org/10.1021/ja202259a.

Richards, P. G., Coles, B., Heptinstall, J., & Walton, D. J. (1994). Electrochemical modification of lysozyme: Anodic reaction of tyrosine residues. *Enzyme and Microbial Technology*, *16*(9), 795–801. https://doi.org/10.1016/0141-0229(94)90038-8.

Rigaud, J. L., Levy, D., Mosser, G., & Lambert, O. (1998). Detergent removal by non-polar polystyrene beads: Applications to membrane protein reconstitution and two-dimensional crystallization. *European Biophysics Journal*, *27*(4), 305–319. https://doi.org/10.1007/s002490050138.

Rypniewski, W. R., Holden, H. M., & Rayment, I. (1993). Structural consequences of reductive methylation of lysine residues in hen egg white lysozyme: An X-ray analysis at 1.8-Å resolution. *Biochemistry*, *32*(37), 9851–9858. https://doi.org/10.1021/bi00088a041.

Shekhawat, S. S., Pham, G. H., Prabakaran, J., & Strieter, E. R. (2014). Simultaneous detection of distinct ubiquitin chain topologies by ^{19}F NMR. *ACS Chemical Biology*, *9*(10), 2229–2236. https://doi.org/10.1021/cb500589c.

Stokes, D. L., & Michael Green, N. (1990). Structure of CaATPase: Electron microscopy of frozen-hydrated crystals at 6 Å resolution in projection. *Journal of Molecular Biology*, *213*(3), 529–538. https://doi.org/10.1016/S0022-2836(05)80213-X.

Strandberg, E., Esteban-Martín, S., Salgado, J., & Ulrich, A. S. (2009). Orientation and dynamics of peptides in membranes calculated from ^{2}H-NMR data. *Biophysical Journal*, *96*(8), 3223–3232. https://doi.org/10.1016/j.bpj.2009.02.040.

Szeverenyi, N. M., Sullivan, M. J., & Maciel, G. E. (1982). Observation of spin exchange by two-dimensional fourier transform ^{13}C cross polarization-magic-angle spinning. *Journal of Magnetic Resonance*, *47*(3), 462–475. https://doi.org/10.1016/0022-2364(82)90213-X. [1969].

Takegoshi, K., Nakamura, S., & Terao, T. (2001). ^{13}C-^{1}H dipolar-assisted rotational resonance in magic-angle spinning NMR. *Chemical Physics Letters*, *344*(5–6), 631–637. https://doi.org/10.1016/S0009-2614(01)00791-6.

Toyoshima, C. (2008). Structural aspects of ion pumping by Ca^{2+}-ATPase of sarcoplasmic reticulum. *Archives of Biochemistry and Biophysics*, *476*(1), 3–11. https://doi.org/10.1016/j.abb.2008.04.017.

Toyoshima, C., Nakasako, M., Nomura, H., & Ogawa, H. (2000). Crystal structure of the calcium pump of sarcoplasmic reticulum at 2.6 A resolution. *Nature*, *405*(6787), 647–655. https://doi.org/10.1038/35015017.

Tugarinov, V., Ollerenshaw, J. E., & Kay, L. E. (2005). Probing side-chain dynamics in high molecular weight proteins by deuterium NMR spin relaxation: An application to an 82-kDa enzyme. *Journal of the American Chemical Society*, *127*(22), 8214–8225. https://doi.org/10.1021/ja0508830.

van Rossum, B. J., Förster, H., & de Groot, H. J. M. (1997). High-field and high-speed CP-MAS ^{13}C NMR heteronuclear dipolar-correlation spectroscopy of solids with frequency-switched Lee–Goldburg homonuclear decoupling. *Journal of Magnetic Resonance*, *124*(2), 516–519. https://doi.org/10.1006/jmre.1996.1089.

Verardi, R., Traaseth, N. J., Masterson, L. R., Vostrikov, V. V., & Veglia, G. (2012). Isotope labeling for solution and solid-state NMR spectroscopy of membrane proteins. In H. S. Atreya (Ed.), *Vol. 992. Isotope labeling in biomolecular NMR* (pp. 35–62). Dordrecht: Springer Netherlands.

Vostrikov, V. V., Gustavsson, M., Gopinath, T., Mullen, D., Dicke, A. A., Truong, V., et al. (2016). Ca(2+) ATPase conformational transitions in lipid bilayers mapped by site-directed ethylation and solid-state NMR. *ACS Chemical Biology*, *11*(2), 329–334. https://doi.org/10.1021/acschembio.5b00953.

Walters, K. J., Ferentz, A. E., Hare, B. J., Hidalgo, P., Jasanoff, A., Matsuo, H., et al. (2001). Characterizing protein-protein complexes and oligomers by nuclear magnetic resonance spectroscopy. In V. D. T. L. James & U. Schmitz (Eds.), *Vol. 339. Nuclear magnetic resonance of biological macromolecules—Part B* (pp. 238–258). Elsevier Masson SAS.

Wang, S., Gopinath, T., & Veglia, G. (2018). Application of paramagnetic relaxation enhancements to accelerate the acquisition of 2D and 3D solid-state NMR spectra of oriented membrane proteins. *Methods*, *138-139*, 54–61. https://doi.org/10.1016/j.ymeth.2017.12.017.

Wang, M., Lu, M., Fritz, M. P., Quinn, C. M., Byeon, I.-J. L., Byeon, C.-H., et al. (2018). Fast magic-angle spinning ^{19}F NMR spectroscopy of HIV-1 capsid protein assemblies. *Angewandte Chemie International Edition*, *57*, 16375–16379. https://doi.org/10.1002/anie.201809060.

Wawrzynow, A., Collins, J. H., & Coan, C. (1993). An iodoacetamide spin-label selectively labels a cysteine side chain in an occluded site on the sarcoplasmic reticulum calcium-ATPase. *Biochemistry*, *32*(40), 10803–10811. https://doi.org/10.1021/bi00091a035.

Weidner, J. P., & Block, S. S. (1964). Alkyl and aryl thiolsulfonates. *Journal of Medicinal Chemistry*, 7(5), 671–673. https://doi.org/10.1021/jm00335a028.

Weingarth, M., Demco, D. E., Bodenhausen, G., & Tekely, P. (2009). Improved magnetization transfer in solid-state NMR with fast magic angle spinning. *Chemical Physics Letters*, *469*(4–6), 342–348. https://doi.org/10.1016/j.cplett.2008.12.084.

White, H. D., & Rayment, I. (1993). Kinetic characterization of reductively methylated myosin subfragment. *Biochemistry*, *32*(37), 9859–9865. https://doi.org/10.1021/bi00088a042.

Wickramasinghe, N. P., Kotecha, M., Samoson, A., Past, J., & Ishii, Y. (2007). Sensitivity enhancement in ^{13}C solid-state NMR of protein microcrystals by use of paramagnetic metal ions for optimizing ^{1}H T_1 relaxation. *Journal of Magnetic Resonance*, *184*(2), 350–356. https://doi.org/10.1016/j.jmr.2006.10.012.

Wu, E. L., Cheng, X., Jo, S., Rui, H., Song, K. C., Davila-Contreras, E. M., et al. (2014). CHARMM-GUI membrane builder toward realistic biological membrane simulations. *Journal of Computational Chemistry*, *35*(27), 1997–2004. https://doi.org/10.1002/jcc.23702.

Xie, Q., Fulton, D. B., & Andreotti, A. H. (2015). A selective NMR probe to monitor the conformational transition from inactive to active kinase. *ACS Chemical Biology*, *10*(1), 262–268. https://doi.org/10.1021/cb5004702.

Xu, R., Ayers, B., Cowburn, D., & Muir, T. W. (1999). Chemical ligation of folded recombinant proteins: Segmental isotopic labeling of domains for NMR studies. *Proceedings of the National Academy of Sciences*, *96*(2), 388–393. https://doi.org/10.1073/pnas.96.2.388.

Ye, L., Larda, S. T., Frank Li, Y. F., Manglik, A., & Prosser, R. S. (2015). A comparison of chemical shift sensitivity of trifluoromethyl tags: Optimizing resolution in ^{19}F NMR studies of proteins. *Journal of Biomolecular NMR*, *62*(1), 97–103. https://doi.org/10.1007/s10858-015-9922-y.

Yokoyama, S. (2003). Protein expression systems for structural genomics and proteomics. *Current Opinion in Chemical Biology*, 7(1), 39–43. https://doi.org/10.1016/S1367-5931(02)00019-4.

Zhang, M., Thulin, E., & Vogel, H. J. (1994). Reductive methylation and pKa determination of the lysine side chains in calbindin D9k. *Journal of Protein Chemistry*, *13*(6), 527–535. https://doi.org/10.1007/BF01901534.

CHAPTER SEVENTEEN

Improved sensitivity and resolution of in-cell NMR spectra

David S. Burz, Leonard Breindel, Alexander Shekhtman*
Department of Chemistry, University at Albany, State University of New York, Albany, NY, United States
*Corresponding author: e-mail address: ashekhtman@albany.edu

Contents

Abstract

In-cell NMR spectroscopy is a powerful tool to study protein structures and interactions under near physiological conditions in both prokaryotic and eukaryotic living cells. The low sensitivity and resolution of in-cell NMR spectra and limited lifetime of cells over the course of an in-cell experiment have presented major hurdles to wide acceptance of the technique, limiting it to a few select systems. These issues are addressed by introducing the use of the CRINEPT pulse sequence to increase the sensitivity and resolution of in-cell NMR spectra and the use of a bioreactor to maintain cell viability for up to 24 h. Application of advanced pulse sequences and bioreactor during in-cell NMR experiments will facilitate the exploration of a wide range of biological processes.

Methods in Enzymology, Volume 621
ISSN 0076-6879
https://doi.org/10.1016/bs.mie.2019.02.029

1. Introduction

In-cell NMR spectroscopy utilizes NMR spectroscopy to study the structure, dynamics and interactions of proteins in living cells. The use of in-cell NMR spectroscopy to study proteins has expanded at a steady pace over the past two decades. Since the initial experiments of overexpressing target proteins in bacterial cells (Serber et al., 2001; Serber & Dotsch, 2001; Wieruszeski, Bohin, Bohin, & Lippens, 2001; Williams, Haggie, & Brindle, 1997), the field has developed a variety of methods for isotopic labeling (Hamatsu et al., 2013; Li et al., 2010; Serber et al., 2004), delivering labeled targets to prokaryotic and eukaryotic cells (Banci et al., 2013; Bertrand, Reverdatto, Burz, Zitomer, & Shekhtman, 2012; Bodart et al., 2008; Hamatsu et al., 2013; Inomata et al., 2009; Ogino et al., 2009; Sakai et al., 2006; Selenko, Serber, Gadea, Ruderman, & Wagner, 2006; Theillet et al., 2016), identifying high and low affinity specific and non-specific protein-protein interactions (Burz, Dutta, Cowburn, & Shekhtman, 2006a, 2006b), determining in-cell atomic resolution structures (Ikeya et al., 2016; Muntener, Haussinger, Selenko, & Theillet, 2016; Sakakibara et al., 2009), studying interactions of the target with the cytosol, mapping the interactions surfaces of target proteins (Danielsson et al., 2015; Kyne & Crowley, 2017; Luh et al., 2013; Majumder, DeMott, Burz, & Shekhtman, 2014; Smith, Zhou, Gorensek, Senske, & Pielak, 2016), detecting targets at physiological concentrations, high throughput drug screening, interaction proteomics, data collection and analysis (Cobbert et al., 2015; DeMott et al., 2018; Ikeya et al., 2010; Theillet et al., 2016; Xie, Thapa, Reverdatto, Burz, & Shekhtman, 2009). Despite these innovations, two major problems continue to plague in-cell NMR experiments: spectral peak broadening and cell viability. In this work we present protocols that help alleviate these difficulties by improving the resolution of in-cell NMR spectra.

1.1 In-cell NMR peak broadening

Multi-dimensional NMR spectroscopy such as heteronuclear single quantum coherence, HSQC, NMR spectroscopy has traditionally been used to investigate target proteins in-cell (Serber & Dotsch, 2001). In-cell spectra are compared to a well-resolved ^{1}H–^{15}N HSQC spectrum of purified isotope-labeled protein in vitro or in cell lysates to assign chemical shifts. However, in-cell, many of the NMR crosspeaks of folded proteins exhibit reduced intensity (broadening) due to a reduced rate of tumbling arising from the increased viscosity of the intracellular medium and interactions with macromolecular components of the cytosol (quinary interactions) that increase the apparent molecular weight of the complex (Crowley, Chow, & Papkovskaia, 2011; Majumder et al., 2015; Ye et al., 2013) (Fig. 1). The contribution from increased viscosity and molecular crowding contributes a comparatively small amount to the peak broadening; the dominant effect arises from quinary interactions (Majumder et al., 2015; Ye et al., 2013). Notable exceptions to this are intrinsically disordered proteins, IDPs, which lack persistent secondary or higher structure, and fail to interact with intracellular constituents; the in-cell spectra of IDPs are much sharper than those typically observed for folded proteins (Pielak et al., 2009). Modifications of traditional NMR pulse sequences (Felli, Gonnelli, & Pierattelli, 2014) have provided major advancements in the ability to resolve crosspeaks that are typically broadened during in-cell NMR experiments.

The increase in apparent molecular weight of a target protein in-cell resulting from quinary interactions leads to a reduced transverse relaxation, T2, which dictates the rate of free induction decay, and results in broader crosspeaks (Wuthrich, 1986). T2 is inversely related to the rotational correlation time, τ_c, of a molecule or molecular complex. τ_c is a time constant that describes the rotation diffusion of a molecule in solution. It is the time

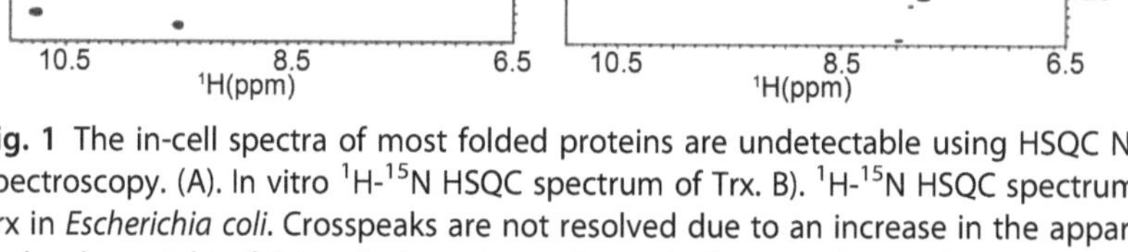

Fig. 1 The in-cell spectra of most folded proteins are undetectable using HSQC NMR spectroscopy. (A). In vitro ^{1}H-^{15}N HSQC spectrum of Trx. B). ^{1}H-^{15}N HSQC spectrum of Trx in *Escherichia coli*. Crosspeaks are not resolved due to an increase in the apparent molecular weight of thioredoxin, only small metabolites are observed.

required for a molecule to rotate by 1 radian and depends on the Stokes radius. Increasing viscosity or molecular weight increases τ_c. The resulting NMR signal from larger molecules decays more rapidly and leads to line broadening. By using transverse relaxation-optimized spectroscopy, TROSY, which suppresses transverse nuclear spin relaxation in heteronuclear NMR experiments during evolution and acquisition cycles (Pervushin, Wider, & Wuthrich, 1998), TROSY provides increased sensitivity and resolution of in-cell NMR peaks for large, >100 kDa, molecular weight species.

The HSQC and HMQC pulse sequences widely used for in-cell NMR spectroscopy use an insensitive nuclei enhanced by polarization transfer, INEPT, pulse sequence to transfer magnetization from protons to heteronuclei, but the efficiency of INEPT deteriorates with increasing τ_c (Riek, Pervushin, & Wuthrich, 2000). Further improvement in resolving the NMR spectra of large molecular species is achieved by using ^{15}N-edited cross relaxation-enhanced polarization transfer, CRINEPT, NMR spectroscopy (Riek, Wider, Pervushin, & Wuthrich, 1999), which increases the efficiency of magnetization transfers between heteronuclei for molecules with τ_c between 50 and 300 ns. The ^{1}H-^{15}N HMQC pulse sequence contains two CRINEPT-like magnetization transfers delay times that can be optimized for molecular species with large rotational correlation times.

The combination of TROSY and CRINEPT creates transverse relaxation optimization during the entire pulse sequence. Importantly, since the efficiency of a CRINEPT polarization transfer increases with the strength of the external magnetic field, the sensitivity of the NMR experiment improves at higher magnetic fields (Riek et al., 2000). Relaxation optimized ^{15}N-edited CRINEPT-heteronuclear multiple quantum coherence, HMQC, transverse relaxation optimized spectroscopy, TROSY, ^{1}H-^{15}N CRINEPT-HMQC-TROSY experiments (Riek et al., 1999) in combination with REDuced PROton density (REDPRO) labeling (Shekhtman, Ghose, Goger, & Cowburn, 2002), which exchanges alpha and beta protons of amino acids for deuterons to minimize proton relaxation, have proven to be useful to detect high molecular weight species in-cell due to superior sensitivity to NMR signals and relative insensitivity to magnetic field inhomogeneity (Wider, 2005). Despite these innovative advances and other improvements such as shortening the data collection and analysis time, in-cell NMR spectroscopic experiments must still confront the dilemma associated with cell viability/stability.

1.2 Cell viability

A major consideration when performing in-cell NMR is verifying the integrity of the cells. Control experiments are routinely performed to verify that cell viability remains at or >90–95% over the course of an experiment. Cells stained with trypan blue, which does not penetrate cells with intact membranes, or colony plated, are quantitatively compared before and after in-cell NMR experiments to assess viability (Burz & Shekhtman, 2010; Serber et al., 2006). In addition, samples are centrifuged following data acquisition and HSQC spectra collected on the supernatants. Target proteins in the supernatant will yield sharp signals indicative of leakage or lysis due to cell death over the course of the experiment.

In-cell NMR experiments often require 2–3 h or more to achieve at least a 3:1 signal to noise ratio.(Banci et al., 2013; Maldonado, Burz, & Shekhtman, 2011; Sakakibara et al., 2009; Serber, Corsini, Durst, & Dotsch, 2005; Theillet et al., 2016). Intracellular changes induced by stimuli, such as cellular adaptation to metabolic (Masters & Stacey, 2007; Wellen & Thompson, 2010) or antibiotic (Galhardo, Hastings, & Rosenberg, 2007) stress, can take hours to fully manifest into detectable signals. During this time the cells are deprived of nutrients and are in essence dying. The metabolic environment is therefore in a state of decay with an ever-changing background. As a result, it is important to ensure that the cells remain in a metabolically active state during in-cell NMR experiments.

To maintain a consistent cellular background during in-cell NMR experiments a bioreactor can be used to ensure that the experiments are conducted with stable, metabolically active cells. In these bioreactors fresh growth medium is continually flowed into the cells while metabolic byproducts are removed. Most bioreactors are complex and require pumps to circulate the medium in a closed system (Inomata, Kamoshida, Ikari, Ito, & Kigawa, 2017; Kubo et al., 2013; Sharaf, Barnes, Charlton, Young, & Pielak, 2010). Current bioreactors can maintain cell viability, as assessed by constant elevated levels of ATP, for up to 12h. Real time, RT, in-cell NMR reactor (Breindel, DeMott, Burz, & Shekhtman, 2018) significantly simplifies previously described flow in-cell NMR methods (Kubo et al., 2013; Sharaf et al., 2010) and maintains cell viability for up to 24h.

In RT-NMR flow is passively sustained by using a gravity siphon. Continuous exchange of medium minimizes cell leakage of labeled proteins. Cells are cast in low melting agarose to keep the sample at the bottom of the NMR tube. To lock on solvent and eliminate the need for deuterated solvent in the medium, an agarose plug in 100% D_2O is located at the bottom of the NMR tube. A horizontal "drip irrigation" system maximizes exposure of the cells to fresh medium and minimizes the upward drag that often dislodges cells from their positions in the NMR tube, which can significantly reduce the quality of the in-cell NMR spectrum.

The metabolic state of the cells is assessed by monitoring the ATP levels. Reduction in the intracellular concentration of ATP is indicative of cell death, whereas constant levels are consistent with a viable metabolically active steady state condition.

1.3 Real time STINT-NMR spectroscopy

Broadening of in-cell spectral peaks arises primarily because of quinary interactions between the target protein and RNA (Majumder et al., 2015; Majumder, DeMott, Reverdatto, Burz, & Shekhtman, 2016). As a result, the in-cell spectrum must be analyzed relative to an *in vitro* or lysate spectrum to identify the interacting surfaces of the target molecule that define the quinary state. Changes in the in-cell target spectrum due to overexpression of an interactor protein or externally administered compounds are, in turn, analyzed relative to the quinary state. Structural interactions NMR spectroscopy, STINT-NMR, is used to identify the interacting surfaces (Burz et al., 2006a; Burz, DeMott, Aldousary, Dansereau, & Shekhtman, 2018; Majumder et al., 2014).

During a STINT-NMR experiment, a series of in-cell spectra are collected and a matrix is created that contains the changes in target protein peak intensities or chemical shifts versus time. Individual signals change at different rates as the concentration of the interactor protein increases. Time dependent degradation of the target protein inside the cell or differences in sample preparations also changes the in-cell spectrum. To distinguish spectral changes that are due to specific binding interactions as the concentration of the interactor or external stimulus increase a rigorous objective analysis, singular value decomposition, SVD, is used (Golub & Van Loan, 2012; Majumder et al., 2014).

SVD analysis discriminates between changes in in-cell spectra due to concentration-dependent and -independent binding interactions (Burz et al., 2018; Majumder et al., 2014). The analysis is simplified by maintaining cells in a steady state. Results of the analysis are presented as a Scree plot (Cattell, 1966) that shows the distribution of singular values that define the relative contribution of each binding mode to the changes in spectral peak intensities or chemical shifts. An abrupt drop in singular values following the first binding mode indicates a change due to specific interactions; a gradual decrease corresponds to random binding of the target protein to components of the cytosol (Majumder et al., 2014).

We describe two techniques for improving the resolution of in-cell NMR experiments. The first describes the application of ^{1}H–^{15}N CRINEPT-HMQC-TROSY spectroscopy to study large slowly tumbling molecular complexes in an effort to improve the resolution and sensitivities of traditional 2D experiments. The second describes the construction of a bioreactor that maintains cells in a steady state for periods of up to 24 h. A third protocol combines the use of ^{1}H–^{15}N CRINEPT-HMQC-TROSY spectroscopy with RT and STINT-NMR, and analysis of the data via SVD, to study the effect of ribosome-binding antibiotics on the quinary structure of *Escherichia coli* thioredoxin.

2. Materials

2.1 Equipment

Shaker incubator
Centrifuge
600 MHz or greater Bruker NMR spectrometer equipped with a cryoprobe
150 cm^2 culture flasks
LB-kanamycin agar plates
2.5 cm diameter × 20 cm cylinder
Polytetrafluoroethylene, PTFE, tubing, 0.5 mm I.D.
Stainless steel dissection needle with 10 μm tip (Roboz Surgical)
Microdissection needle holder (Roboz Surgical)
Low pressure tee (Idex Peek)
10 mL syringe
5 mm NMR tube
5 mm screw-cap NMR tube
Teflon membrane/PTFE/silicone septum (New Era)
Inlet reservoir flask
Outlet reservoir flask
Sonicator
Lyophilizer
Inversion microscope equipped with digital camera

Topspin data analysis software
CARA analysis software
MATLAB® software
EXCEL software

2.2 Chemicals

pRSF-Trx expression plasmid
E. coli strain BL21(DE3) codon+
HeLa cells
Luria broth, LB, medium
Kanamycin, Kn
M9 (minimal) medium
Deuterated M9 medium
$^{15}NH_4Cl$
Glucose
1 M isopropyl β-D-1-thiogalactopyranoside, IPTG
NMR buffer, 10 mM potassium phosphate buffer pH 6.5
D_2O
LB-Kn agar plates
Dulbecco's modified Eagle medium, DMEM (Gibco)
Fetal bovine serum, FBS (Gibco)
Trypsin/EDTA (Sigma-Aldrich)
Phosphate buffered saline, PBS
Cambrex Seaprep agarose
35% perchloric acid
2 M KOH
0.5 M EDTA, pH 8
Trypan blue (Hyclone)
Streptomycin

3. Methods

3.1 In-cell ^1H-^{15}N CRINEPT-HMQC-TROSY NMR spectroscopy

For this protocol *E. coli* thioredoxinA, *trxA*, was cloned into pRSF (Novagen) (Majumder et al., 2015). pRSF-Trx confers kanamycin resistance, codes for lac repressor and expresses N-terminal His-tagged bacterial Trx from the T7 promoter/lac operon, which is induced by IPTG in *E. coli* strain BL21(DE3) codon+. The His-tag was not removed from the protein since it does interfere with the experiments (Majumder et al., 2015). NMR peak assignments for the target protein are required to complete the analysis.

3.1.1 Over-expression and labeling of Trx

1. Transform *E. coli* BL21(DE3) codon+ with plasmid pRSF-Trx.
2. Inoculate 10 mL of LB containing 35 μg/mL of kanamycin, LB-Kn, with a single colony of transformed cells.
3. Grow the culture overnight at 37 °C with shaking at 275 rpm.
4. Dilute the overnight culture to OD_{600} ~0.07 in 50 mL of LB-Kn and grow to an OD_{600} of 0.9–1.0 at 37 °C, 275 rpm.
5. Centrifuge the cells at 200 × *g* for 15 min at room temperature.
6. Wash the cells twice with 50 mL of minimal, M9, medium.
7. Resuspend the cells in 200 mL of deuterated M9 supplemented with 1.0 g/L of $^{15}NH_4Cl$ and 0.2% glucose for uniform [*U*- ^{2}D,^{15}N] labeling.
8. Incubate the culture for 20 min at 37 °C.
9. Induce protein overexpression by adding by adding 0.1% culture volume of 1 M IPTG.
10. Allow overexpression of REDPRO-labeled target protein to proceed for up to 36 h at 37 °C, 275 rpm.

3.1.2 Sample preparation for in-cell NMR spectroscopy

1. Remove a 100 mL aliquot from the overexpressing cell culture.
2. Centrifuge the cells at 200 × *g* for 15 min at room temperature.
3. Wash the cells twice with NMR buffer, and re-suspend in 450 μL of NMR buffer supplemented with 60 μL of D_2O.
4. Remove 10 μL of the sample for a cell viability assay.
5. Prepare serial dilutions at 10^{-4}, 10^{-5} and 10^{-6}.
6. Plate 100 μL of each dilution onto LB-Kn plates in triplicate.
7. Incubate the plates overnight at 37 °C and count the resulting colonies.
8. Transfer the remaining sample to a standard 5 mm NMR tube.

3.1.3 In-cell NMR spectroscopy

NMR spectra were recorded at 298 K on a 700 MHz Avance II NMR spectrometer (Bruker) equipped with a TXI z-gradient cryoprobe.

1. Shim the sample manually using the z, z2 and z3 shims.
2. Select/create the ^{1}H-^{15}N CRINEPT-HMQC-TROSY experiment as described in Riek et al. (1999), Fig. 3B.

3. Apply water suppression by using water selective pulses along the $+z$-axis for the entire sequence with a recycle delay of 300 ms.
4. Set the number of transients to 512.
5. Set the spectral widths to 12 ppm and 30 ppm in the ^{1}H and ^{15}N dimensions, respectively.
6. Collect 1024 and 128 points in the ^{1}H and ^{15}N dimensions, respectively.
7. Use the first block of ^{1}H-^{15}N CRINEPT-HMQC-TROSY to optimize the CRINEPT transfer delay time by increasing the CRIPT transfer time by 0.2–0.3 ms starting from 1 ms.
8. Monitor the amide envelope and tryptophan imide peaks to determine the transfer time that produces maximum peak volume, i.e., T_{opt}.
9. Set T_{opt} to the optimal value determined in step 7 (Fig. 2).
10. Collect a ^{1}H-^{15}N CRINEPT-HMQC-TROSY spectrum of the cells (Fig. 2).
11. When data collection is complete, remove a 10 μL aliquot for cell viability assays (Section 3.1.2). A final colony count within 90% of the original count is considered viable.
12. Centrifuge the remaining sample at 4500 × g for 10 min at RT.
13. Transfer the supernatant to a standard 5 mm NMR tube and test for cell leakage by collecting a ^{1}H-^{15}N HSQC spectrum.
14. Re-suspend the cell pellet in 0.3 mL of NMR buffer.

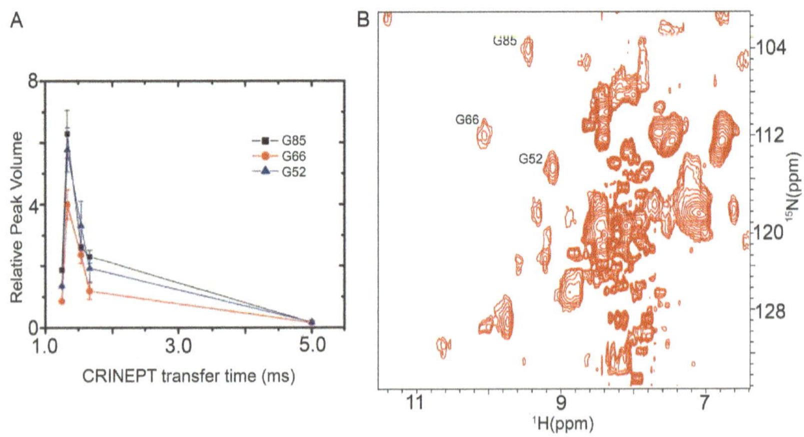

Fig. 2 CRINEPT NMR spectroscopy resolves in-cell crosspeaks for proteins with high apparent molecular weights due to quinary interactions. (A) The relative volumes of G52, G66 and G85 crosspeaks are plotted against the CRINEPT transfer delay times. An endogenous tryptophan indole amide peak in the in-cell spectra is used as a reference. (B) ^{1}H–^{15}N CRINEPT-HMQC-TROSY spectrum of REDPRO-labeled Trx in *E. coli*. The G52, G66 and G85 crosspeaks used for optimizing the CRINEPT transfer delay time are indicated.

15. Freeze-thaw (−80 °C-RT) five times to lyse the cells.
16. Centrifuge the sample at 18000 × *g* for 15 min at RT.
17. Collect a ^1H-^{15}N CRINEPT-HMQC-TROSY spectrum on the clarified lysate using the optimized value for T_{opt}.

3.2 Bioreactor for cell viability

This protocol describes the preparation of *E. coli* and HeLa cells for viability testing. *E. coli* are transformed with expression vector to mimic the cellular background used in in-cell NMR experiments.

3.2.1 Preparation of E. coli *cells*

1. Transform *E. coli* BL21(DE3) codon + with plasmid pRSF-Trx.
2. Inoculate 10 mL of LB medium supplemented with 35 µg/mL of kanamycin with a single colony of pRSF-Trx and grow overnight at 37 °C and 225 rpm in a shaker incubator.
3. Transfer the overnight culture to 200 mL of LB medium supplemented with 35 µg/mL of kanamycin and grow overnight at 37 °C and 225 rpm to an OD_{600} of 0.7–1.0.
4. Centrifuge the culture at 200 × *g* for 20 min and wash the cells twice with M9 medium.

3.2.2 Preparation of HeLa cells

To assess cell viability two sets of cast threads are prepared, one for NMR spectroscopy and one for trypan blue staining (Section 3.2.7).

1. Seed HeLa cells onto four 150 cm^2 flasks in complete medium, low-glucose DMEM supplemented with 10% FBS, and culture in a 5% CO_2 incubator at 37 °C.
2. Culture the cells to 80% confluence (2×10^7 cells/flask).
3. Treat the plates with 0.25% trypsin/EDTA for 5 min at 37 °C to harvest the cells.
4. Neutralize the trypsin by diluting 5 × with complete medium.
5. Centrifuge the cells at 200 × *g* for 10 min at 25 °C and wash with 15 mL of pre-warmed (37 °C) PBS.

3.2.3 Cell casting

The same procedure is used for *E. coli* and HeLa cells.

1. Centrifuge the cells at 200 × *g* for 20 min at 4 °C.
2. Re-suspend the pellet 1:1 (v/v) with 3.0% Cambrex Seaprep agarose in M9 medium (for *E. coli*) or complete medium (HeLa cells) at 37 °C so that the final concentration of the gel is 1.5%.

3. Inject the 37 °C mixture into 2 m of PTFE tubing. Wrap the tubing around a 2.5 cm diameter cylinder and place in an ice bath for 15 min to allow the gel to solidify.
4. Insert a 135 μL plug of 3.0% agarose in D_2O in the bottom of a 5 mm screw cap NMR tube. This is for the sample lock.
5. Extrude the immobilized cells into the NMR tube by injecting sterile M9 medium into the tubing.

3.2.4 The perfusion system

1. Thermally seal one end of a 2 m length of PTFE inlet line.
2. Make eight 50 μm orifices over a 3 cm length at the sealed end of the inlet line using a stainless steel dissection needle with 10 μm tip.
3. Prepare a 3 m length of PTFE outlet line.
4. Attach a low-pressure tee 4 in. from the unsealed end of the inlet line.
5. Insert the two lines through the septum in the screw cap of the NMR tube with the orifices of the inlet line inside the tube.
6. Attach a 10 mL syringe to the low-pressure tee.
7. Insert the inlet and outlet lines into an NMR tube that contains cast cell threads. Set the end of the inlet tubing just above the D_2O agarose plug and the outlet tubing ~2 cm above the cell threads (Fig. 3).

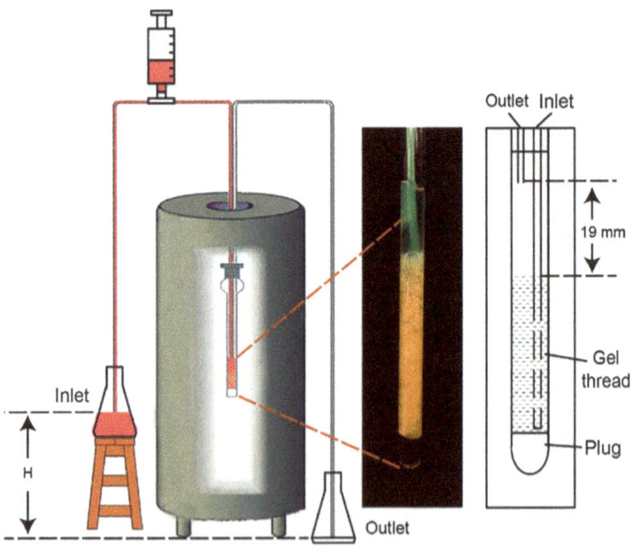

Fig. 3 RT in-cell NMR. A gravity siphon drives the continuous flow of fresh medium through the cells.

8. Insert the free end of the inlet line into a reservoir of M9 medium.
9. Insert the free end of the outlet line into a waste reservoir.
10. Use the syringe to prime the inlet line with fresh M9 medium.
11. Initiate the flow of medium, using the syringe if necessary.
12. Adjust the height difference between the reservoirs to maintain a constant flow rate of 100 μL/min. In this experiment the height difference was 0.86 m.
13. Equilibrate the system to for 30 min to eliminate any air from the system.

3.2.5 Cell viability-I: Intracellular ATP levels

This experiment was performed using a Bruker 600 MHz Avance III NMR spectrometer equipped with a QCI-P cryoprobe,

1. Incubate the sample containing either *E. coli* or Hela cells for 30 min at RT before positioning it into the NMR spectrometer.
2. Collect a proton-decoupled ^{31}P spectrum.
3. Continue collecting spectra at 3 h intervals over a 24 h period.
4. Repeat the experiment without the perfusion system (Fig. 4).
5. Integrate the ^{31}P peak intensities at −11.5 ppm using three different ranges to calculate a mean standard error for each spectrum.
6. Plot the average intensity versus time to assess the cell viability of cells in the perfusion system.

3.2.6 Cell viability-II: Prokaryotic cells

In this procedure cellular metabolites are extracted from *E. coli* as a baseline control and to identify in-cell peaks.

1. Suspend 3 g of *E. coli* cells (Section 3.2.1) in 5 mL of H_2O and 0.5 mL of 35% (v/v) perchloric acid.
2. Sonicate the cells for 5 min with a 50% duty cycle.
3. Centrifuge at 13,300 × *g* for 1 h at 4 °C.
4. Neutralize the supernatant by using 2 M KOH. Monitor the process with pH indicator paper.
5. Centrifuge at 13,300 × *g* for 1 h at 4 °C.
6. Lyophilize the supernatant and re-dissolve in 1 mL of 90% H_2O/10% D_2O supplemented with 5 mM EDTA from a 0.5 M EDTA, pH 8 stock.
7. Transfer to an NMR tube and collect a proton-decoupled ^{31}P spectrum.

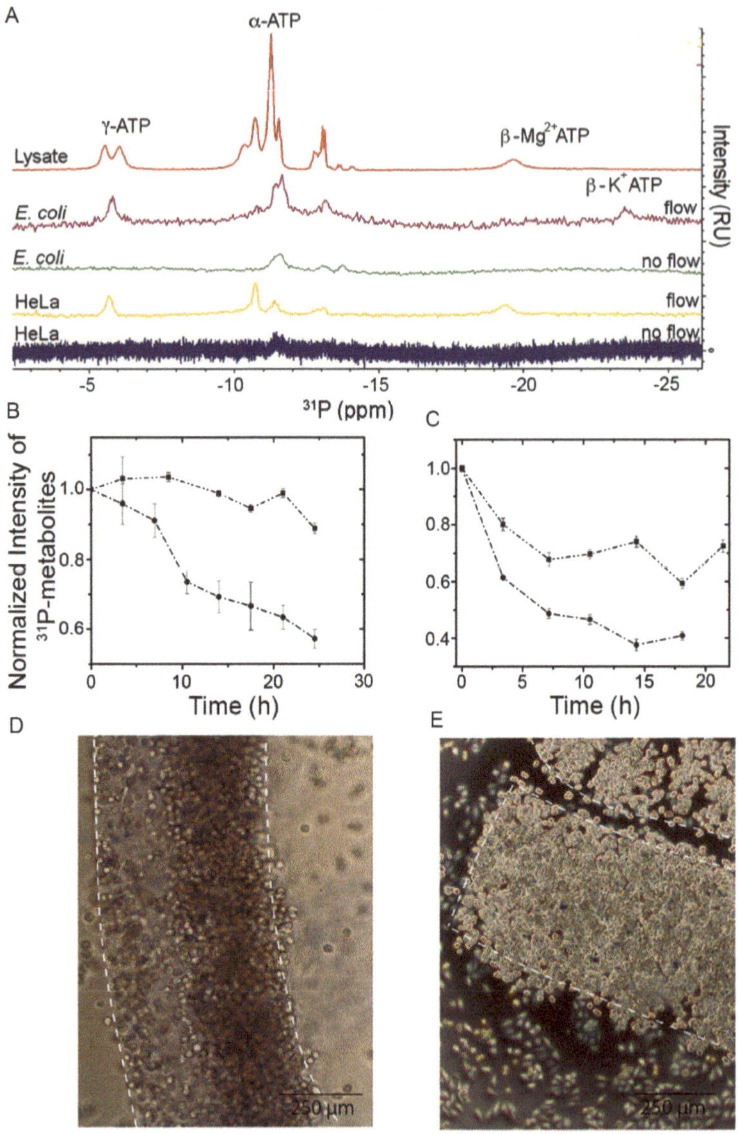

Fig. 4 Cell viability is maintained for up to 24 h in a bioreactor. (A) ^{31}P NMR spectra showing α-, β- and γ-ATP levels after 6 h for *E. coli* and HeLa cells in the bioreactor and in *E. coli* cell lysate. (B) Signal intensity of ^{31}P-metabolites over time in *E. coli* with (squares) and without (circles) M9 flow. (C) Intensity of ^{31}P-metabolites over time in HeLa cells with (squares) and without (circles) DMEM flow. (D) Trypan blue staining of cast HeLa cells before in-cell spectroscopy. (E) Trypan blue staining of cast HeLa cells 24 h after casting.

3.2.7 *Cell viability-III: Eukaryotic cells*

1. Stained cells were imaged using a Nikon Eclipse TS100 inversion microscope equipped with a Canon EOS digital camera.
2. Divide the HeLa cells cast in agarose threads (Section 3.2.3) into two samples.
3. Incubate the first sample in a 5% CO_2 incubator at 37 °C for 24 h.
4. Suspend the second sample in 200 μL of DMEM at 37 °C and add 200 μL of Trypan blue stock solution.
5. Allow the cells to stain for ~10 min and examine the cells under a microscope.
6. After 24 h, stain the second aliquot with trypan blue and examine under a microscope.
7. Compare the extent of cell staining to assess cell viability (Fig. 4). A final cell count within 90% of the original count is considered viable.

3.3 Real time in-cell STINT-NMR spectroscopy

For this procedure, Trx is overexpressed and labeled in *E. coli*, and the effects of externally administered streptomycin on the interaction surface of Trx is monitored in real time.

3.3.1 *NMR spectroscopy*

1. Overexpress and label Trx in *E. coli* (Section 3.1.1).
2. Cast the cells into agarose threads (Section 3.2.3).
3. Set up the perfusion system for the bioreactor (Section 3.2.4).
4. Set T_{opt} to 1.3 ms and collect a ^1H-^{15}N CRINEPT-HMQC-TROSY in-cell basis spectrum (Section 3.1.3).
5. Add streptomycin to a final concentration of 50 μg/mL to the inlet reservoir.
6. Collect five consecutive ^1H-^{15}N CRINEPT-HMQC-TROSY interaction spectra over the next 18 h (Fig. 5).

3.3.2 *Data processing*

1. Process the data with Topspin 3.0 (Bruker).
2. Upload in vitro assignments for Trx from the BMRB databank and open the Trx ^1H-^{15}N CRINEPT-HMQC -TROSY spectrum using CARA software (Masse & Keller, 2005).
3. Assign in-cell and lysate chemical shifts based on the proximity of each crosspeak to the in vitro assignments. Disregard residues with chemical shifts that differ by >0.1 ppm between the lysate and purified protein.

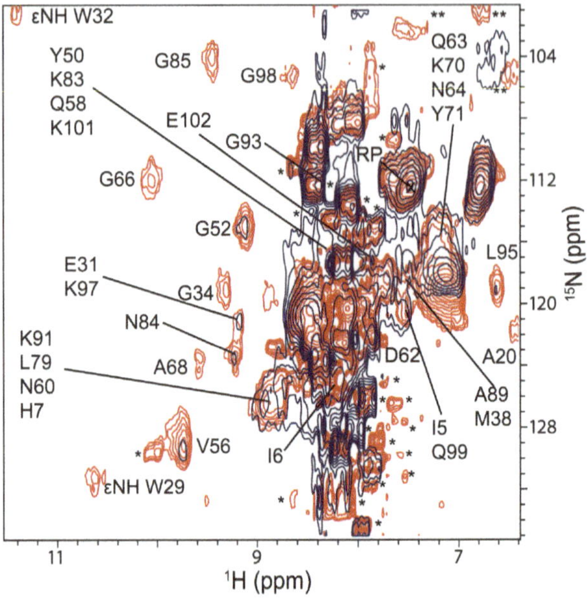

Fig. 5 Binding of streptomycin to ribosomes changes the quinary structure of Trx in *E. coli*. Overlay of in-cell ^{1}H-^{15}N CRINEPT-HMQC-TROSY spectra of Trx without (red) and with (blue) streptomycin (final spectrum). Residues that broaden in the presence of streptomycin are indicated. Single and double asterisks indicate peaks from metabolites and unassigned side chain protons, respectively. The spectra are shown at the same contour levels. The reference peak used for intensity normalization is indicated by RP.

4. Integrate crosspeak intensities by using base rectangle intensity summing.
5. Normalize intensities between the in-cell and lysate spectra using an amino proton peak from either glutamine or asparagine side chain that does not change chemical shift between the two spectra. For Trx a glutamine amide side chain at 7.49 and 112.4 ppm in the proton and nitrogen dimensions, respectively, was used.
6. Export the chemical shifts and peak integrations from CARA as a .TXT file to EXCEL.
7. Calculate changes in peak intensities for each amino acid residue, ΔI, using

$$\Delta I = \left(I_{lysate} - I_{in-cell}\right) / I_{in-cell}$$

where I_{lysate} and $I_{in-cell}$ are normalized, integrated peak intensities from lysate and in-cell spectra.

8. Calculate changes in chemical shifts for each amino acid residue, $\Delta\delta$, using

$$\Delta\delta = \left(\delta H^2 + (\delta N/4)^2\right)^{1/2}.$$

where δ_H and δN are the changes in hydrogen and nitrogen chemical shifts, respectively, between the lysate and in-cell spectra.
9. Assemble the data into matrices with residue numbers in columns and the corresponding changes in chemical shift or peak intensity values in rows.
10. Save the matrices as ASCII text files

3.3.3 Data processing of interaction spectra

1. Process the interaction spectra data with Topspin 3.0 (Bruker).
2. Upload the in-cell assignments of Trx from Section 3.3.2 using CARA software.
3. Assign in-cell chemical shifts for each spectrum based on the proximity of each crosspeak to the in-cell assignments.
4. Integrate crosspeak intensities by using base rectangle intensity summing.
5. Normalize intensities between the in-cell interaction spectra and in-cell basis spectrum using an amino proton peak from either a glutamine or asparagine side chain that does not change chemical shift between the two spectra.
6. Export the chemical shifts and peak integrations from CARA as a .TXT file to EXCEL.
7. Calculate changes in peak intensities for each amino acid residue, ΔI, using

$$\Delta I = \left(I_{basis} - I_{interaction}\right)/I_{interaction}$$

where I_{basis} and $I_{interaction}$ are normalized, integrated peak intensities from the in-cell Trx spectra in the absence and presence of streptomycin.
8. Calculate changes in chemical shifts for each amino acid residue, $\Delta\delta$, using

$$\Delta\delta = \left(\delta H^2 + (\delta N/4)^2\right)^{1/2}$$

where δ_H and δN are the changes in hydrogen and nitrogen chemical shifts, respectively, between the basis and interaction in-cell spectra.

9. Assemble the data into matrices with residue numbers in columns and the corresponding changes in chemical shift or peak intensity values in rows.
10. Save the matrices as ASCII text files.

3.3.4 SVD analysis of interaction spectra

1. Start MATLAB® (Mathworks).
2. Load the chemical shift or intensity matrix text file into MATLAB®: e.g., **LOAD TRXIntensitySVD.txt.**
3. Define the uploaded matrix, M: **M = TRXIntensitySVD**.
4. Initiate SVD analysis of matrix M: **[U,S,V] = svd(M).**
5. Save the S matrix: **save ('S.txt', 'S','-ascii')**.
6. Calculate the first binding mode: **B1 = S(1,1)*U(:,1).**
7. Save the first binding mode: **save ('B1.txt', 'B1','-ascii').**
8. Calculate the second binding mode: **B2 = S(2,2)*U(:,2).**
9. Save the second binding mode: **save ('B2.txt', 'B2','-ascii').**
10. Open MATLAB® results in EXCEL.
11. Create a Scree (bar) plot of singular values versus the binding mode.
12. Fit the singular values using linear regression to calculate the R^2 value (Fig. 6).
13. Bar plot the first and the second binding modes for the protein residue sequence.
14. Set a threshold line at 1.5 times the maximum contribution of the second binding mode (Fig. 6).
15. Identify interacting residues as those whose first binding mode exceeds the threshold.
16. Map broadened residues onto the protein surface based on PDB file(s) using SWISS-PDB Viewer (Fig. 6).

4. Results

Typical results for selected protocols are presented.

4.1 In-cell ^{1}H-^{15}N CRINEPT-HMQC-TROSY NMR spectroscopy of Trx

Optimizing the CRINEPT transfer delay for in-cell ^{1}H-^{15}N CRINEPT-HMQC-TROSY spectra maximizes resolution. For Trx the optimum

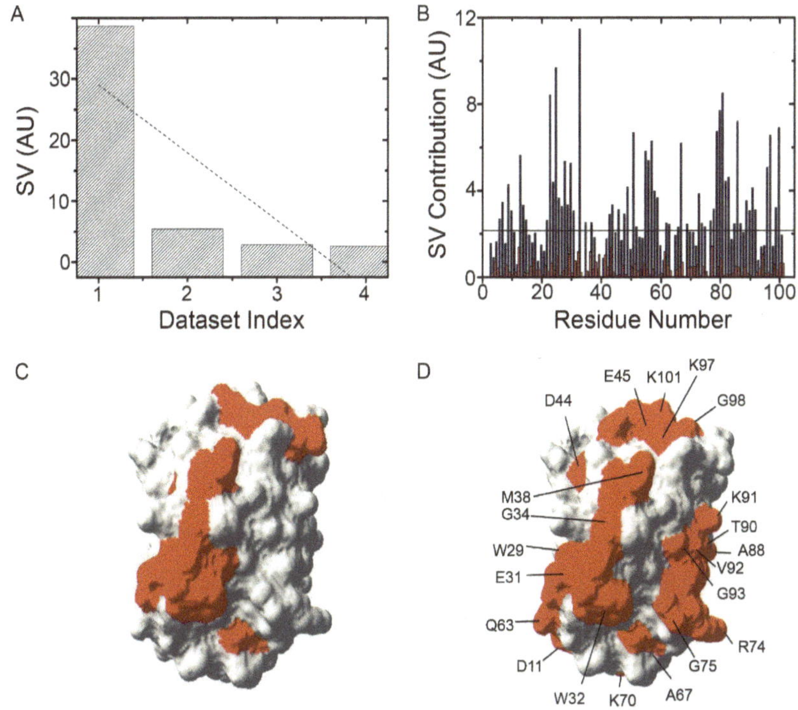

Fig. 6 SVD analysis of Trx RT STINT NMR spectra. (A) Distribution of singular values, SV, of each data set index (binding mode) for Trx residues in the presence of streptomycin. (B) The contribution of each residue in response to adding streptomycin for the first (blue) and second (red) binding modes. (C) Surface map of Trx (Protein Data Bank entry 1X0B) showing the quinary interaction surface (red) in the absence of antibiotics. (D) Surface map of Trx showing the quinary interaction surface (red) in the presence of streptomycin. Residues that comprise the interaction surface are indicated.

CRINEPT transfer delay was ~1.3 ms, which corresponds to an apparent molecular weight of ~1.1 MDa (Fig. 2A). The resulting spectrum shows crosspeaks that were not visible using HSQC NMR spectroscopy (Fig. 2B).

4.2 Bioreactor for cell viability

A schematic for RT-NMR is shown in Fig. 3. The height difference between the reservoirs provides the hydrodynamic force to drive the flow and allows external stimuli to be introduced into the cells from the inlet reservoir. The inlet line terminates at the bottom of the gel thread of cast cells. The outlet line terminates above the cast cells thread. An agarose plug in D_2O at the base of the NMR tube is used to lock the sample.

4.3 Cell viability

Cell viability is assessed by measuring ATP levels during in-cell experiments using ^{31}P NMR spectroscopy. In the presence of flow the phosphate levels are maintained. In the absence of flow, phosphate levels are rapidly depleted. In *E. coli* the intracellular level of Mg^{2+} is ~1 mM, which is lower than the concentration of ATP, 3–10 mM, the charge is mostly counterbalanced by K^+ ions (Fig. 4A). Flow of medium maintains cellular levels of ATP, ADP, NAD^+, and NAD(H), at ≥90% in *E. coli* and ≥75% in HeLa cells for up to 24 h (Fig. 4B and C). Staining HeLa cells with trypan blue shows that the cells remain 99±1% viable over 24 h. An increase in cell density is consistent with ongoing metabolic activity (Fig. 4D and E).

4.4 Real time in-cell STINT-NMR spectroscopy

Changes in the in-cell NMR spectra resulting from the introduction of external stimulus reveal corresponding changes in quinary structure for Trx (Fig. 5).

4.5 SVD analysis of interaction spectra

Residues that form the quinary interacting surface are identified through SVD analysis of in-cell NMR spectra (Fig. 6A). The sharp drop and poor linear fit of the singular values ($r^2=0.66$) are consistent with a specific change in quinary structure. Amino acids whose contribution to individual singular values exceed a threshold value make up the quinary interaction surface (Fig. 6B). The threshold line in Fig. 6B represents the average noise level of the NMR spectra. The amino acids identified in the SVD analysis are mapped onto the surface of the target molecule (Fig. 6C and D). Trx engages in quinary interactions in-cell that change when the antibiotic streptomycin, which binds to the 30S ribosome particle but not to Trx, is introduced into the cells during the course of an RT-STINT NMR experiment.

5. Summary

The protocols presented addressed the major limitations facing most in-cell NMR experiments: spectral sensitivity and resolution, and cell viability. The introduction of CRINEPT-based NMR spectroscopy allowed for observation of target proteins that were previously invisible by conventional HSQC NMR. The gravity flow-based bioreactor provided a simple solution to maintain metabolically active cells over the time course of the experiment. Because of these modifications the field of in-cell NMR is now well positioned to study the complex interplay between cytosolic components, cellular processes and protein structural interactions.

Acknowledgment

This work was supported by NIH R01 GM085006.

References

Banci, L., Barbieri, L., Bertini, I., Luchinat, E., Secci, E., Zhao, Y., et al. (2013). Atomic-resolution monitoring of protein maturation in live human cells by NMR. *Nature Chemical Biology*, *9*(5), 297–299. https://doi.org/10.1038/nchembio.1202.

Bertrand, K., Reverdatto, S., Burz, D. S., Zitomer, R., & Shekhtman, A. (2012). Structure of proteins in eukaryotic compartments. *Journal of the American Chemical Society*, *134*(30), 12798–12806. https://doi.org/10.1021/ja304809s.

Bodart, J. F., Wieruszeski, J. M., Amniai, L., Leroy, A., Landrieu, I., Rousseau-Lescuyer, A., et al. (2008). NMR observation of Tau in Xenopus oocytes. *Journal of Magnetic Resonance*, *192*(2), 252–257. (doi:S1090-7807(08)00089-X [pii) 10.1016/j.jmr.2008.03.006.

Breindel, L., DeMott, C., Burz, D. S., & Shekhtman, A. (2018). Real-time in-cell nuclear magnetic resonance: Ribosome-targeted antibiotics modulate Quinary protein interactions. *Biochemistry*, *57*, 540–546. https://doi.org/10.1021/acs.biochem.7b00938.

Burz, D. S., DeMott, C. M., Aldousary, A., Dansereau, S., & Shekhtman, A. (2018). Quantitative determination of interacting protein surfaces in prokaryotes and eukaryotes by using in-cell NMR spectroscopy. *Methods in Molecular Biology*, *1688*, 423–444. https://doi.org/10.1007/978-1-4939-7386-6_20.

Burz, D. S., Dutta, K., Cowburn, D., & Shekhtman, A. (2006a). In-cell NMR for protein-protein interactions (STINT-NMR). *Nature Protocols*, *1*(1), 146–152.

Burz, D. S., Dutta, K., Cowburn, D., & Shekhtman, A. (2006b). Mapping structural interactions using in-cell NMR spectroscopy (STINT-NMR). *Nature Methods*, *3*(2), 91–93.

Burz, D. S., & Shekhtman, A. (2010). The STINT-NMR method for studying in-cell protein-protein interactions. *Current Protocols in Protein Science*, *61*, 17.11.1–17.11.15. Chapter 17, Unit 17 11. https://doi.org/10.1002/0471140864.ps1711s61.

Cattell, R. B. (1966). The scree test for the number of factors. *Multivariate Behavioral Research*, *1*, 245–276.

Cobbert, J. D., DeMott, C., Majumder, S., Smith, E. A., Reverdatto, S., Burz, D. S., et al. (2015). Caught in action: Selecting peptide aptamers against intrinsically disordered proteins in live cells. *Scientific Reports*, *5*, 9402. https://doi.org/10.1038/srep09402.

Crowley, P. B., Chow, E., & Papkovskaia, T. (2011). Protein interactions in the Escherichia coli cytosol: An impediment to in-cell NMR spectroscopy. *Chembiochem*, *12*(7), 1043–1048. https://doi.org/10.1002/cbic.201100063.

Danielsson, J., Mu, X., Lang, L., Wang, H., Binolfi, A., Theillet, F. X., et al. (2015). Thermodynamics of protein destabilization in live cells. *Proceedings of the National Academy of Sciences of the United States of America*, *112*(40), 12402–12407. https://doi.org/10.1073/pnas.1511308112.

DeMott, C. M., Girardin, R., Cobbert, J., Reverdatto, S., Burz, D. S., McDonough, K., et al. (2018). Potent inhibitors of Mycobacterium tuberculosis growth identified by using in-cell NMR-based screening. *ACS Chemical Biology*, *13*, 733–741. https://doi.org/10.1021/acschembio.7b00879.

Felli, I. C., Gonnelli, L., & Pierattelli, R. (2014). In-cell (1)(3)C NMR spectroscopy for the study of intrinsically disordered proteins. *Nature Protocols*, *9*(9), 2005–2016. https://doi.org/10.1038/nprot.2014.124.

Galhardo, R. S., Hastings, P. J., & Rosenberg, S. M. (2007). Mutation as a stress response and the regulation of evolvability. *Critical Reviews in Biochemistry and Molecular Biology*, *42*(5), 399–435. https://doi.org/10.1080/10409230701648502.

Golub, G. H., & Van Loan, C. F. (2012). *Matrix computations* (4th ed.). Baltimore, USA: The Johns Hopkins University Press.

Hamatsu, J., O'Donovan, D., Tanaka, T., Shirai, T., Hourai, Y., Mikawa, T., et al. (2013). High-resolution heteronuclear multidimensional NMR of proteins in living insect cells using a baculovirus protein expression system. *Journal of the American Chemical Society*, *135*(5), 1688–1691. https://doi.org/10.1021/ja310928u.

Ikeya, T., Hanashima, T., Hosoya, S., Shimazaki, M., Ikeda, S., Mishima, M., et al. (2016). Improved in-cell structure determination of proteins at near-physiological concentration. *Scientific Reports*, *6*, 38312. https://doi.org/10.1038/srep38312.

Ikeya, T., Sasaki, A., Sakakibara, D., Shigemitsu, Y., Hamatsu, J., Hanashima, T., et al. (2010). NMR protein structure determination in living E. coli cells using nonlinear sampling. *Nature Protocols*, *5*(6), 1051–1060. https://doi.org/10.1038/nprot.2010.69.

Inomata, K., Kamoshida, H., Ikari, M., Ito, Y., & Kigawa, T. (2017). Impact of cellular health conditions on the protein folding state in mammalian cells. *Chemical Communications (Cambridge, England)*, *53*(81), 11245–11248. https://doi.org/10.1039/c7cc06004a.

Inomata, K., Ohno, A., Tochio, H., Isogai, S., Tenno, T., Nakase, I., et al. (2009). High-resolution multi-dimensional NMR spectroscopy of proteins in human cells. *Nature*, *458*(7234), 106–109. nature07839 [pii]. https://doi.org/10.1038/nature07839.

Kubo, S., Nishida, N., Udagawa, Y., Takarada, O., Ogino, S., & Shimada, I. (2013). A gel-encapsulated bioreactor system for NMR studies of protein-protein interactions in living mammalian cells. *Angewandte Chemie (International Ed. in English)*, *52*(4), 1208–1211. https://doi.org/10.1002/anie.201207243.

Kyne, C., & Crowley, P. B. (2017). Short arginine motifs drive protein stickiness in the Escherichia coli cytoplasm. *Biochemistry*, *56*(37), 5026–5032. https://doi.org/10.1021/acs.biochem.7b00731.

Li, C., Wang, G. F., Wang, Y., Creager-Allen, R., Lutz, E. A., Scronce, H., et al. (2010). Protein (19)F NMR in Escherichia coli. *Journal of the American Chemical Society*, *132*(1), 321–327. https://doi.org/10.1021/ja907966n.

Luh, L. M., Hansel, R., Lohr, F., Kirchner, D. K., Krauskopf, K., Pitzius, S., et al. (2013). Molecular crowding drives active pin1 into nonspecific complexes with endogenous proteins prior to substrate recognition. *Journal of the American Chemical Society*, *135*(37), 13796–13803. https://doi.org/10.1021/ja405244v.

Majumder, S., DeMott, C. M., Burz, D. S., & Shekhtman, A. (2014). Using singular value decomposition to characterize protein-protein interactions by in-cell NMR spectroscopy. *Chembiochem*, *15*(7), 929–933. https://doi.org/10.1002/cbic.201400030.

Majumder, S., DeMott, C. M., Reverdatto, S., Burz, D. S., & Shekhtman, A. (2016). Total cellular RNA modulates protein activity. *Biochemistry*, *55*(32), 4568–4573. https://doi.org/10.1021/acs.biochem.6b00330.

Majumder, S., Xue, J., DeMott, C. M., Reverdatto, S., Burz, D. S., & Shekhtman, A. (2015). Probing protein quinary interactions by in-cell nuclear magnetic resonance spectroscopy. *Biochemistry*, *54*(17), 2727–2738. https://doi.org/10.1021/acs.biochem.5b00036.

Maldonado, A. Y., Burz, D. S., & Shekhtman, A. (2011). In-cell NMR spectroscopy. *Progress in Nuclear Magnetic Resonance Spectroscopy*, *59*(3), 197–212. doi:S0079-6565(10)00112-3 [pii]. https://doi.org/10.1016/j.pnmrs.2010.11.002.

Masse, J. E., & Keller, R. (2005). AutoLink: Automated sequential resonance assignment of biopolymers from NMR data by relative-hypothesis-prioritization-based simulated logic. *Journal of Magnetic Resonance*, *174*(1), 133–151.

Masters, J. R., & Stacey, G. N. (2007). Changing medium and passaging cell lines. *Nature Protocols*, *2*(9), 2276–2284. https://doi.org/10.1038/nprot.2007.319.

Muntener, T., Haussinger, D., Selenko, P., & Theillet, F. X. (2016). In-cell protein structures from 2D NMR experiments. *Journal of Physical Chemistry Letters*, 7(14), 2821–2825. https://doi.org/10.1021/acs.jpclett.6b01074.

Ogino, S., Kubo, S., Umemoto, R., Huang, S., Nishida, N., & Shimada, I. (2009). Observation of NMR signals from proteins introduced into living mammalian cells by reversible membrane permeabilization using a pore-forming toxin, streptolysin O. *Journal of the American Chemical Society*, *131*(31), 10834–10835. https://doi.org/10.1021/ja904407w.

Pervushin, K. V., Wider, G., & Wuthrich, K. (1998). Single transition-to-single transition polarization transfer (ST2-PT) in [15N,1H]-TROSY. *Journal of Biomolecular NMR*, *12*(2), 345–348. https://doi.org/10.1023/A:1008268930690.

Pielak, G. J., Li, C., Miklos, A. C., Schlesinger, A. P., Slade, K. M., Wang, G. F., et al. (2009). Protein nuclear magnetic resonance under physiological conditions. *Biochemistry*, *48*(2), 226–234.

Riek, R., Pervushin, K., & Wuthrich, K. (2000). TROSY and CRINEPT: NMR with large molecular and supramolecular structures in solution. *Trends in Biochemical Sciences*, *25*(10), 462–468.

Riek, R., Wider, G., Pervushin, K., & Wuthrich, K. (1999). Polarization transfer by cross-correlated relaxation in solution NMR with very large molecules. *Proceedings of the National Academy of Sciences of the United States of America*, *96*(9), 4918–4923.

Sakai, T., Tochio, H., Tenno, T., Ito, Y., Kokubo, T., Hiroaki, H., et al. (2006). In-cell NMR spectroscopy of proteins inside Xenopus laevis oocytes. *Journal of Biomolecular NMR*, *36*(3), 179–188. https://doi.org/10.1007/s10858-006-9079-9.

Sakakibara, D., Sasaki, A., Ikeya, T., Hamatsu, J., Hanashima, T., Mishima, M., et al. (2009). Protein structure determination in living cells by in-cell NMR spectroscopy. *Nature*, *458*(7234), 102–105. nature07814 [pii]. https://doi.org/10.1038/nature07814.

Selenko, P., Serber, Z., Gadea, B., Ruderman, J., & Wagner, G. (2006). Quantitative NMR analysis of the protein G B1 domain in Xenopus laevis egg extracts and intact oocytes. *Proceedings of the National Academy of Sciences of the United States of America*, *103*(32), 11904–11909.

Serber, Z., Corsini, L., Durst, F., & Dotsch, V. (2005). In-cell NMR spectroscopy. *Methods in Enzymology*, *394*, 17–41.

Serber, Z., & Dotsch, V. (2001). In-cell NMR spectroscopy. *Biochemistry*, *40*(48), 14317–14323.

Serber, Z., Keatinge-Clay, A. T., Ledwidge, R., Kelly, A. E., Miller, S. M., & Dotsch, V. (2001). High-resolution macromolecular NMR spectroscopy inside living cells. *Journal of the American Chemical Society*, *123*(10), 2446–2447.

Serber, Z., Selenko, P., Hansel, R., Reckel, S., Lohr, F., Ferrell, J. E., Jr., et al. (2006). Investigating macromolecules inside cultured and injected cells by in-cell NMR spectroscopy. *Nature Protocols*, *1*(6), 2701–2709.

Serber, Z., Straub, W., Corsini, L., Nomura, A. M., Shimba, N., Craik, C. S., et al. (2004). Methyl groups as probes for proteins and complexes in in-cell NMR experiments. *Journal of the American Chemical Society*, *126*(22), 7119–7125.

Sharaf, N. G., Barnes, C. O., Charlton, L. M., Young, G. B., & Pielak, G. J. (2010). A bioreactor for in-cell protein NMR. *Journal of Magnetic Resonance*, *202*(2), 140–146. https://doi.org/10.1016/j.jmr.2009.10.008.

Shekhtman, A., Ghose, R., Goger, M., & Cowburn, D. (2002). NMR structure determination and investigation using a reduced proton (REDPRO) labeling strategy for proteins. *FEBS Letters*, *524*(1–3), 177–182.

Smith, A. E., Zhou, L. Z., Gorensek, A. H., Senske, M., & Pielak, G. J. (2016). In-cell thermodynamics and a new role for protein surfaces. *Proceedings of the National Academy of Sciences of the United States of America*, *113*(7), 1725–1730. https://doi.org/10.1073/pnas.1518620113.

Theillet, F. X., Binolfi, A., Bekei, B., Martorana, A., Rose, H. M., Stuiver, M., et al. (2016). Structural disorder of monomeric alpha-synuclein persists in mammalian cells. *Nature*, *530*(7588), 45–50. https://doi.org/10.1038/nature16531.

Wellen, K. E., & Thompson, C. B. (2010). Cellular metabolic stress: Considering how cells respond to nutrient excess. *Molecular Cell*, *40*(2), 323–332. https://doi.org/10.1016/j.molcel.2010.10.004.

Wider, G. (2005). NMR techniques used with very large biological macromolecules in solution. *Methods in Enzymology*, *394*, 382–398. https://doi.org/10.1016/S0076-6879(05)94015-9.

Wieruszeski, J. M., Bohin, A., Bohin, J. P., & Lippens, G. (2001). In vivo detection of the cyclic osmoregulated periplasmic glucan of Ralstonia solanacearum by high-resolution magic angle spinning NMR. *Journal of Magnetic Resonance*, *151*(1), 118–123. https://doi.org/10.1006/jmre.2001.2348.

Williams, S. P., Haggie, P. M., & Brindle, K. M. (1997). 19F NMR measurements of the rotational mobility of proteins in vivo. *Biophysical Journal*, *72*(1), 490–498. https://doi.org/10.1016/S0006-3495(97)78690-9.

Wuthrich, K. (1986). *NMR of proteins and nucleic acids*. New York: John Wiley&Sons.

Xie, J., Thapa, R., Reverdatto, S., Burz, D. S., & Shekhtman, A. (2009). Screening of small molecule interactor library by using in-cell NMR spectroscopy (SMILI-NMR). *Journal of Medicinal Chemistry*, *52*(11), 3516–3522. https://doi.org/10.1021/jm9000743.

Ye, Y., Liu, X., Zhang, Z., Wu, Q., Jiang, B., Jiang, L., et al. (2013). (19) F NMR spectroscopy as a probe of cytoplasmic viscosity and weak protein interactions in living cells. *Chemistry*, *19*(38), 12705–12710. https://doi.org/10.1002/chem.201301657.

CHAPTER EIGHTEEN

High-throughput methods in aptamer discovery and analysis

Kyle H. Cole[a], **Andrej Lupták**[a,b,c,*]
[a]Department of Molecular Biology and Biochemistry, University of California, Irvine, CA, United States
[b]Department of Pharmaceutical Sciences, University of California, Irvine, CA, United States
[c]Department of Chemistry, University of California, Irvine, CA, United States
*Corresponding author: e-mail address: aluptak@uci.edu

Contents

Abstract

Aptamers are small, functional nucleic acids that bind a variety of targets, often with high specificity and affinity. Genomic aptamers constitute the ligand-binding domains of riboswitches, whereas synthetic aptamers find applications as diagnostic and therapeutic tools, and as ligand-binding domains of regulatory RNAs in synthetic biology. Discovery and characterization of aptamers has been limited by a lack of high-throughput approaches that uncover the target-binding domains and the biochemical properties of individual sequences. With the advent of high-throughput sequencing, large-scale analysis of *in vitro* selected populations of aptamers (and catalytic nucleic acids, such as ribozymes and DNAzmes) became possible. In recent years the development of new experimental approaches and software tools has led to significant streamlining of the selection–pool analysis. This article provides an overview of post-selection data analysis and describes high–throughput methods that facilitate rapid discovery and biochemical characterization of aptamers.

Methods in Enzymology, Volume 621
ISSN 0076-6879
https://doi.org/10.1016/bs.mie.2019.02.009

1. Introduction

Aptamers are nucleic acids that can recognize ligands ranging from ions and small molecules, to more complex targets such as proteins and whole cells (Wu & Kwon, 2016). Because of their versatility and relative ease of isolation from high-diversity libraries, aptamers have found applications in diagnostics and drug development. A classic example of an FDA-approved therapeutic aptamer is the anti-vascular endothelial growth factor (anti-VEGF) aptamer used in the treatment of ocular vascular disease (Ng et al., 2006). Aptamer-based therapeutics compete within the niche currently filled by biologics such as monoclonal antibodies, but unlike monoclonal antibodies, they do not suffer from batch-to-batch differentiation (Shaughnessy, 2012). *In vitro* selection of nucleic-acid aptamers has gained popularity due to these applications; however, aptamer discovery is limited by the necessity of post-selection biochemical analysis (Jijakli et al., 2016). High-throughput aptamer discovery methods aim to resolve these issues through combinatorial approaches to identify ligand-targeting sequences while simultaneously determining biochemical and structurally-relevant information.

A traditional *in vitro* selection (SELEX) experiment used target molecules conjugated to solid media to separate bound molecules from highly-diverse pools of sequences. After successive rounds of selection, enriched pools were collected and cloned into *Escherichia coli* (*E. coli*) plasmids. To identify individual aptamers, the plasmids were sequenced by chain-termination methods, such as Sanger sequencing (Griffin, Toole, & Leung, 1993; Morris, Jensen, Julin, Weil, & Gold, 1998). This strategy proved laborious, because a relatively high number of individual colonies had to be sequenced to determine the distribution of selected motifs. If a specific sequence or structural motif dominated the enriched pool, the likelihood of identifying less-abundant motifs was low, due to the relatively low-throughput of this sequencing approach.

As the cost of high-throughput sequencing (HTS) decreased, deep sequencing was introduced to *in vitro* selection protocols, with the intent of improving aptamer discovery and increasing data validation through improved sequencing depth (Schütze et al., 2011). The first applications of HTS to *in vitro* selections aimed to identify new binding targets of transcription factors (Jolma et al., 2010; Slattery et al., 2011; Zhao & Stormo, 2011; Zykovich, Korf, & Segal, 2009). At the same time, Zimmerman et al.

developed a HTS method of identifying naturally-occurring RNA aptamers encoded in *E. coli* genomic DNA (Zimmermann, Gesell, Chen, Lorenz, & Schroeder, 2010), and other laboratories built methodology to select for growth-factor-specific DNA aptamers (Cho et al., 2010) and RNA inhibitors for HIV reverse transcriptase (Ditzler et al., 2013). With increased sequencing depth and computational methods, the fitness landscape of an *in vitro* selection could be visualized (Pitt & Ferré-D'Amaré, 2010), and evolutionary steps tracked between rounds of selection toward functional RNAs (Jiménez, Xulvi-Brunet, Campbell, Turk-MacLeod, & Chen, 2013; Neveu, Kim, & Benner, 2013). While these methods incorporated HTS for more expedient aptamer discovery or ribozyme sequence analysis, recent high-throughput discovery methods have combined aptamer discovery with determination of rate or equilibrium constants, as well as probing of ligand-induced conformational changes, to streamline the biochemical analysis necessary for aptamer motif definition and activity validation (Fig. 1). The objective of the following sections is to provide an overview of post-selection sequence analysis methods and high-throughput aptamer discovery methods.

2. Post-selection sequence analysis

Post-selection analysis of deep-sequenced pools can become overwhelming due to the sheer quantity of available sequences, presumably enriched for functional molecules. Software development aimed at tackling this issue has progressed significantly over the last decade, with tools intended to reduce the burden of deep sequencing data manipulation by providing analysis of sequence enrichment and sorting of aptamer sequences based on genotypic familiarity. Early high-throughput sequencing adopters used analysis packages designed for genomic sequencing data manipulation, in addition to homebrewed UNIX scripts utilizing the *sed*, *awk*, and *grep* commands (Hoon, Zhou, Janda, Brenner, & Scolnick, 2011). The availability of these custom-built scripts was limited, and scripts were typically built to answer specific experimental questions and were not intended for general enriched-pool workup. More recently, analysis of sequencing data for functional aptamers has greatly improved and the benefits of HTS have become realized; for example, for SELEX-specific constant region trimming and determination of aptamer frequencies in enriched pools (Hamada, 2018).

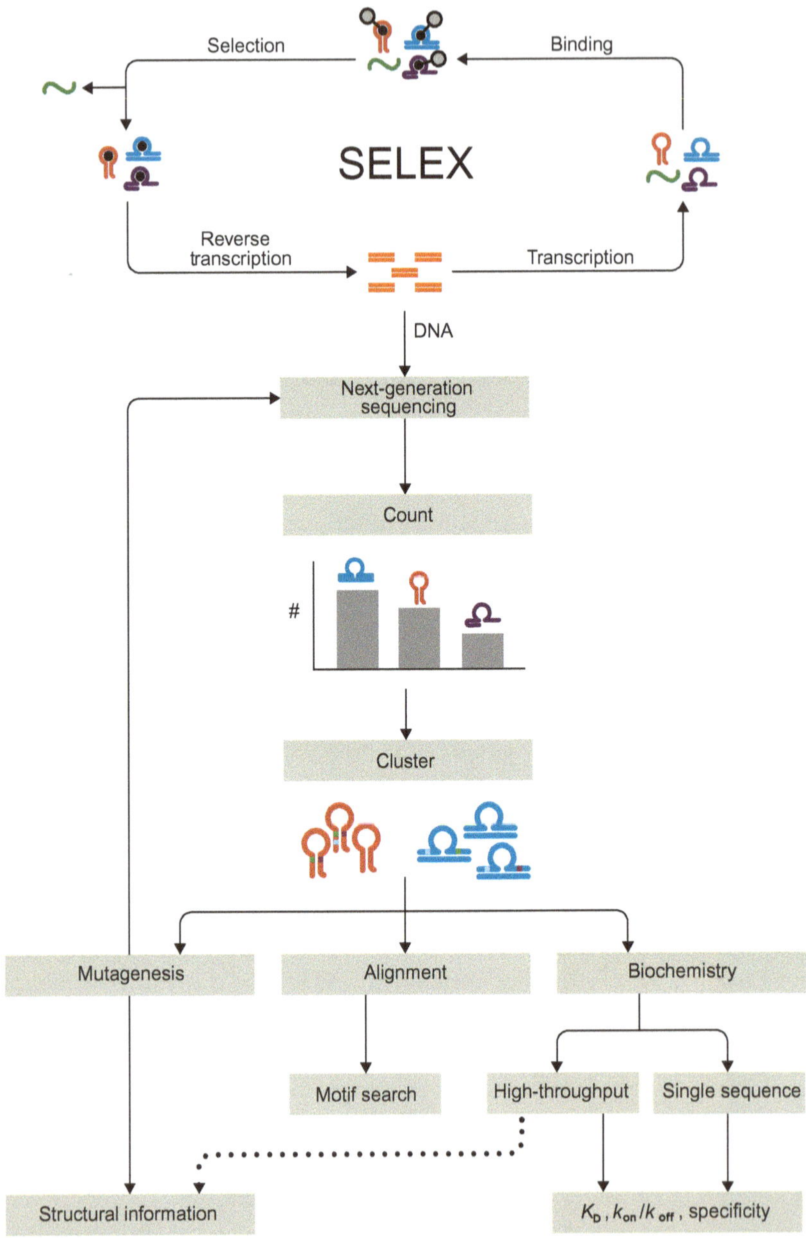

Fig. 1 Generalized workflow of *in vitro* selection (SELEX) of RNA aptamers and post-selection analysis. After several rounds of selection, the enriched pool is amplified and submitted for high-throughput sequencing. After sequencing, the reads are counted and clustered. Clustered sequences provide the basis for downstream analysis, such as mutagenesis, alignment and sequence-motif discovery, and biochemistry. *(Continued)*

2.1 Clustering

In silico parsing of aptamer sequences provides the foundation of biochemical analysis by identifying sequence families enriched through selective pressure. The analysis typically begins by minimizing and sorting redundant sequencing data by identifying unique sequences and sorting them based on shared sequence characteristics. One of the initial high-throughput sequencing analysis packages specific for *in vitro* selection was Sequence Evolution With Adaptive Landscape (SEWAL), which assesses the overall fitness landscape of *in vitro* selection pools for the purpose of performing functional genotypic analysis (Pitt, Rajapakse, & Ferré-D'Amaré, 2010). SEWAL produces 3D frequency plots using a sorting algorithm to determine sequence patterns and visualizes changes in the sequencing space due to selective pressure across consecutive rounds. Identification of changes due to selective pressure can be used to predict evolutionary paths that individual sequences follow toward improved fitness.

Similarly, FASTAptamer and AptaCluster compress pool data into nonredundant sequence sets (count) and sort sequences based on similarities in genotype (clustering) (Alam, Chang, & Burke, 2015; Hoinka, Berezhnoy, Sauna, Gilboa, & Przytycka, 2014). The FASTAptamer package of scripts provides a simple, command-line data workup package that can be installed on any system capable of running Perl binaries, whereas AptaCluster requires a C++ compiler for installation, which can be difficult for inexperienced users. Both packages use counting to rank sequences based upon frequency, aiding in the identification of sequences enriched throughout the selection to assist in determining the complexity of the enriched pool.

As mentioned above, clustering sorts sequences into families of sequences which share a similar genotype. Clustering begins by separating abundant unique sequences into individual clusters. Less abundant sequences that are similar to the seed sequence of a given cluster, but vary by mutations (including insertions and deletions; indels), are then sorted. Alignments of individual clusters, using tools such as MUSCLE or MAFFT,

---

Fig. 1—Cont'd Mutagenesis of enriched pools, followed by reselection and next-generation sequencing, reveals functionally critical (immutable) positions, as well as structural information in the form of sequence covariation of base-paired positions. Finally, enriched sequences identified through counting and clustering can be individually tested for target binding specificity, and determination of binding constants, such as K_D and k_{on}/k_{off}. High-throughput analysis methods yield these kinetic and thermodynamic constants for many sequences simultaneously, with some methods, such as Apta-Seq (Abdelsayed et al., 2017), revealing structural information as well.

provide visualization of allowed mutations and the output of these tools is useful in determining sequence consensus in addition to potential evolutionary relationships among clustered sequences derived from different selection rounds (Ameta, Winz, Previti, & Jäschke, 2014; Edgar, 2004; Katoh, Misawa, Kuma, & Miyata, 2002). And while clustering leads to increased understanding of the genotypic variation that exists within families of aptamers, confirmation of aptamer activity relies on further biochemical analysis.

2.2 Motif searching

Identification of motif sequences from enriched pools provides insight in to functional binding targets of aptamers. Methods aimed at identifying transcription factor binding sites, such as ChIP-seq, contributed to the development of DNA motif analysis software packages, such as the MEME family of tools, which identify reoccurring sequence motifs enriched throughout genomic sequencing data (Bailey et al., 2009). However, identification of motifs based on sequence alone does not necessarily coincide with target binding. BEEML, and similar methods, account for binding by utilizing a position weight matrix (PWM) to determine motifs based on energy models which determine the contribution of each base-pair to transcription factor binding (Jolma et al., 2010; Zhao & Stormo, 2011). These computational methods identify motifs based on the energetic contributions of each nucleotide position, improving the likelihood of identifying binding sequences.

RNA aptamers, on the other hand, maintain complex structures, requiring motif identification software that incorporates secondary structure prediction.

To address this issue, Backofen and coworkers developed MEME in RNAs Including secondary Structure (MEMERIS), which built on MEME by incorporating high throughput analysis capabilities in conjunction with secondary structure prediction, but is restricted to motifs within predicted single-stranded loop and bulge regions of RNA aptamer sequences (Hiller, Pudimat, Busch, & Backofen, 2006). RNAcontext improved upon the motif analysis landscape by including loops, bulges, and stems into its predictions (Kazan, Ray, Chan, Hughes, & Morris, 2010). Because RNAcontext distinguishes between these secondary structures, it can also predict the preferred conformation of the identified motif within the aptamer.

Similarly, AptaMotif and APTANI, which use a minimum free energy (MFE) approach for structure prediction in conjunction with iterative sampling and sequence alignments, have been developed to determine potential motifs across multiple input files (i.e., selection rounds) (Caroli, Taccioli, De La Fuente, Serafini, & Bicciato, 2016; Hoinka, Zotenko, Friedman, Sauna, & Przytycka, 2012). However, both of these methods are limited to identification of dominant motifs—an outcome that may not be desirable in selections for complex targets that are expected to yield multiple motifs, or in genomic SELEX, in which an exhaustive mapping of all aptamers is often preferred. AptaTrace overcomes this problem by tracing motifs through multiple rounds of selection data and testing whether these regions undergo selection toward secondary structure, such as hairpin loops or bulges (Dao et al., 2016). More recently, AptaSuite was developed as an all-encompassing software package, which includes AptaTrace and AptaCluster, and is the first open-source package designed for selection schemes to feature a graphical user interface (GUI) (Hoinka, Backofen, & Przytycka, 2018). AptaSuite can be installed on any system that runs Java Runtime Environment (JRE), providing ease-of-use to less-experienced users and removing the hassle of installing packages for adapter trimming and motif prediction.

The above-mentioned high-throughput strategies predict motifs based on the frequency of reoccurrence between rounds of selection. Alternatively, covariance model (CM) based algorithms predict secondary-structures and identify motifs based on the dependent variability of nucleotides in a given sequence (covariance). One example is CMFinder, which applies a covariance-probabilistic model to predict motifs from sequences that are dissimilar and unaligned (Yao, Weinberg, & Ruzzo, 2006). CMFinder can align and predict motifs from any input of sequences, whether *in vitro* evolved or genomic, and the output is used in the identification of sequence homology. Similarly, the Infernal package constructs a CM based on an alignment of RNAs and searches for genomic RNA homologs based on primary sequence and, most importantly, conserved-secondary structure (Nawrocki & Eddy, 2013). The output of Infernal can be applied to CMFinder for improved motif prediction in the case of genomic aptamers. In contrast, RNArobo uses a context-based motif searching algorithm for identification of novel motifs when no known homology has been previously annotated (Rampášek, Jimenez, Lupták, Vinař, & Brejová, 2016). RNArobo parses sequence searches based upon an input descriptor that defines a simplified motif map and individual structural elements; for example, a combination of secondary structure and binding-loop sequence. The methods employed by RNArobo are expedient in determining sequences containing difficult-to-predict structural elements, such as complex pseudoknots, adding to its effectiveness in discovering genomic aptamers from *in vitro* selections.

In the case of genomic selections, sequence data require further manipulation to identify conservation and homology, in addition to aptamer *loci*. Initial sequence alignment to reference genomes can be accomplished through multiple methods. For example, Bowtie2 is beneficial for alignments of short reads obtained from selections to a reference genome (Langmead & Salzberg, 2012), while Integrated Genome Browser (IGB) provides a user interface for alignments to annotated-reference genomes, and provides additional tools including sequence pileups for visualization of consensus sequences (Freese, Norris, & Loraine, 2016). These basic alignment methods are useful when comparing HTS data to single-reference genomes but are less efficient when comparing multiple species. Packages such as the Basic Local Alignment Search Tool (BLAST) or HMMER3, allow users to query custom- or downloadable- sequence databases for more advanced conservation and homology searches (Camacho et al., 2009; Eddy, 2011). BLAST provides a powerful webserver and command-line based software package that delivers a well-rounded source for alignments, conservation, and homology, but can be nonintuitive to use. HMMER3 is an alternative homology identification package that uses a hidden Markov model (HMM) to improve upon sequence identification, with improved efficiency compared to BLAST. Information obtained from reference-genome alignments, conservation, and homology provides a basis for downstream biological characterization of genomic aptamers, such as expression analysis, *in vivo* target binding, and potential regulatory functions.

2.3 Mutagenesis

Candidate aptamer sequences that dominate a selected pool bias it toward enriched genotypes and this genotypic bias restricts possible nucleotide variation across the pool, which limits structurally-relevant information. Upon initial selection, aptamer sequences may be constrained within their given sequence space, because mutations introduced by the polymerase enzymes during the *in vitro* selection process may not be frequent enough to thoroughly sample the permissive sequence variation around a founding aptamers sequence. Mutagenesis of an enriched pool introduces such sequence variation and subsequent selection provides alternative genotypic outcomes (Jiménez et al., 2013). Consequently, mutagenesis followed by reselection, is a useful strategy to overcome the lack of sequence diversity, allowing a more thorough characterization of the functional sequence space, in addition to potentially revealing functionally superior variants of the selected aptamers.

Mutagenesis of an enriched pool can be accomplished through mutagenic PCR or error-prone PCR methods (EP-PCR). Mutagenic PCR is facilitated by nucleoside analogs, such as 8-oxo-dG and dPTP, to introduce templated mismatches. 8-oxo-dG and dPTP can be utilized by *Taq* polymerase allowing for introduction of the analogs into standard PCR reaction mixtures in excess, and the rate at which mutations are introduced can be tuned by the ratio of the analogs to standard dNTPs (Zaccolo, Williams, Brown, & Gherardi, 1996). EP-PCR takes advantage of the low fidelity of *Taq* polymerase to introduce mutations through doubling of the pool (Wilson & Keefe, 2001). PCR is inherently error-prone and contributes to the genotypic variation of *in vitro* selection; however, by increasing the concentration of Mg^{2+} and introducing Mn^{2+} metal ions, the error rate of *Taq* polymerase increases from 0.02% to 0.066% per nucleotide position when coupled with disproportionate dNTP concentrations (Cadwell & Joyce, 1992).

Sequence covariation is a sought-after outcome of mutagenic selection and provides benefit to motif and secondary structure prediction. Due to the lack of sequence diversity of selected pools, covariation of sequences that form the secondary structure elements of functional nucleic acids tends be low. As mentioned previously, software such as CMFinder and Infernal utilize covariance models for motif and homology prediction (Nawrocki & Eddy, 2013; Yao et al., 2006). Each tool utilizes covariance in secondary structure prediction strategies by identifying nucleotide positions which mutate dependently. Compensatory evolution of nucleotide positions is most likely the result of structural motifs which rely on the interaction of the two positions to maintain functionality (Parsch, Braverman, & Stephan, 2000). Secondary-structure prediction based on covariation data is accomplished with software such as ViennaRNA's RNAalifold, which can take several hundred aligned sequences as input through the webserver and command-line versions (Bernhart, Hofacker, Will, Gruber, & Stadler, 2008). Sequence covariation can be also revealed through alignment of homologous sequences using resources like the BLAST-like alignment tool (BLAT) (Kent, 2002), but mutagenic selection has the potential to provide more diversity, which is ideal for structural and motif prediction.

3. High-throughput characterization methods

While deep sequencing has increased the diversity of potential aptamers from selections by aiding in the identification of less abundant sequences, validation relies on biochemical confirmation of ligand binding. Given the potentially high number of functional sequences in an enriched pool, binding assays of aptamers to immobilized ligands can prove tedious, unless the pool is dominated by a small number of motifs (Sassanfar & Szostak, 1993). Moreover, traditional binding assays do not provide insights into conformational dynamics upon ligand binding or the location of binding domains. High-throughput methods of aptamer discovery aim to address these issues through combinatorial measurements of binding kinetics or thermodynamics, as well as structural probing, followed by HTS. These methods intend to alleviate the strain of *in vitro* and *in silico* selection workup by identifying ligand-associated motifs and determining binding or rate constants.

Sequence enrichment does not necessarily equate to stronger or more specific binding; therefore, identification of high-affinity binders in a pool traditionally requires binding or rate constant measurements of individual sequences. To increase the experimental throughput of these measurements, one approach characterizes of functional RNAs by probing their dynamic properties against a titration of ligand. The technique, Apta-Seq, utilizes a combinatorial, multiplexed approach focused on RNA chemical modification and deep HTS to evaluate structural interaction of aptamers with their target ligand (Abdelsayed et al., 2017). Apta-Seq utilizes selective 2′-hydroxyl acylation with primer extension (SHAPE) to determine whether a 2′-hydroxyl is solvent-accessible to allow acylation by a SHAPE reagent, such as 2-(azidomethyl)nicotinic acid acyl imidazole (NAI-N_3), under variable ligand concentrations (Merino, Wilkinson, Coughlan, & Weeks, 2005; Spitale et al., 2015). Apta-Seq is based on an established workflow known as SHAPE-seq to map HTS data of reverse transcriptase (RT) stops caused by acylation of 2′-hydroxyls of pool RNAs, leading to truncated cDNAs with heterogenous 3′ ends (Lucks et al., 2011). The heterogenous 3′ ends of the cDNAs are resolved by introducing an RT primer with a 5′ overhang and self-ligating the cDNA by CircLigase (Lucigen). Together with the RT primer sequence, the introduced 5′ overhang can be subsequently used for priming a PCR amplification reaction, and submitted to HTS. RT stop sites are mapped using ShapeFinder, a package used to quantitatively map RT stops at nucleotide resolution and when a ligand titration is used, the stops uncover RNA conformational changes due ligand binding (Vasa, Guex, Wilkinson, Weeks, & Giddings, 2008). When this method was applied to a human-genomic *in vitro* selection for ATP-binders, it revealed three novel human-genome-derived ATP-binding RNA aptamers, and confirmed the existence of two previously identified aptamers (Abdelsayed et al., 2017; Vu et al., 2012). These results demonstrated the robustness of the method to perform structural analysis in a high-throughput pipeline and allowed determination of dissociation rate constants, K_Ds, for selected pools while eliminating single-clone biochemical assessment. This general workflow provides a unique framework which can be modified for all nucleic acids, including DNA and XNA aptamers, if base-specific modifications such as dimethyl sulfate (DMS), 1-ethyl-3-(3-dimethylaminopropyl) carbodiimide (EDC), or nicotinoyl azide (NAz) probing are introduced (Feng et al., 2018; Kwok, Ding, Tang, Assmann, & Bevilacqua, 2013; Mitchell et al., 2018; Wang, Sexton, Culligan, & Simon, 2018; Zinshteyn et al., 2018).

Identification of structural motifs is critical for the determination of ligand binding domains, assisting in the minimization of aptamer sequences derived from a given pool. Previously, natural RNA aptamers, the ligand-binding domains of riboswitches, have mostly been discovered via computational investigation for sequence and structural homology (Barrick & Breaker, 2007). However, this approach limits the discovery of novel, naturally occuring aptamers to those which share consensus with characterized sequences or sequence models, such as those modeled with CMFinder. To alleviate this limitation, Parallel Analysis of RNA Conformations Exposed to Ligand Binding (PARCEL) was developed as a high-throughput method of identifying RNA-ligand interactions in whole transcriptomes (Tapsin et al., 2018). The PARCEL workflow uses three methods of RNA-ligand interaction assessment in parallel, followed by next-generation sequencing. RNA-footprinting with double-strand specific RNase V1 and single-strand specific S1 nuclease is performed in parallel in the presence or absence of a target metabolite. Additionally, the protocol applies partial hydrolysis of transcripts, in-line probing, to determine transcript flexibility (Regulski & Breaker, 2008; Soukup & Breaker, 1999). The methodology relies on the assumption that the target ligand provides protection to the RNA when bound. Transcripts treated in the presence or absence of metabolite are then sequenced and mapped to a reference genome. Reads are normalized to ligand conditions and while PARCEL utilizes nuclease cleavage and in-line probing, both methods could be applied in parallel with acylation in Apta-Seq to provide additional structural insight and K_Ds.

A different approach to determining the strength of aptamer-ligand interactions is based on the measurement of binding rate constants. High-throughput sequencing kinetics (HTSK), is a an efficient method for the determination of k_{on} and k_{off} rate constants for pools of mRNA-peptide fusions (Jalali-Yazdi, Lai, Takahashi, & Roberts, 2016; Roberts & Szostak, 1997). HTSK provides k_{on}/k_{off} rates for the majority of sequences in a given pool through a time-point analysis protocol. To determine the association rate constants, k_{on}, an enriched pool is added to ligands immobilized on magnetic beads and at various time points, bead fractions are removed, washed, and the isolated sequences are amplified and prepared for next-generation sequencing. The fractional composition of a given sequence can be determined by measuring the frequency of that sequence at a given time point, and the amount of bound aptamer can be determined by multiplying the fractional composition by the total counts of the bound pool at each time point. Dissociation rate constants, k_{off}, are determined following the on-rate time points by washing the beads in the presence of excess ligand and collecting fractions at various time-points. Fractional composition of sequences is determined similarly to the on-rates and K_Ds can be calculated from the rate constants, assuming a simple binding model.

RNA Bind-n-Seq is a similar approach to HTSK, used to identify RNA-protein interactions by combining high-throughput *in vitro* selection of RNA with rate constant measurements (Lambert et al., 2014). A randomized pool of RNA is incubated with varying concentrations of streptavidin-tagged RNA binding proteins (RBPs), which can be pulled down using biotinylated beads, allowing the bound RNA to be eluted and sequenced. Association and dissociation rate constants can be determined from these sequenced data, assuming complete pull-down of the tagged RBPs; however, this method is limited to proteins or target molecules that can be tagged for affinity-binding assays.

3.1 On-chip analysis

The Illumina sequencing platform has been utilized for deep sequencing of whole genomes, single-cell RNA-seq, and for understanding epigenetics through bisulphite sequencing, among many other applications

(Lippert et al., 2017; Raine, Manlig, Wahlberg, Syvänen, & Nordlund, 2017; Ziegenhain et al., 2017). After sequencing, the DNA-displayed chips can be used for further experiments that utilize the strong fluorescent signal, good physical separation, and known sequences of the DNA polonies generated by the sequencing method. The Illumina platform's versatility for aptamer discovery was first illustrated with high-throughput sequencing fluorescent ligand interaction profiling (HITS-FLIP) (Nutiu et al., 2011). HITS-FLIP quantitatively measures DNA-protein affinity through changes in fluorescent ligand associations of millions of DNA clusters bound to the Illumina flowcell. On/off rates can be determined by varying the protein concentrations introduced across the flowcell and measuring subsequent changes in fluorescence.

Similarly, the chip-hybridized associated-mapping platform (CHAMP) takes advantage of used MiSeq chips with linked DNA clusters to determine DNA-protein interactions (Jung et al., 2017). The DNA clusters are reamplified to remove associated fluorescent nucleotides and a fluorescent-oligonucleotide probe is hybridized to the clusters and used as a reference marker for the DNA sequences. Fluorescent proteins are incubated in varying concentrations, imaged, and associations between the binding experiment and the sequences of individual colonies are made to the DNA sequence through imaging and software analysis. In another example of this approach, two groups built on HITS-FLIP for high-throughput RNA aptamer discovery (Buenrostro et al., 2014; Tome et al., 2014). Both methods use diverse single-stranded DNA libraries which contain T7 promoters to initiate transcription on the sequencer flow cell, and employ an approach to halt transcription with the intent on maintaining the association of the transcript with the DNA cluster. Fluorescently-labeled proteins that bind the halted transcripts can be detected through fluorescent imaging of the flowcell. This approach allows identification of RNA aptamers that bind the protein ligand through the sequence of the associated DNA. Additionally, genotypic variations can be directly compared, decreasing the need for downstream mutagenic analysis and thus providing covariation information for structure prediction. These methods are currently limited to ligands which can be fluorescently labeled, such as target proteins labeled with fluorescent dyes or fused to fluorescent proteins. On the other hand, many targets of SELEX are modified for bead-binding during *in vitro* selection experiments, and this chemistry can often be exploited for flourescent labeling allowing for optical detection of binding on HTS chips.

4. Conclusions

Increased interest in nucleic-acid aptamers has, in recent years, led to the development of improved high-throughput methods of aptamer discovery. High-throughput methods that incorporate biochemical assessment of individual sequences decrease the time requirement of aptamer identification and increases the likelihood of uncovering larger diversity of functional sequences. Current methods have been used to uncover human ATP-binding RNA aptamers, analyze MS2 aptamers, and identify CRISPR-Cas complex specificity (Abdelsayed et al., 2017; Buenrostro et al., 2014; Jung et al., 2017). These techniques are valuable in the advancement of *in vitro* selection and aid in understanding of the evolution of functional RNAs from random sequence space. Continued improvement to these methods will support the growth of aptamer-based drug development and should fill the shortcomings in current therapeutic biologics.

References

Abdelsayed, M. M., Ho, B. T., Vu, M. M. K., Polanco, J., Spitale, R. C., & Lupták, A. (2017). Multiplex aptamer discovery through apta-seq and its application to ATP aptamers derived from human-genomic SELEX. *ACS Chemical Biology*, *12*(8), 2149–2156. https://doi.org/10.1021/acschembio.7b00001.

Alam, K. K., Chang, J. L., & Burke, D. H. (2015). FASTAptamer: A bioinformatic toolkit for high-throughput sequence analysis of combinatorial selections. *Molecular Therapy Nucleic Acids*, *4*(3), e230. https://doi.org/10.1038/mtna.2015.4.

Ameta, S., Winz, M.-L., Previti, C., & Jäschke, A. (2014). Next-generation sequencing reveals how RNA catalysts evolve from random space. *Nucleic Acids Research*, *42*(2), 1303–1310. https://doi.org/10.1093/nar/gkt949.

Bailey, T. L., Boden, M., Buske, F. A., Frith, M., Grant, C. E., Clementi, L., et al. (2009). MEME suite: Tools for motif discovery and searching. *Nucleic Acids Research*, *37*(1), W202–W208. https://doi.org/10.1093/nar/gkp335.

Barrick, J. E., & Breaker, R. R. (2007). The distributions, mechanisms, and structures of metabolite-binding riboswitches. *Genome Biology*, *8*(11), R239. https://doi.org/10.1186/gb-2007-8-11-r239.

Bernhart, S. H., Hofacker, I. L., Will, S., Gruber, A. R., & Stadler, P. F. (2008). RNAalifold: Improved consensus structure prediction for RNA alignments. *BMC Bioinformatics*, *9*(1), 474. https://doi.org/10.1186/1471-2105-9-474.

Buenrostro, J. D., Araya, C. L., Chircus, L. M., Layton, C. J., Chang, H. Y., Snyder, M. P., et al. (2014). Quantitative analysis of RNA-protein interactions on a massively parallel array reveals biophysical and evolutionary landscapes. *Nature Biotechnology*, *32*, 562. https://doi.org/10.1038/nbt.2880.

Cadwell, R. C., & Joyce, G. F. (1992). Randomization of genes by PCR mutagenesis. *PCR Methods and Applications*, *2*(1), 28–33. https://doi.org/10.1101/gr.2.1.28.

Camacho, C., Coulouris, G., Avagyan, V., Ma, N., Papadopoulos, J., Bealer, K., et al. (2009). BLAST+: Architecture and applications. *BMC Bioinformatics*, *10*, 421. https://doi.org/10.1186/1471-2105-10-421.

Caroli, J., Taccioli, C., De La Fuente, A., Serafini, P., & Bicciato, S. (2016). APTANI: A computational tool to select aptamers through sequence-structure motif analysis of HT-SELEX data. *Bioinformatics*, *32*(2), 161–164. https://doi.org/10.1093/bioinformatics/btv545.

Cho, M., Xiao, Y., Nie, J., Stewart, R., Csordas, A. T., Oh, S. S., et al. (2010). Quantitative selection of DNA aptamers through microfluidic selection and high-throughput sequencing. *Proceedings of the National Academy of Sciences of the United States of America*, *107*(35), 15373–15378. https://doi.org/10.1073/pnas.1009331107.

Dao, P., Hoinka, J., Takahashi, M., Zhou, J., Ho, M., Wang, Y., et al. (2016). AptaTRACE elucidates RNA sequence-structure motifs from selection trends in HT-SELEX experiments. *Cell Systems*, *3*(1), 62–70. https://doi.org/10.1016/j.cels.2016.07.003.

Ditzler, M. A., Lange, M. J., Bose, D., Bottoms, C. A., Virkler, K. F., Sawyer, A. W., et al. (2013). High-throughput sequence analysis reveals structural diversity and improved potency among RNA inhibitors of HIV reverse transcriptase. *Nucleic Acids Research*, *41*(3), 1873–1884. https://doi.org/10.1093/nar/gks1190.

Eddy, S. R. (2011). Accelerated Profile HMM Searches. *PLoS Computational Biology*, 7(10), e1002195. https://doi.org/10.1371/journal.pcbi.1002195.

Edgar, R. C. (2004). MUSCLE: Multiple sequence alignment with high accuracy and high throughput. *Nucleic Acids Research*, *32*(5), 1792–1797. https://doi.org/10.1093/nar/gkh340.

Feng, C., Chan, D., Joseph, J., Muuronen, M., Coldren, W. H., Dai, N., et al. (2018). Light-activated chemical probing of nucleobase solvent accessibility inside cells. *Nature Chemical Biology*, *14*, 276. https://doi.org/10.1038/nchembio.2548.

Freese, N. H., Norris, D. C., & Loraine, A. E. (2016). Integrated genome browser: Visual analytics platform for genomics. *Bioinformatics*, *32*(14), 2089–2095. https://doi.org/10.1093/bioinformatics/btw069.

Griffin, L. C., Toole, J. J., & Leung, L. L. K. (1993). The discovery and characterization of a novel nucleotide-based thrombin inhibitor. *Gene*, *137*(1), 25–31. https://doi.org/10.1016/0378-1119(93)90247-Z.

Hamada, M. (2018). In silico approaches to RNA aptamer design. *Biochimie*, *145*, 8–14. https://doi.org/10.1016/j.biochi.2017.10.005.

Hiller, M., Pudimat, R., Busch, A., & Backofen, R. (2006). Using RNA secondary structures to guide sequence motif finding towards single-stranded regions. *Nucleic Acids Research*, *34*(17), e117. https://doi.org/10.1093/nar/gkl544.

Hoinka, J., Backofen, R., & Przytycka, T. M. (2018). AptaSUITE: A full-featured bioinformatics framework for the comprehensive analysis of aptamers from HT-SELEX experiments. *Molecular Therapy—Nucleic Acids*, *11*, 515–517. https://doi.org/10.1016/j.omtn.2018.04.006.

Hoinka, J., Berezhnoy, A., Sauna, Z. E., Gilboa, E., & Przytycka, T. M. (2014). AptaCluster—A method to cluster HT-SELEX aptamer pools and lessons from its application. *Research in Computational Molecular Biology*, *8394*, 115–128. https://doi.org/10.1007/978-3-319-05269-4_9.

Hoinka, J., Zotenko, E., Friedman, A., Sauna, Z. E., & Przytycka, T. M. (2012). Identification of sequence-structure RNA binding motifs for SELEX-derived aptamers. *Bioinformatics*, *28*(12), i215–i223. https://doi.org/10.1093/bioinformatics/bts210.

Hoon, S., Zhou, B., Janda, K. D., Brenner, S., & Scolnick, J. (2011). Aptamer selection by high-throughput sequencing and informatic analysis. *BioTechniques*, *51*(6), 413–416. https://doi.org/10.2144/000113786.

Jalali-Yazdi, F., Lai, L. H., Takahashi, T. T., & Roberts, R. W. (2016). High-throughput measurement of binding kinetics by mRNA display and next-generation sequencing. *Angewandte Chemie*, *55*(12), 4007–4010. https://doi.org/10.1002/anie.201600077.

Jijakli, K., Khraiwesh, B., Fu, W., Luo, L., Alzahmi, A., Koussa, J., et al. (2016). The in vitro selection world. *Methods, 106*, 3–13. https://doi.org/10.1016/j.ymeth.2016.06.003.

Jiménez, J. I., Xulvi-Brunet, R., Campbell, G. W., Turk-MacLeod, R., & Chen, I. A. (2013). Comprehensive experimental fitness landscape and evolutionary network for small RNA. *Proceedings of the National Academy of Sciences of the United States of America, 110*(37), 14984–14989. https://doi.org/10.1073/pnas.1307604110.

Jolma, A., Kivioja, T., Toivonen, J., Cheng, L., Wei, G., Enge, M., et al. (2010). Multiplexed massively parallel SELEX for characterization of human transcription factor binding specificities. *Genome Research, 20*(6), 861–873. https://doi.org/10.1101/gr.100552.109.

Jung, C., Hawkins, J. A., Jones, S. K., Jr., Xiao, Y., Rybarski, J. R., Dillard, K. E., et al. (2017). Massively parallel biophysical analysis of CRISPR-cas complexes on next generation sequencing chips. *Cell, 170*(1), 35–47.e13. https://doi.org/10.1016/j.cell.2017.05.044.

Katoh, K., Misawa, K., Kuma, K.-i., & Miyata, T. (2002). MAFFT: A novel method for rapid multiple sequence alignment based on fast Fourier transform. *Nucleic Acids Research, 30*(14), 3059–3066. https://doi.org/10.1093/nar/gkf436.

Kazan, H., Ray, D., Chan, E. T., Hughes, T. R., & Morris, Q. (2010). RNAcontext: A new method for learning the sequence and structure binding preferences of RNA-binding proteins. *PLoS Computational Biology, 6*(7), e1000832. https://doi.org/10.1371/journal.pcbi.1000832.

Kent, W. J. (2002). BLAT—The blast-like alignment tool. *Genome Research, 12*(4), 656–664. https://doi.org/10.1101/gr.229202.

Kwok, C. K., Ding, Y., Tang, Y., Assmann, S. M., & Bevilacqua, P. C. (2013). Determination of in vivo RNA structure in low-abundance transcripts. *Nature Communications, 4*, 2971. https://doi.org/10.1038/ncomms3971.

Lambert, N., Robertson, A., Jangi, M., McGeary, S., Sharp, P. A., & Burge, C. B. (2014). RNA bind-n-seq: Quantitative assessment of the sequence and structural binding specificity of RNA binding proteins. *Molecular Cell, 54*(5), 887–900. https://doi.org/10.1016/j.molcel.2014.04.016.

Langmead, B., & Salzberg, S. L. (2012). Fast gapped-read alignment with Bowtie 2. *Nature Methods, 9*, 357. https://doi.org/10.1038/nmeth.1923.

Lippert, C., Sabatini, R., Maher, M. C., Kang, E. Y., Lee, S., Arikan, O., et al. (2017). Identification of individuals by trait prediction using whole-genome sequencing data. *Proceedings of the National Academy of Sciences of the United States of America, 114*(38), 10166–10171. https://doi.org/10.1073/pnas.1711125114.

Lucks, J. B., Mortimer, S. A., Trapnell, C., Luo, S., Aviran, S., Schroth, G. P., et al. (2011). Multiplexed RNA structure characterization with selective 2′-hydroxyl acylation analyzed by primer extension sequencing (SHAPE-Seq). *Proceedings of the National Academy of Sciences of the United States of America, 108*(27), 11063–11068. https://doi.org/10.1073/pnas.1106501108.

Merino, E. J., Wilkinson, K. A., Coughlan, J. L., & Weeks, K. M. (2005). RNA structure analysis at single nucleotide resolution by selective 2‘-hydroxyl acylation and primer extension (SHAPE). *Journal of the American Chemical Society, 127*(12), 4223–4231. https://doi.org/10.1021/ja043822v.

Mitchell, D., Renda, A. J., Douds, C. A., Babitzke, P., Assmann, S. M., & Bevilacqua, P. C. (2018). In vivo RNA structural probing of uracil and guanine base pairing by 1-ethyl-3-(3-dimethylaminopropyl)carbodiimide (EDC). *RNA, 25*, 147–157. https://doi.org/10.1261/rna.067868.118. Advanced Online Publication.

Morris, K. N., Jensen, K. B., Julin, C. M., Weil, M., & Gold, L. (1998). High affinity ligands from in vitro selection: Complex targets. *Proceedings of the National Academy of Sciences of the United States of America, 95*(6), 2902–2907. https://doi.org/10.1073/pnas.95.6.2902.

Nawrocki, E. P., & Eddy, S. R. (2013). Infernal 1.1: 100-fold faster RNA homology searches. *Bioinformatics*, *29*(22), 2933–2935. https://doi.org/10.1093/bioinformatics/btt509.

Neveu, M., Kim, H.-J., & Benner, S. A. (2013). The "strong" RNA world hypothesis: Fifty years old. *Astrobiology*, *13*(4), 391–403. https://doi.org/10.1089/ast.2012.0868.

Ng, E. W. M., Shima, D. T., Calias, P., Cunningham, E. T., Jr., Guyer, D. R., & Adamis, A. P. (2006). Pegaptanib, a targeted anti-VEGF aptamer for ocular vascular disease. *Nature Reviews Drug Discovery*, *5*, 123. https://doi.org/10.1038/nrd1955.

Nutiu, R., Friedman, R. C., Luo, S., Khrebtukova, I., Silva, D., Li, R., et al. (2011). Direct measurement of DNA affinity landscapes on a high-throughput sequencing instrument. *Nature Biotechnology*, *29*, 659. https://doi.org/10.1038/nbt.1882.

Parsch, J., Braverman, J. M., & Stephan, W. (2000). Comparative sequence analysis and patterns of covariation in RNA secondary structures. *Genetics*, *154*(2), 909–921.

Pitt, J. N., & Ferré-D'Amaré, A. R. (2010). Rapid construction of empirical RNA fitness landscapes. *Science*, *330*(6002), 376–379. https://doi.org/10.1126/science.1192001.

Pitt, J. N., Rajapakse, I., & Ferré-D'Amaré, A. R. (2010). SEWAL: An open-source platform for next-generation sequence analysis and visualization. *Nucleic Acids Research*, *38*(22), 7908–7915. https://doi.org/10.1093/nar/gkq661.

Raine, A., Manlig, E., Wahlberg, P., Syvänen, A.-C., & Nordlund, J. (2017). SPlinted ligation adapter tagging (SPLAT), a novel library preparation method for whole genome bisulphite sequencing. *Nucleic Acids Research*, *45*(6), e36. https://doi.org/10.1093/nar/gkw1110.

Rampášek, L., Jimenez, R. M., Lupták, A., Vinař, T., & Brejová, B. (2016). RNA motif search with data-driven element ordering. *BMC Bioinformatics*, *17*(1), 216. https://doi.org/10.1186/s12859-016-1074-x.

Regulski, E. E., & Breaker, R. R. (2008). In-line probing analysis of riboswitches. In J. Wilusz (Ed.), *Post-transcriptional gene regulation* (pp. 53–67). Totowa, NJ: Humana Press. https://doi.org/10.1007/978-1-59745-033-1_4.

Roberts, R. W., & Szostak, J. W. (1997). RNA-peptide fusions for the in vitro selection of peptides and proteins. *Proceedings of the National Academy of Sciences of the United States of America*, *94*(23), 12297–12302. https://doi.org/10.1073/pnas.94.23.12297.

Sassanfar, M., & Szostak, J. W. (1993). An RNA motif that binds ATP. *Nature*, *364*, 550. https://doi.org/10.1038/364550a0.

Schütze, T., Wilhelm, B., Greiner, N., Braun, H., Peter, F., Mörl, M., et al. (2011). Probing the SELEX process with next-generation sequencing. *PLoS One*, *6*(12), e29604. https://doi.org/10.1371/journal.pone.0029604.

Shaughnessy, A. F. (2012). Monoclonal antibodies: Magic bullets with a hefty price tag. *BMJ345*, e8346. https://doi.org/10.1136/bmj.e8346.

Slattery, M., Riley, T., Liu, P., Abe, N., Gomez-Alcala, P., Dror, I., et al. (2011). Cofactor binding evokes latent differences in DNA binding specificity between hox proteins. *Cell*, *147*(6), 1270–1282. https://doi.org/10.1016/j.cell.2011.10.053.

Soukup, G. A., & Breaker, R. R. (1999). Relationship between internucleotide linkage geometry and the stability of RNA. *RNA*, *5*(10), 1308–1325. https://doi.org/10.1017/S1355838299990891.

Spitale, R. C., Flynn, R. A., Zhang, Q. C., Crisalli, P., Lee, B., Jung, J.-W., et al. (2015). Structural imprints in vivo decode RNA regulatory mechanisms. *Nature*, *519*(7544), 486–490. https://doi.org/10.1038/nature14263.

Tapsin, S., Sun, M., Shen, Y., Zhang, H., Lim, X. N., Susanto, T. T., et al. (2018). Genome-wide identification of natural RNA aptamers in prokaryotes and eukaryotes. *Nature Communications*, *9*(1), 1289. https://doi.org/10.1038/s41467-018-03675-1.

Tome, J. M., Ozer, A., Pagano, J. M., Gheba, D., Schroth, G. P., & Lis, J. T. (2014). Comprehensive analysis of RNA-protein interactions by high-throughput sequencing-RNA affinity profiling. *Nature Methods*, *11*(6), 683–688. https://doi.org/10.1038/nmeth.2970.

Vasa, S. M., Guex, N., Wilkinson, K. A., Weeks, K. M., & Giddings, M. C. (2008). ShapeFinder: A software system for high-throughput quantitative analysis of nucleic acid reactivity information resolved by capillary electrophoresis. *RNA*, *14*(10), 1979–1990. https://doi.org/10.1261/rna.1166808.

Vu, M. M. K., Jameson, N. E., Masuda, S. J., Lin, D., Larralde-Ridaura, R., & Lupták, A. (2012). Convergent evolution of adenosine aptamers spanning bacterial, human, and random sequences revealed by structure-based bioinformatics and genomic SELEX. *Chemistry & Biology*, *19*(10), 1247–1254. https://doi.org/10.1016/j.chembiol.2012.08.010.

Wang, P. Y., Sexton, A. N., Culligan, W. J., & Simon, M. D. (2018). Carbodiimide reagents for the chemical probing of RNA structure in cells. *RNA*, *25*, 135–146. https://doi.org/10.1261/rna.067561.118. Advanced Online Publication.

Wilson, D. S., & Keefe, A. D. (2001). Random mutagenesis by PCR. *Current Protocols in Molecular Biology*, *51*(1), 8.3.1–8.3.9. https://doi.org/10.1002/0471142727.mb0803s51.

Wu, Y. X., & Kwon, Y. J. (2016). Aptamers: The "evolution" of SELEX. *Methods*, *106*, 21–28. https://doi.org/10.1016/j.ymeth.2016.04.020.

Yao, Z., Weinberg, Z., & Ruzzo, W. L. (2006). CMfinder—A covariance model based RNA motif finding algorithm. *Bioinformatics*, *22*(4), 445–452. https://doi.org/10.1093/bioinformatics/btk008.

Zaccolo, M., Williams, D. M., Brown, D. M., & Gherardi, E. (1996). An approach to random mutagenesis of DNA using mixtures of triphosphate derivatives of nucleoside analogues. *Journal of Molecular Biology*, *255*(4), 589–603. https://doi.org/10.1006/jmbi.1996.0049.

Zhao, Y., & Stormo, G. D. (2011). Quantitative analysis demonstrates most transcription factors require only simple models of specificity. *Nature Biotechnology*, *29*(6), 480–483. https://doi.org/10.1038/nbt.1893.

Ziegenhain, C., Vieth, B., Parekh, S., Reinius, B., Guillaumet-Adkins, A., Smets, M., et al. (2017). Comparative analysis of single-cell RNA sequencing methods. *Molecular Cell*, *65*(4), 631–643. https://doi.org/10.1016/j.molcel.2017.01.023.

Zimmermann, B., Gesell, T., Chen, D., Lorenz, C., & Schroeder, R. (2010). Monitoring genomic sequences during SELEX using high-throughput sequencing: Neutral SELEX. *PLoS One*, *5*(2), e9169. https://doi.org/10.1371/journal.pone.0009169.

Zinshteyn, B., Chan, D., England, W., Feng, C., Green, R., & Spitale, R. C. (2018). Assaying RNA structure with LASER-seq. *Nucleic Acids Research*, *47*, 43–55. gky1172. https://doi.org/10.1093/nar/gky1172.

Zykovich, A., Korf, I., & Segal, D. J. (2009). Bind-n-seq: High-throughput analysis of in vitro protein–DNA interactions using massively parallel sequencing. *Nucleic Acids Research*, *37*(22), e151. https://doi.org/10.1093/nar/gkp802.

CHAPTER NINETEEN

Drop-in-well chamber for droplet interface bilayer with built-in electrodes

Kazuhiro Urakubo, Masayuki Iwamoto, Shigetoshi Oiki*
Department of Molecular Physiology and Biophysics, University of Fukui, Fukui, Japan
*Corresponding author: e-mail address: oiki-fki@umin.ac.jp

Contents

Methods in Enzymology, Volume 621
ISSN 0076-6879
https://doi.org/10.1016/bs.mie.2019.02.012

Abstract

Various methods have been developed for the formation of planar lipid bilayers, and recent techniques using water-in-oil droplets, such as droplet interface bilayer (DIB) and contact bubble bilayer (CBB) methods, allow the ready formation of bilayers with arbitrary lipid compositions. Here, we developed a simple and portable DIB system using drop-in-wells, shaping two merging wells for settling electrolyte droplets. An aliquot of the electrolyte solution (1 μL) is dropped into an organic solvent, and the droplet sinks to the drop-in-well at the bottom, where two monolayer-lined droplets come in contact to form the bilayer. Pre-installed electrodes allow electrophysiological measurements. The detailed drop-in-well method is presented, and some variations of the method, such as the use of microelectrodes and a sheet with a small hole for low-noise recordings, are extended. Examples of single channel current recordings of the KcsA potassium channel are demonstrated.

1. Introduction

Lipid bilayers, which include those in liposomes (Heginbotham, Kolmakova-Partensky, & Miller, 1998; Kasahara & Hinkle, 1976), nanodiscs (Banerjee & Nimigean, 2011), cell-sized liposomes (Yanagisawa, Iwamoto, Kato, Yoshikawa, & Oiki, 2011), and planar lipid bilayers (Oiki, 2012; Wonderlin, Finkel, & French, 1990; Ehrlich, 1992; Oiki, Koeppe, & Andersen, 1995; Favre, Sun, & Moczydlowski, 1999; Gutsmann, Heimburg, Keyser, Mahendran & Winterhalter, 2015), have been used for reconstitution and functional examination of membrane proteins. In liposomes, the flux of substances across the membrane has been measured (Kasahara & Hinkle, 1976; Nimigean, 2006). In planar lipid bilayers (PLBs), channel proteins have been examined for single-channel current recordings under applied membrane potentials (Oiki, 2015). Membrane potential is an inherent property of all cell membranes, and ready applicability of the membrane potential strengthens the PLBs. To form PLB, methods using water-in-oil droplets lined with lipid monolayer have been extensively applied, which renovated the lipid bilayer experiments. The original method used a microfluidic system as well as a small oval chamber for settling the droplets (Funakoshi, Suzuki, & Takeuchi, 2006). Many variants of these methods, i.e., the droplet interface bilayer (DIB) (Bayley et al., 2008) and contact bubble bilayer (CBB) (Iwamoto & Oiki, 2015; Oiki & Iwamoto, 2018) methods, have been developed progressively (Najem et al., 2015; Iwamoto & Oiki, 2018). In the conventional DIB method, electrolyte droplets were suspended around the tip of AgCl electrodes, and then, the droplets in the oil phase were brought into contact by manipulating the electrodes (Bayley et al., 2008). In the CBB method, the electrolyte solution is blown from a glass pipette into an oil phase (bubble), and two bubbles are docked via pipette manipulation (Iwamoto & Oiki, 2019a). This method allows various maneuvers to the bilayer, not only to form asymmetric bilayers (Hwang, Chen, Cronin, Holden, & Bayley, 2008; Iwamoto & Oiki, 2013) but also to alter the membrane compositions (membrane perfusion) (Iwamoto & Oiki, 2017) and tension via mechanical manipulation of the bubbles (Iwamoto & Oiki, 2018).

Among a spectrum of water-in-oil droplet methods, a simple and portable method for lipid bilayer formation is presented herein. The procedure involves usual biochemical pipetting works. One microliter of electrolyte solution is dropped in an organic solvent (hexadecane) filled in a chamber, having two merging wells at the bottom. The droplet slips along the slope of the plastic surface down the bottom (drop-in-wells), while the monolayer is formed at the oil-water interface of a droplet. When the droplet reaches the drop-in-wells, the pre-installed AgCl electrode at the bottom secures electrical connectivity with the droplet electrolyte solution. Two droplets spontaneously contact each other to form a DIB without inspection of microscopic observations.

We present the basic procedure and some variations of the drop-in-well method, such as the use of glass microelectrodes for preventing electrode interference by droplet constituents, and the use of a thin sheet with a small hole for low-noise recordings. Recently, the effects of membrane lipids to channels have been elucidated (Iwamoto & Oiki, 2013, 2018), and the drop-in-well method allows examination of channel-membrane interplays under various lipid compositions.

2. DIB formation in the drop-in-wells

2.1 Fabrication of the drop-in-wells

The drop-in-well chamber was designed as follows (Fig. 1). There was a concaved well of 16 mm in diameter with a shallow slope (the depth was 3.3 mm). At the bottom of the well, there were two small dimples (the diameter was 1.6 mm and the depth was 0.8 mm), which partly coalesced. Materials for the chamber should be chemically resistant to organic solvents, and the chamber was carved from an acryl block using programmable cutting machine.

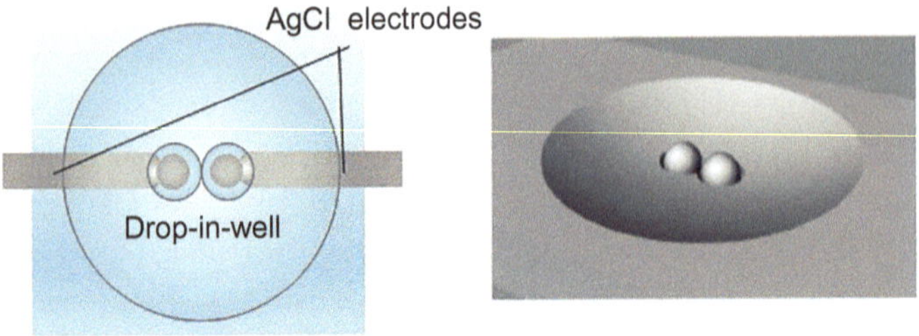

Fig. 1 Drop-in-well chamber and droplet interface bilayer. The design of the drop-in-well chamber. At the bottom of a shallow concave well, there were two small concave dimples (~1 mm in diameter and ~0.2 mm in depth), which were partly merging (drop-in-well) (*Left*). The drop-in-well chamber was carved from an acryl block using a programmable cutting machine. At the bottom of each drop-in-well, AgCl ink was painted as a built-in electrode (gray). Two droplets settled on the drop-in-wells (*Right*). The chamber was filled with hexadecane (100 μL), into which 1 μL of electrolyte solution was dropped. The droplet sank down along the slope and reached one of the drop-in-well. Two droplets came into contact in the drop-in-well to form bilayer spontaneously.

2.2 Lipid species for bilayer formation

Appropriate lipid species for bilayer formation must be selected. Some lipids do not form bilayers because of their high phase transition temperature above the temperature during experiments. Some lipids form the bilayer but the membrane resistance was too low to perform single-channel current recordings. Diphytanoylphosphatidylcholine has been used frequently (Iwamoto, Matsunaga, & Oiki, 2014; Iwamoto, Shimizu, Muramatsu, & Oiki, 2010; Matsuki et al., 2016), as well as azolectin, mixed phospholipids predominated with phosphatidylcholine.

2.3 Lipid addition: Lipid-in vs. lipid-out

Phospholipids for the formation of monolayers at the oil-water interface were served from either the organic solvent phase or the droplet electrolyte solution (Fig. 2A). The former is called the lipid-out method, and the latter is called the lipid-in method. In the lipid-out method, phospholipids were dispersed in hexadecane. Azolectin can be dispersed in hexadecane up to 100 mg/mL, but other lipids, such as PC, were dispersed at much lower concentrations. In the lipid-in method, liposomes with the desired lipid compositions were added in the electrolyte solutions for droplets in both sides. In either case, amphipathic phospholipids transfer spontaneously to the oil-water interface. Lipid transfer toward the interface was faster in the lipid-in method than the lipid-out method (Venkatesan et al., 2015).

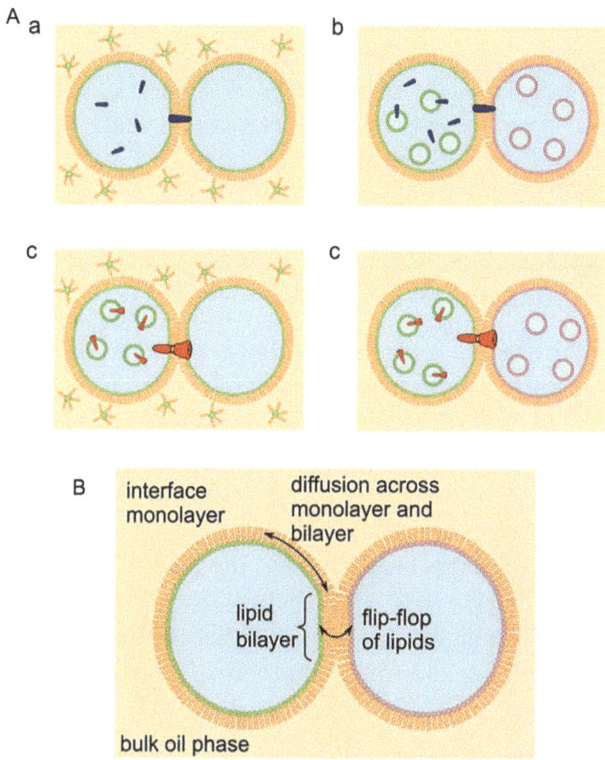

Fig. 2 Monolayer and bilayer in droplet interface bilayer (DIB). (A) Methods for serving phospholipids to the oil-water interface and channel reconstitution. Phospholipids were added in either the oil phase (a, c) or the aqueous phase (b, d). For channel molecules, water-soluble channels, such as channel forming toxins (gramicidin A, polytheonamide, amphidinol 3, α-hemolysin), were added in the droplet electrolyte solution (a, b). For channel proteins, they were reconstituted into liposomes (c, d). (B) The phases in DIB. An asymmetric bilayer is shown. A leaflet of the bilayer was continuous to the relevant monolayer, but the flip-flop of lipids across the leaflets is generally slow. *Adapted from Iwamoto, M., & Oiki, S. (2019a). Lipid bilayer experiments with contact bubble bilayers for patch-clampers.* Journal of Visualized Experiments *(143), e58840.*

Furthermore, membrane electrical resistance was higher in the lipid-in method for unknown reasons. An asymmetric membrane was formed exclusively with the lipid-in method (Iwamoto & Oiki, 2015).

2.4 Flux measurements for charge-neutral substances

DIB is a simple system for flux measurements (Braziel, Sullivan, & Lee, 2018; Elfaramawy et al., 2018; Fischer, Holden, Pentelute, & Collier, 2011; Huang, Lein, Gunderson, & Holden, 2011; Szabo & Wallace, 2016).

After the formation of DIB and incorporation of the transporting membrane proteins, substances were transported across the membrane, and the amount of the transported substance can be cropped by picking up a downstream (*trans*) droplet using the micropipette. The DIB is stable, and the flux of substances for hours can be examined.

3. Electrical measurements

One of the advantages of the DIB method is the ability to perform electrophysiological measurements. In addition to the current recordings through channels under the voltage clump, the electrophysiological system allows for measurements of the membrane capacitance and resistance (Gillis, 1995). From measurements of electrical capacitance and the membrane area, the bilayer thickness can be evaluated. Hexadecane is currently used as an organic solvent, and the bilayer specific capacitance was $0.65\,\mu F/cm^2$. This was smaller than that of the cell membrane ($1.0\,\mu F/cm^2$), indicating that the bilayers in hexadecane were thicker than the cell membrane with its hydrophobic core of 3 nm in thickness. Decane has been frequently used for the conventional PLB method, but when used in DIB, decane evaporates rapidly from the surface. The membrane was thicker, and with a specific capacitance of $0.3\,\mu F/cm^2$. Electrical resistances of bilayers vary substantially with the lipid compositions. For instance, dioleoylphosphatidylcholine bilayers have low electrical resistances (Venkatesan et al., 2015).

3.1 Drop-in-wells with built-in electrodes

To establish a simple "test-tube"-like experiment, AgCl electrodes were built into the bottom of the drop-in-well. AgCl ink is commercially available, and was used to paint the bottom of the wells. The surface of the AgCl electrode was wetted first by electrolyte solution, which prevents wetting of the electrode surface by oil and secures the electrical connectivity with the droplet electrolyte solution.

3.1.1 *A portable setup of* drop-in-wells

For portable applications, a compact electrical device such as Pico2 can be used (Kawano et al., 2014). This system allows ready current recordings outside the electrophysiological room.

3.2 Drop-in-wells with glass microelectrodes

The drop-in-well method with the built-in electrode is a convenient method, but in some experiments, continuous electrical measurements are not necessary. In some cases, the built-in electrodes can cause problems. The surface of the built-in electrode was large, and one may concern the adhesion of some components in the aqueous solution to the surface. This may interfere with the electrode potential, leading to erroneous readings of the membrane potential. In these cases, electrical measurements should be performed using glass microelectrodes (Iwamoto, Elfaramawy, Yamatake, Matsuura, & Oiki, 2018, Iwamoto & Oiki, 2019b).

3.2.1 Microelectrode method

Glass microelectrodes filled with 3M KCl were inserted into the droplets after the formation of DIB. The tip of the electrode was 1–2 μm in diameter, and the pipette was filled with 3M KCl. The electrode resistance was ≪1 MΩ. Microelectrodes provide accurate evaluation of membrane voltage irrespective of the solution compositions. Furthermore, the application of the membrane potential and the measurements of the single-channel currents were performed on demand by inserting the microelectrodes into the droplets (Fig. 3). The electrodes

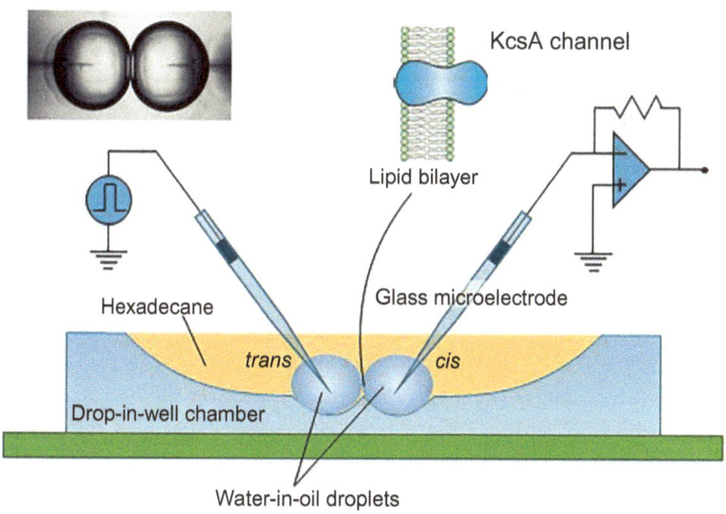

Fig. 3 The drop-in-well method with glass electrodes. Two glass microelectrodes filled with 3 M KCL were inserted into the droplets, and electrophysiological measurements were performed under a two-electrode voltage clamp configuration. *Adopted from Iwamoto, M., Elfaramawy, M.A., Yamatake, M., Matsuura, T., & Oiki, S. (2018). Concurrent in vitro synthesis and functional detection of nascent activity of the KcsA channel under a membrane potential,* ACS Synthetic Biology, 7 *(4), 1003–1011.*

offsets can be adjusted by inserting two electrodes into a droplet, and the liquid junction potential need not be of concern as far as concentrated KCl is used.

4. Thin-walled sheet for recordings with low background noise

4.1 Thin-sheet sheet

A drawback of using DIB for the electrophysiological measurements is high background noise based on its large bilayer area. Moreover, the shape of the droplets deforms over time because of gravity, and the bilayer area increases. The bilayer size is as large as several hundred micrometers in diameter, hindering recordings of single-channel currents of the order of several pA. To circumvent the increased noise, the contact area should be limited by a small hole bored in a sheet separating two droplets. A thin sheet was set at the center of the drop-in-wells before the electrolyte solutions were dropped into the wells (Fig. 4).

Technically, the thickness of the sheet is important. When a thick sheet (>100 μm) is used, a hole with larger diameter must be used for the following reason: two monolayers from both droplets must advance into the midst of the hole length (or sheet thickness) to come into contact with each other, but the organic solvent intercalated between the two monolayers prevents the contact. A thinner sheet allows for the formation of a smaller bilayer, but the thinner sheet (<10 μm in thickness) itself added electrical capacitance with its large surface area from the contact of the droplets, even though the bilayer area was small (Kawano et al., 2014). Consequently, the use of a thin sheet results in high electrical background noise. We used a Teflon sheet of 50 μm in the thickness and a hole with 200 μm in diameter, which compromises for a low background noise. To facilitate the contact, the hole shape was conical rather than cylindrical to allow for intercalated organic solvents to escape (Wonderlin et al., 1990).

The thin sheet has additional benefits. For settling the thin sheet, a slit is grooved at the two merging wells in the drop-in-well chamber. The sheet was set in the slit such that the hole was placed at the center of the two merging wells, where the two droplets come into contact. In our chamber, the sheet can be shifted along the slit, and the hole in the sheet can be displaced from the center, by which two droplets cannot come into contact with each other. This on-and-off contact maneuver provides a switchable system between a completely isolated two-droplet system and an open system,

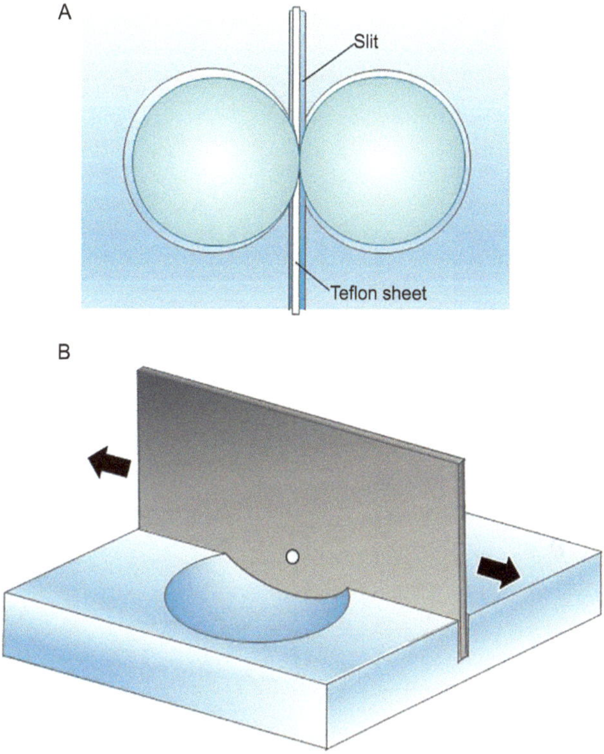

Fig. 4 Low-noise recordings using a thin sheet. At the point where the two drop-in-wells merge, a thin slit was grooved. A small hole (~200 μm in diameter) was bored in a thin-walled Teflon sheet (50 μm; see Protocol). This sheet was placed in the shallow slit such that the hole was located at the height where the two droplets come into contact. Top view (A) and perspective view (B).

allowing for transport between the two droplets. In the isolated system, chemical reactions in each droplet proceeds independently, while in the open system, transport between droplets is allowed and measured.

5. Materials

5.1 Equipment

1. Patch-clamp amplifier (Axopatch 200B, Molecular Devices, Sunnyvale, CA)
2. AD converter (Digidata1550A, Molecular Devices, Sunnyvale, CA)
3. Micropipette puller (P-87, Sutter Instrument, Novato, CA, USA)
4. Manipulator (EMM-3NV, Narishige, Tokyo, Japan)

5. Thin wall borosilicate capillary glass (length; 100 mm, OD/ID; 1.5/1.1 mm)
6. Pipette holder × 2
7. AgCl wire electrode × 2
8. Rotary evaporator
9. N_2 gas cylinder
10. Vacuum pump
11. Desiccator
12. Eggplant flask (5–10 mL)
13. Tuberculin syringe (1 mL)
14. Drop-in-well chamber (handmade)
15. Teflon sheet (thickness, 50 μm; diameter, 200 μm)
16. A stainless-steel rod with its tip having a conical shape with a smooth surface
17. A gas burner

5.2 Chemicals

1. Azolectin (P3644, Sigma-Aldrich, St. Louis, MO)
2. Hexadecane
3. Potassium chloride
4. AgCl ink (Ag/AgCl ink 011464, BAS Inc., Tokyo, Japan)
5. Distilled water

5.3 Drop-in-well chamber

The drop-in-well chamber was fabricated from an acryl block. For low noise recordings, the drop-in-well chamber with a slit at the midst of the merging wells for settling the thin sheet was fabricated. For the built-in electrode, AgCl ink was painted at the bottoms of each drop-in-well, and dried for several hours. Care should be taken to prevent merging of the AgCl-painted region in both wells. Silver wires were embedded in the AgCl paint for electrical leads.

5.4 Channel molecules

5.4.1 Channel forming toxins

Channel forming toxins are typical samples to test the electrophysiological measurements. These molecules are designed to transfer from aqueous solution to the target membrane. Accordingly, they are simply added to one or both of the droplets (Fig. 2Aa and b), leading to spontaneous insertion and single-channel current recordings (Iwamoto et al., 2010, 2017). Gramicidin A is a most frequently used channel (Oiki, Koeppe, & Andersen, 1994).

5.4.2 Channel proteins

The KcsA channel is a prototypical potassium channel from *Streptomyces lividans*.

6. Protocol

The protocol for the built-in electrode chamber is presented first. Both lipid-in and lipid-out methods were demonstrated in parallel.

6.1 Preparation of lipid solution

1 Disperse azolectin (e.g., 10 mg in powder) in chloroform at a desired concentration (e.g., 50 mg/mL).
2 Place the azolectin solution in an eggplant flask and set it on a rotary evaporator connected to a N_2 gas cylinder. Rotate the flask under N_2 flow at room temperature until a thin phospholipid film appears (20–40 min).
3 Place the flask into a desiccator connected to a vacuum pump. Aspirate the inside of the desiccator using the vacuum pump for several hours to evaporate the chloroform thoroughly.
4 For the lipid-in method, add an appropriate volume of electrolyte solution to the flask to obtain a 2 mg/mL lipid suspension. Sonicate the suspension for 10–60 s using a bath sonicator to obtain a multilayered vesicle (MLV) suspension. For preparation of channel-containing vesicles, add a channel protein solution (channel proteins are solubilized using appropriate detergents) to the MLV suspension (2% volume is the maximum) and sonicate for several seconds using a bath sonicator.

   For the lipid-out method, add an appropriate volume of hexadecane to the flask to obtain a 100 mg/mL azolectin solution.

6.2 Electrophysiological recordings

1 Connect the wirings of the chamber to the head stage of the patch-clamp amplifier and to the ground, respectively. Fill the drop-in-well chamber with an appropriate electrolyte solution to electrically short-circuit two AgCl electrodes at the bottom of the chamber. Adjust the electrode offset potential of the patch-clamp amplifier. Discard the electrolyte solution after the adjustment.

2 Fill the drop-in-well chamber with 150 μL hexadecane. Check the leak current between electrodes. If the leak current is substantial, clean the chamber. Drop 1 μL of the MLV suspension and the channel-containing MLV suspension, respectively, above the drop-in-wells by immersing the tip of the pipette beneath the hexadecane surface (Fig. 5). Wait for the two droplets to come into contact (formation of the DIB).

For the lipid-out method, add 150 μL of the hexadecane solution of lipids into the drop-in well chamber. Drop one μL of the electrolyte solution and channel-containing electrolyte solution, respectively, on the drop-in-wells. The channel-containing electrolyte solution should be prepared just prior to the experiment by diluting the channel protein

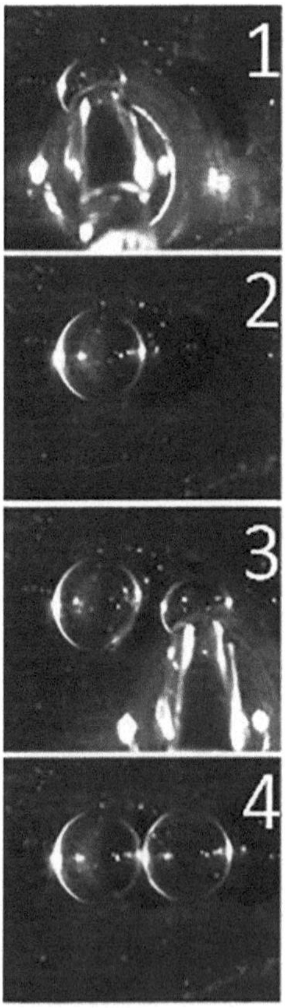

Fig. 5 Drop-in procedure. The tip of the micropipette should be immersed under the surface of the organic phase. A droplet sinks down to the drop-in-well, where monolayer-lined droplets contact to form DIB.

solution (channel proteins are solubilized using appropriate detergents) by the electrolyte solution >50 times. Wait for the formation of the DIB.

3 Set the membrane voltage to the desired value using the patch-clamp amplifier. Wait for the insertion of the channel proteins into the DIB, which is confirmed by the emergence of current signal. Currents were recorded using an amplifier through the cut-off frequency of 1 kHz sampled by 5 kHz using an AD converter.

4 In cases where asymmetric electrolyte solutions are used, correct the liquid junction potential, such that the calculated value is added to the applied membrane potential for the true membrane potential.

5 After breaking the bilayer, zero electrode offset potential should be confirmed.

6.3 Electrophysiological recordings using glass micropipettes

1 Set a glass capillary to a micropipette puller. Pull the capillary by two steps to fabricate a pair of micropipettes with fine tapered tip (diameter, 1–2 μm).

2 Fill 3 M KCl solution into the micropipettes using the tuberculin syringe and the injection needle. Connect the AgCl wire electrodes to the pipette holders equipped with a pressure port. Mount the micropipette to the pipette holders. Confirm that the AgCl wire electrode is soaked into the 3 M KCl solution. Connect one of the holders to the head stage of the patch-clamp amplifier and the other one to the electrical ground.

3 Soak the tip of both micropipettes into the same solution of 3 M KCl by manipulating the micromanipulator. Adjust the electrode offset potential of the patch-clamp amplifier.

4 Prepare the DIB either by using the lipid-in or lipid-out method.

5 Insert the tip of the glass microelectrodes to the respective droplet. Set the membrane voltage to the desired value by the patch-clamp amplifier.

6.4 Thin sheet for recordings with low background noise

1 Fabricate a small hole on a Teflon sheet as follows.

Heat a stainless-steel rod (diameter, 3 mm) by a gas burner. Make a slight bulge (<1 mm) on the surface of a Teflon sheet (thickness, 50 μm) by pushing the conical tip of the heated rod. Shave the slight bulge with a razor, making a hole of approximately 200 μm in diameter (the shape of the hole is conical with a sharp edge). Cut the sheet such that the hole is located 1 mm from an edge of the sheet. Wash the Teflon sheet thoroughly with chloroform/methanol.

2 Set the sheet in the drop-in-well chamber on a slit, such that the position of the hole is off-centered from the droplet-merging site.
3 Drop electrolyte solutions in both sides of the sheet either by lipid-in or lipid-out method. Slide the sheet such that the small hole on the sheet is placed just between the droplets, resulting in the forming the DIB. Perform the electrophysiological recording.

7. Single-channel current recordings

Background noise was reduced to ~1/8 with the Teflon sheet. Single-channel KcsA currents are shown in Fig. 6, representing signature activities with fast gating (Fig. 6) (Chakrapani, Cordero-Morales, & Perozo, 2007; Iwamoto et al., 2006; LeMasurier, Heginbotham, & Miller, 2001).

8. Summary

The drop-in-well chamber allows for the formation of a lipid bilayer using a water-in-oil droplet bilayer. The procedure is as simple as the addition of an aliquot of electrolyte solution into an organic solvent. The bilayer is automatically formed without visual inspection and manipulation. Variations in the drop-in-well method, such as the use of glass microelectrodes and a thin sheet for recordings with low-background noise, helped in widening the scope of its application. For example, *in vitro* transcription/translation of the KcsA channel protein was performed inside the droplet, followed by spontaneous insertion of the channel into the membrane (Iwamoto et al., 2018).

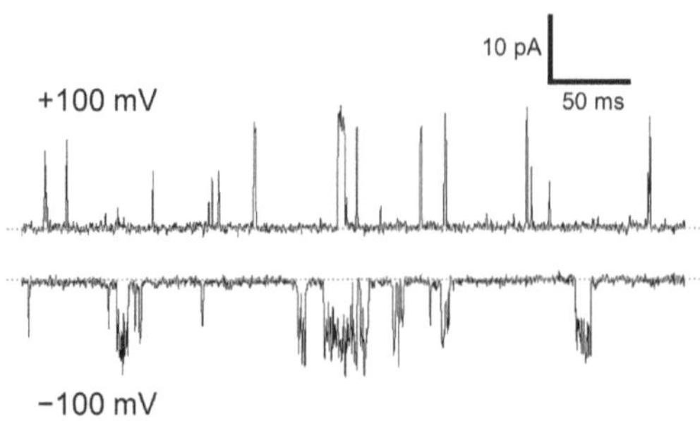

Fig. 6 Representative single-channel current of the KcsA potassium channel. The currents were recorded at the membrane potential of +100 mV through a 1 kHz filter. The *cis* and *trans* droplets contained 200 mM KCl and were asymmetrically buffered to pH 4.0 and 7.5, respectively.

Membrane effects on channel molecules include direct chemical interactions and physical perturbations. For example, cholesterol is a prerequisite for some channels to exhibit channel activities (Iwamoto et al., 2017). On the other hand, for the KcsA potassium channel, sterols, including cholesterol, attenuated the channel activity, and the effect of sterol was mediated by decreasing the bilayer tension rather than by direct chemical actions (Iwamoto & Oiki, 2018). Lipid bilayers with their varieties of methods would open a frontier of the study of channel-membrane interplays.

Acknowledgment

This work was supported in part by KAKENHI Grant numbers 25440067, 17K07360 (M.I.), and 16H00759, 17H04017 (S.O.).

References

Banerjee, S., & Nimigean, C. M. (2011). Non-vesicular transfer of membrane proteins from nanoparticles to lipid bilayers. *The Journal of General Physiology*, *137*(2), 217–223.

Bayley, H., Cronin, B., Heron, A., Holden, M. A., Hwang, W. L., Syeda, R., et al. (2008). Droplet interface bilayers. *Molecular BioSystems*, *4*(12), 1191–1208.

Braziel, S., Sullivan, K., & Lee, S. (2018). Quantitative Raman microspectroscopy for water permeability parameters at a droplet interface bilayer. *Analyst*, *143*(3), 747–755.

Chakrapani, S., Cordero-Morales, J. F., & Perozo, E. (2007). A quantitative description of KcsA gating II: Single-channel currents. *The Journal of General Physiology*, *130*(5), 479–496.

Ehrlich, B. E. (1992). Planar lipid bilayers on patch pipettes: Bilayer formation and ion channel incorporation. *Methods in Enzymology*, *207*, 463–470.

Elfaramawy, M. A., Fujii, S., Uyeda, A., Osaki, T., Takeuchi, S., Kato, Y., et al. (2018). Quantitative analysis of cell-free synthesized membrane proteins at the stabilized droplet interface bilayer. *Chemical Communications*, *54*(86), 12226–12229.

Favre, I., Sun, Y. M., & Moczydlowski, E. (1999). Reconstitution of native and cloned channels into planar bilayers. *Methods in Enzymology*, *294*, 287–304.

Fischer, A., Holden, M. A., Pentelute, B. L., & Collier, R. J. (2011). Ultrasensitive detection of protein translocated through toxin pores in droplet-interface bilayers. *Proceedings of the National Academy of Sciences of the United States of America*, *108*(40), 16577–16581.

Funakoshi, K., Suzuki, H., & Takeuchi, S. (2006). Lipid bilayer formation by contacting monolayers in a microfluidic device for membrane protein analysis. *Analytical Chemistry*, *78*(24), 8169–8174.

Gillis, K. D. (1995). Techniques for membrane capacitance measurements. In B. Sakman & E. Neher (Eds.), *Single-channel recording* (pp. 155–198). New York: Plenum.

Gutsmann, T., Heimburg, T., Keyser, U., Mahendran, K. R., & Winterhalter, M. (2015). Protein reconstitution into freestanding planar lipid membranes for electrophysiological characterization. *Nature Protocols*, *10*(**1**), 188–198.

Heginbotham, L., Kolmakova-Partensky, L., & Miller, C. (1998). Functional reconstitution of a prokaryotic K+ channel. *The Journal of General Physiology*, *111*(6), 741–749.

Huang, J., Lein, M., Gunderson, C., & Holden, M. A. (2011). Direct quantitation of peptide-mediated protein transport across a droplet-interface bilayer. *Journal of the American Chemical Society*, *133*(4), 15818–15821.

Hwang, W. L., Chen, M., Cronin, B., Holden, M. A., & Bayley, H. (2008). Asymmetric droplet interface bilayers. *Journal of the American Chemical Society*, *130*(18), 5878–5879.
Iwamoto, M., Elfaramawy, M. A., Yamatake, M., Matsuura, T., & Oiki, S. (2018). Concurrent in vitro synthesis and functional detection of nascent activity of the KcsA channel under a membrane potential. *ACS Synthetic Biology*, 7(4), 1004–1011.
Iwamoto, M., Matsunaga, S., & Oiki, S. (2014). Paradoxical one-ion pore behavior of the long β-helical peptide of marine cytotoxic polytheonamide B. *Scientific Reports*, *4*, 3636.
Iwamoto, M., & Oiki, S. (2013). Amphipathic antenna of an inward rectifier K^+ channel responds to changes in the inner membrane leaflet. *Proceedings of the National Academy of Sciences of the United States of America*, *110*(2), 749–754.
Iwamoto, M., & Oiki, S. (2015). Contact bubble bilayers with flush drainage. *Scientific Reports*, *5*, 9110.
Iwamoto, M., & Oiki, S. (2017). Membrane perfusion of hydrophobic substances around channels embedded in the contact bubble bilayer. *Scientific Reports*, 7(1), 6857.
Iwamoto, M., & Oiki, S. (2018). Constitutive boost of a K^+ channel via inherent bilayer tension and a unique tension-dependent modality. *Proceedings of the National Academy of Sciences of the United States of America*, *115*(51), 13117–13122.
Iwamoto, M., & Oiki, S. (2019a). Lipid bilayer experiments with contact bubble bilayers for patch-clampers. *Journal of Visualized Experiments*, *143*, e58840.
Iwamoto, M., & Oiki, S. (2019b). In bulla functional channel expression systems that mimic bacterial synthetic membranes. *Methods in Enzymology*. in press.
Iwamoto, M., Shimizu, H., Inoue, F., Konno, T., Sasaki, Y. C., & Oiki, S. (2006). Surface structure and its dynamic rearrangements of the KcsA potassium channel upon gating and tetrabutylammonium blocking. *The Journal of Biological Chemistry*, *281*(38), 28379–28386.
Iwamoto, M., Shimizu, H., Muramatsu, I., & Oiki, S. (2010). A cytotoxic peptide from a marine sponge exhibits ion channel activity through vectorial-insertion into the membrane. *FEBS Letters*, *584*(18), 3995–3999.
Iwamoto, M., Sumino, A., Shimada, E., Kinoshita, M., Matsumori, N., & Oiki, S. (2017). Channel formation and membrane deformation via sterol-aided polymorphism of amphidinol. *Scientific Reports*, 7, 10782.
Kasahara, M., & Hinkle, P. C. (1976). Reconstitution of D-glucose transport catalyzed by a protein fraction from human erythrocytes in sonicated liposomes. *Proceedings of the National Academy of Sciences of the United States of America*, 73(2), 396–400.
Kawano, R., Tsuji, Y., Kamiya, K., Kodama, T., Osaki, T., Miki, N., et al. (2014). A portable lipid bilayer system for environmental sensing with a transmembrane protein. *PLoS One*, *9*(7), e102427.
LeMasurier, M., Heginbotham, L., & Miller, C. (2001). KcsA: It's a potassium channel. *The Journal of General Physiology*, *118*(3), 303–314.
Matsuki, Y., Iwamoto, M., Mita, K., Shigemi, K., Matsunaga, S., & Oiki, S. (2016). Rectified proton grotthuss conduction across a long water-wire in the test nanotube of the polytheonamide B channel. *Journal of the American Chemical Society*, *138*(12), 4168–4177.
Najem, J. S., Dunlap, M. D., Yasmann, A., Freeman, E. C., Grant, J. W., Sukharev, S., et al. (2015). Multifunctional, micropipette-based method for incorporation and stimulation of bacterial mechanosensitive ion channels in droplet interface bilayers. *Journal of Visualized Experiments*, (105), e53362.
Nimigean, C. M. (2006). A radioactive uptake assay to measure ion transport across ion channel–containing liposomes. *Nature Protocols*, *1*(3), 1207–1212.
Oiki, S. (2012). Planar lipid bilayer method for studying channel molecules. In Y. Okada (Ed.), *Patch clamp techniques: From beginning to advanced protocols* (pp. 229–275). Tokyo: Springer protocols handbooks, Springer-Verlag.

Oiki, S. (2015). Channel function reconstitution and re-animation: A single-channel strategy in the postcrystal age. *The Journal of Physiology*, *593*(12), 2553–2573.

Oiki, S., & Iwamoto, M. (2018). Lipid bilayers manipulated through monolayer technologies for studies of channel-membrane interplay. *Biological and Pharmaceutical Bulletin*, *41*(3), 303–311.

Oiki, S., Koeppe, R. E., & Andersen, O. S. (1994). Asymmetric gramicidin channels: Heterodimeric channels with a single F6Val1 residue. *Biophysical Journal*, *66*(6), 1823–1832.

Oiki, S., Koeppe, R. E., & Andersen, O. S. (1995). Voltage-dependent gating of an asymmetric gramicidin channel. *Proceedings of the National Academy of Sciences of the United States of America*, *92*(6), 2121–2125.

Szabo, M., & Wallace, M. I. (2016). Imaging potassium-flux through individual electropores in droplet interface bilayers. *Biochimica et Biophysica Acta—Biomembranes*, *1858*(3), 613–617.

Venkatesan, G. A., Lee, J., Farimani, A. B., Heiranian, M., Collier, C. P., Aluru, N. R., et al. (2015). Adsorption kinetics dictate monolayer self-assembly for both lipid-in and lipid-out approaches to droplet interface bilayer formation. *Langmuir*, *31*(47), 12883–12893.

Wonderlin, W. F., Finkel, A., & French, R. J. (1990). Optimizing planar lipid bilayer single-channel recordings for high resolution with rapid voltage steps. *Biophysical Journal*, *58*(2), 289–297.

Yanagisawa, M., Iwamoto, M., Kato, A., Yoshikawa, K., & Oiki, S. (2011). Oriented reconstitution of a membrane protein in a giant unilamellar vesicle: Experimental verification with the potassium channel KcsA. *Journal of the American Chemical Society*, *133*(30), 11774–11779.

FSC
www.fsc.org
MIX
Papier | Fördert gute Waldnutzung
FSC® C083411

Kolibri 360 GmbH
Ferdinand-Jühlke-Straße 7
99095 Erfurt, Deutschland
produktsicherheit@kolibri360.de